AF785235

Generalized Fractional Brownian Motion

Series Editor
Nikolaos Limnios

Generalized Fractional Brownian Motion

Mounir Zili

ISTE

WILEY

First published 2026 in Great Britain and the United States by ISTE Ltd and John Wiley & Sons, Inc.

ISTE Ltd
27-37 St George's Road
London SW19 4EU
UK

www.iste.co.uk

John Wiley & Sons, Inc.
111 River Street
Hoboken, NJ 07030
USA

www.wiley.com

Library of Congress Control Number: 2025949679

British Library Cataloguing-in-Publication Data
A CIP record for this book is available from the British Library
ISBN 978-1-78630-963-1

The manufacturer's authorized representative according to the EU General Product Safety Regulation is Wiley-VCH GmbH, Boschstr. 12, 69469 Weinheim, Germany, e-mail: Product_Safety@wiley.com.

Contents

Introduction

The theory of stochastic processes provides a powerful framework for modeling phenomena across a vast range of disciplines, from financial markets to telecommunications and hydrology. A central challenge in this field is developing versatile models that capture complex behaviors like long-range dependence, self-similarity and nonstationary increments, which are frequently observed in real-world data. Among the many classes of stochastic processes, Gaussian processes have gained particular attention, serving as a foundation for numerous successful modeling tools. For a comprehensive overview of surveys and the complete literature on these processes, see Aloès et al. (2001); Aloès and Nualart (2003); Karatzas and Shreve (1998); Lifshits (2012); Revuz and Yor (1999), and references therein.

Within the realm of Gaussian processes, the *fractional Brownian motion* (fBm) stands as a significant generalization of classical Brownian motion. It is widely used for its ability to model phenomena exhibiting long-range dependence, self-similarity and, importantly, stationary increments. In parallel, the *sub-fractional Brownian motion* (sfBm) represents a distinct class of processes that preserves many of these characteristics but notably lacks stationary increments, positioning it as a model for phenomena with a different underlying structure. For the foundational work presented in this book, we will also utilize a version of the fBm defined on the entire real line, known as the *two-sided fractional Brownian motion* (tsfBm), which is essential for our central subject. For more information on these foundational processes, see Bojdecki et al. (2004a); Mandelbrot and Van Ness (1968); Mishura (2008); Tudor (2007, 2008), and references therein.

While numerous extensions of fBm and sfBm have emerged in recent decades, including multifractional Brownian motion (Peltier and Lévy Véhel 1995), mixed fBm (Zili 2006), bifractional Brownian motion (Houdré and Villa 2003), mixed sub-fractional Brownian motion (El-Nouty and Zili 2015; Mishura and Zili 2018; Zili 2014) and generalized sub-fractional Brownian motion (Sghir 2013), these extensions are typically limited to generalizing either fBm or sfBm individually. This forces researchers to select a model based on whether increments are stationary, a constraint that limits the flexibility and scope of their analysis.

In contrast, *generalized fractional Brownian motion*, the main topic of this book, offers a unique approach by simultaneously extending both fBm and sfBm. To avoid confusion with recent work by Pang and Taqqu (2019), who also used the term "generalized fractional Brownian motion" for a different extension, we will consistently refer to the specific generalization presented herein as the *Zili generalized fractional Brownian motion (ZgfBm)*, first introduced by Mounir Zili in 2017 (Zili 2017). This versatile Gaussian process is defined as a linear combination of a tsfBm on the entire real line, B^H, and its time-reversed counterpart. By adjusting its three parameters (H, a and b), ZgfBm can capture complex systems with heterogeneous time-dependent characteristics, offering greater flexibility compared to traditional fBm and sfBm models, which are limited by a single Hurst parameter. This versatility allows for modeling a wider range of random physical phenomena, including both stationary and nonstationary behaviors. For further details on ZgfBm, please refer to Zili (2017, 2018, 2024, 2025).

ZgfBm provides a unified framework for a broad spectrum of behaviors. When $a = 1$ and $b = 0$, the process reduces to a standard fBm with stationary increments. When $a = b = 1/\sqrt{2}$, it becomes a sub-fractional Brownian motion with nonstationary increments. This inherent flexibility allows for the modeling of phenomena with both stationary and nonstationary behaviors using a single mathematical object. Moreover, the dependence on the additional parameters a and b enables researchers to fine-tune the level of correlation between increments, a feature absent in simpler models. This ability to construct more adequate and specialized models is a key motivation for the in-depth study of ZgfBm.

This book represents a pivotal step in the mathematical theory of stochastic processes. Our primary purpose is to lay the first comprehensive milestones of a rigorous stochastic analysis for ZgfBm, thereby establishing a solid mathematical foundation that will enable its widespread use as a powerful and versatile modeling tool. To achieve this, we embark on a journey that begins with foundational theory and systematically progresses to advanced topics, each chapter building upon the last to reveal the intricate structure of this remarkable process.

This monograph is crafted to be an invaluable resource for a diverse audience. It is intended for researchers and practitioners in stochastic analysis, mathematical finance and other fields who require a deep understanding of advanced Gaussian processes. However, its value extends significantly to Master's and PhD. students who are studying standard fractional and sub-fractional Brownian motions. By thoroughly exploring ZgfBm as a general form, the book provides a unified perspective that illuminates the underlying principles and shared structures of these more specific processes. This approach offers a richer and more complete understanding of the entire class of fractional Gaussian processes, making the book a cornerstone for both advanced study and future research.

This book is a carefully structured guide, designed to be both self-contained and accessible as it takes the reader from foundational concepts to advanced results. Our journey begins with a review of the theoretical groundwork essential for understanding the processes that follow.

In *Chapter 1*, we lay the essential groundwork for the entire book. We begin by exploring the fundamental properties of Gaussian random variables and vectors, including moments and quadratic forms. We then move to the general theory of stochastic processes, defining Gaussian processes and providing the conditions for their existence. This chapter also introduces the concepts of Gaussian random measures and Wiener integration, including the crucial space-time white noise and its associated integral. This comprehensive review provides the necessary theoretical tools for understanding the more advanced concepts that follow.

Building upon this, *Chapter 2* delves into the properties of key fractional processes, including fBm and its odd and even variants, as well as the sfBm. This allows us to establish a common language and understanding of the processes that ZgfBm will generalize.

With these foundational concepts established, *Chapter 3* formally introduces the central subject of this book, investigating its unique properties, including its covariance structure and relationship to its components. This provides the reader with a deep understanding of its characteristics before moving on to its applications.

The following chapters then explore the practical and analytical power of ZgfBm. *Chapter 4* presents a complete methodology for estimating the three parameters of the process (H, a and b), a crucial step for real-world applications. We then develop a rigorous stochastic calculus for ZgfBm in *Chapter 5*, deriving generalized Itô formulas that unify and extend prior work. The book culminates in *Chapter 6* with an investigation into the regularity of solutions to a generalized stochastic heat equation, demonstrating the process' utility in modeling complex physical phenomena.

To help readers master the material, each chapter concludes with a set of carefully designed exercises, and a final appendix ensures the book is fully self-contained by providing a collection of important technical results. Through this comprehensive and systematic exposition, this book aims to establish ZgfBm as a new cornerstone in the field of stochastic analysis.

1

Gaussian Processes

Gaussian processes are a powerful and versatile class of stochastic processes fundamental to modern statistics, machine learning, and various scientific and engineering disciplines. Their inherent strength lies in their ability to model complex functions and time-dependent phenomena probabilistically, offering not only predictions but also a robust quantification of uncertainty.

Defined by the property that any finite collection of their random variables follows a multivariate Gaussian distribution, these processes are exceptionally well suited for capturing intricate dependencies over time or space, making them invaluable for tasks ranging from regression and classification to the analysis of complex dynamic systems.

To prepare for our study of advanced processes like the generalized fractional Brownian motion, we must first establish the foundational building blocks of this field. This chapter is dedicated to this essential groundwork.

We begin by exploring the fundamental properties of Gaussian random variables and vectors and their distributions. We then move to a comprehensive study of Gaussian processes, defining them and discussing the conditions for their existence. As a specific and important example, we will define the Wiener process (Brownian motion) as a particular case of a Gaussian process, establish its existence and recall its fundamental properties. The chapter will conclude with an introduction to Gaussian random measures and the core principles of Wiener integration, providing the necessary theoretical toolkit for the chapters to come.

1.1. Gaussian variables

To start, we will define what makes a random variable Gaussian.

DEFINITION 1.1.– *A random variable X is called a Gaussian variable (or normal variable) if its probability density function (PDF) with respect to the Lebesgue measure is given by:*

$$f_X(x) = \frac{1}{\sigma\sqrt{2\pi}} \exp\left(-\frac{(x-m)^2}{2\sigma^2}\right).$$

Here, $m \in \mathbb{R}$ represents the mean and $\sigma > 0$ represents the standard deviation. We denote this as $X \sim \mathcal{N}(m, \sigma^2)$. In the special case where $m = 0$ and $\sigma = 1$, X is called a standard Gaussian variable, denoted $X \sim \mathcal{N}(0,1)$.

Among the many properties satisfied by such variables, we will cite only those which will be particularly useful in this book.

LEMMA 1.1.– *If $X \sim \mathcal{N}(m, \sigma^2)$, then:*

1) The expected value is $\mathbb{E}[X] = m$.

2) The variance is $\mathrm{Var}[X] = \sigma^2$.

3) The characteristic function is given by:

$$\varphi_X(t) = \mathbb{E}[\exp(itX)] = \exp\left(imt - \frac{\sigma^2 t^2}{2}\right).$$

4) If $Y \sim \mathcal{N}(m_2, \sigma_2^2)$ is independent of X, then their sum $X + Y$ is also a Gaussian random variable:

$$X + Y \sim \mathcal{N}(m + m_2, \sigma^2 + \sigma_2^2).$$

5) For any $(\alpha, \beta) \in (\mathbb{R} \setminus \{0\}) \times \mathbb{R}$, the linear transformation $\alpha X + \beta$ is also a Gaussian random variable:

$$\alpha X + \beta \sim \mathcal{N}(\alpha m + \beta, \alpha^2 \sigma^2).$$

Building on these fundamental characteristics, we proceed to a vital property concerning the *moments of Gaussian variables.*

LEMMA 1.2.– *If $X \sim \mathcal{N}(m, \sigma^2)$, then its central moments are:*

– $\mathbb{E}[(X-m)^n] = 0$ if n is an odd positive integer.

– $\mathbb{E}[(X-m)^n] = \sigma^n (n-1)!!$ if n is an even positive integer, where the double factorial is given by:

$$(n-1)!! = (n-1)(n-3)\cdots 1 = \frac{n!}{2^{n/2}(n/2)!}.$$

In particular, for a standard Gaussian variable $Z \sim \mathcal{N}(0,1)$:

– $\mathbb{E}[Z^n] = 0$ *if n is an odd positive integer.*

– $\mathbb{E}[Z^n] = (n-1)!!$ *if n is an even positive integer.*

Proof. The proof of this lemma is standard; however, its derivation provides significant understanding and is therefore given as Exercise 1.2. □

1.1.1. *Chi-squared distribution*

Building upon the concept of Gaussian variables, we now introduce a distribution fundamental to statistical inference that arises directly from sums of squared Gaussian variables: the chi-squared distribution.

DEFINITION 1.2.– *Let $Z_1, Z_2, \ldots, Z_k$ be k independent and identically distributed (i.i.d.) standard Gaussian random variables, that is, $Z_i \sim \mathcal{N}(0,1)$ for $i = 1, \ldots, k$. A random variable Y defined as the sum of their squares:*

$$Y = \sum_{i=1}^{k} Z_i^2$$

is said to follow a chi-squared distribution with k degrees of freedom, denoted as $Y \sim \chi^2(k)$.

Its PDF for $y > 0$ is given by:

$$f_Y(y;k) = \frac{1}{2^{k/2}\Gamma(k/2)} y^{(k/2)-1} e^{-y/2},$$

where $\Gamma(\cdot)$ is the Gamma function, defined for real numbers $x > 0$ by the integral:

$$\Gamma(x) = \int_0^{\infty} t^{x-1} e^{-t} dt.$$

The chi-squared distribution possesses several key characteristics important for its application.

LEMMA 1.3.– *If $Y \sim \chi^2(k)$ with k degrees of freedom, then:*

1) The expected value (mean) is $\mathbb{E}[Y] = k$.

2) The variance is $\mathrm{Var}[Y] = 2k$.

3) The characteristic function is given by: $\varphi_Y(t) = (1-2it)^{-k/2}$.

1.2. Gaussian vectors

Having defined Gaussian variables, we now move to multivariate settings by introducing the definition of a Gaussian vector.

DEFINITION 1.3.– *Let* $\langle x, y\rangle = x_1y_1 + ... + x_ny_n$ *denote the Euclidean inner product in* $\mathbb{R}^n$ *for* $x, y \in \mathbb{R}^n$. *A random vector* $X = (X_1, X_2, ..., X_n)'$ *is called a Gaussian vector if any linear combination of its components,*

$$\langle \lambda, X\rangle = \lambda_1 X_1 + \lambda_2 X_2 + ... + \lambda_n X_n,$$

is a Gaussian random variable for any vector $\lambda = (\lambda_1, \lambda_2, ..., \lambda_n)' \in \mathbb{R}^n$. *Here, the prime notation* $(.)'$ *denotes the transpose of the vector.*

If X is a Gaussian vector, its *expectation vector* is:

$$M_X = \mathbb{E}[X] = (\mathbb{E}[X_1], ..., \mathbb{E}[X_n])',$$

and its *covariance matrix* is:

$$K_X = \text{Cov}(X) = (\text{Cov}(X_i, X_j))_{1\leq i,j\leq n}\,.$$

We denote this as $X \sim \mathcal{N}_n(M_X, K_X)$. A special case is the *standard Gaussian vector*, where $M_X = \mathbf{0}$ (the zero vector) and $K_X = I_n$ (the identity matrix).

For Gaussian vectors, we will focus on their most relevant properties for this book.

LEMMA 1.4.– *If* $X \sim \mathcal{N}_n(M_X, K_X)$, *its characteristic function is given by:*

$$\varphi_X(\lambda) = \mathbb{E}\left[\exp\{i\lambda' X\}\right] = \exp\left(i\langle\lambda, M_X\rangle - \frac{1}{2}\langle\lambda, K_X\lambda\rangle\right).$$

This can also be written in matrix notation as:

$$\varphi_X(\lambda) = \exp\left(i\lambda' M_X - \frac{1}{2}\lambda' K_X \lambda\right)$$

for any $\lambda \in \mathbb{R}^n$.

A crucial property of Gaussian vectors is their behavior under linear transformations. Specifically, a linear transformation of a Gaussian vector remains a Gaussian vector, a fact we formally state and prove below.

LEMMA 1.5.– *If* $\mathbf{Y}$ *is a Gaussian vector,* $\mathbf{Y} \sim \mathcal{N}_n(\boldsymbol{\mu}, \boldsymbol{\Sigma})$, *and* $S = \mathbf{b}'\mathbf{Y}$ *where* $\mathbf{b}$ *is a constant vector, then* S *is a Gaussian random variable, specifically:*

$$S = \mathbf{b}'\mathbf{Y} \sim \mathcal{N}(\mathbf{b}'\boldsymbol{\mu}, \mathbf{b}'\boldsymbol{\Sigma}\mathbf{b}).$$

Proof. Since the scalar $S = \mathbf{b}'\mathbf{Y} = \sum_{j=1}^{n} b_j Y_j$ is a linear combination of the components of a Gaussian vector, S must be a Gaussian random variable by definition. To determine its specific distribution, we need to find its mean and variance.

For the mean of S, using the linearity of expectation:

$$\mathbb{E}[S] = \mathbb{E}[\mathbf{b}'\mathbf{Y}] = \mathbf{b}'\mathbb{E}[\mathbf{Y}] = \mathbf{b}'\boldsymbol{\mu}.$$

Now, the variance of a random variable S is $\text{Var}(S) = \mathbb{E}[(S - \mathbb{E}[S])^2]$. Substituting $S = \mathbf{b}'\mathbf{Y}$ and $\mathbb{E}[S] = \mathbf{b}'\boldsymbol{\mu}$:

$$\begin{aligned}\text{Var}(S) &= \mathbb{E}[(\mathbf{b}'\mathbf{Y} - \mathbf{b}'\boldsymbol{\mu})^2] \\ &= \mathbb{E}[(\mathbf{b}'(\mathbf{Y} - \boldsymbol{\mu}))^2] \\ &= \mathbb{E}[(\mathbf{b}'(\mathbf{Y} - \boldsymbol{\mu}))(\mathbf{Y} - \boldsymbol{\mu})'\mathbf{b}] \\ &= \mathbf{b}'\mathbb{E}[(\mathbf{Y} - \boldsymbol{\mu})(\mathbf{Y} - \boldsymbol{\mu})']\mathbf{b}.\end{aligned}$$

By definition, $\mathbb{E}[(\mathbf{Y} - \boldsymbol{\mu})(\mathbf{Y} - \boldsymbol{\mu})']$ is the covariance matrix of $\mathbf{Y}$, which is $\boldsymbol{\Sigma}$. Therefore,

$$\text{Var}(S) = \mathbf{b}'\boldsymbol{\Sigma}\mathbf{b}.$$

Thus, S follows a Gaussian distribution with mean $\mathbf{b}'\boldsymbol{\mu}$ and variance $\mathbf{b}'\boldsymbol{\Sigma}\mathbf{b}$. □

Extending this concept to transformations resulting in a vector, the following proposition demonstrates that any linear transformation of a Gaussian vector produces another Gaussian vector.

PROPOSITION 1.1.– *If* $\mathbf{Z}$ *is a Gaussian vector,* $\mathbf{Z} \sim \mathcal{N}_n(\mathbf{0}, \Gamma)$, *and* $\mathbf{V} = P'\mathbf{Z}$ *where* P *is a constant* $n \times n$ *matrix and* P' *is its transpose, then* $\mathbf{V}$ *is also a Gaussian vector, specifically:*

$$\mathbf{V} = P'\mathbf{Z} \sim \mathcal{N}_n\left(\mathbf{0}, P'\Gamma P\right).$$

Proof. The proof proceeds by determining the mean vector and covariance matrix of $\mathbf{V}$ and then showing that its characteristic function has the Gaussian form.

The mean of $\mathbf{V}$ is given by the expectation of $\mathbf{V}$: $\mathbb{E}[\mathbf{V}] = \mathbb{E}[P'\mathbf{Z}]$.

Since P' is a constant matrix, we can take it out of the expectation:

$$\mathbb{E}[\mathbf{V}] = P'\mathbb{E}[\mathbf{Z}] = 0,$$

since $\mathbf{Z} \sim \mathcal{N}_n(\mathbf{0}, \Gamma)$. Thus, the mean of $\mathbf{V}$ is also the zero vector.

The covariance matrix of $\mathbf{V}$, denoted by $\Sigma_{\mathbf{V}}$, is defined as:

$$\Sigma_{\mathbf{V}} = \mathbb{E}\left[(\mathbf{V} - \mathbb{E}[\mathbf{V}])(\mathbf{V} - \mathbb{E}[\mathbf{V}])'\right].$$

Since $\mathbb{E}[\mathbf{V}] = \mathbf{0}$, this simplifies to:

$$\Sigma_{\mathbf{V}} = \mathbb{E}\left[\mathbf{V}\mathbf{V}'\right] = \mathbb{E}\left[(P'\mathbf{Z})(P'\mathbf{Z})'\right].$$

Using the properties $(AB)' = B'A'$ and $(A')' = A$, we have:

$$(P'\mathbf{Z})' = \mathbf{Z}'(P')' = \mathbf{Z}'P.$$

Thus,

$$\Sigma_{\mathbf{V}} = \mathbb{E}\left[P'\mathbf{Z}\mathbf{Z}'P\right].$$

Since P' and P are constant matrices, we can take them out of the expectation:

$$\Sigma_{\mathbf{V}} = P'\mathbb{E}\left[\mathbf{Z}\mathbf{Z}'\right]P.$$

The covariance matrix of $\mathbf{Z}$, denoted by $\Sigma_{\mathbf{Z}}$, is given by:

$$\Sigma_{\mathbf{Z}} = \mathbb{E}\left[(\mathbf{Z} - \mathbb{E}[\mathbf{Z}])(\mathbf{Z} - \mathbb{E}[\mathbf{Z}])'\right] = \mathbb{E}\left[\mathbf{Z}\mathbf{Z}'\right] = \Gamma.$$

Substituting this back into the expression for $\Sigma_{\mathbf{V}}$: $\Sigma_{\mathbf{V}} = P'\Gamma P$.

The characteristic function of $\mathbf{V}$ can be found by substituting its definition into the characteristic function of a random vector:

$$\phi_{\mathbf{V}}(\mathbf{t}) = \mathbb{E}\left[e^{it'\mathbf{V}}\right] = \mathbb{E}\left[e^{it'(P'\mathbf{Z})}\right].$$

Using the property $\mathbf{t}'(P'\mathbf{Z}) = (P\mathbf{t})'\mathbf{Z}$ and letting $\mathbf{s} = P\mathbf{t}$, we can express this in terms of the characteristic function of $\mathbf{Z}$:

$$\phi_{\mathbf{V}}(\mathbf{t}) = \mathbb{E}(e^{i\mathbf{s}'\mathbf{Z}}) = \phi_{\mathbf{Z}}(\mathbf{s}).$$

Since $\mathbf{Z}$ is a Gaussian vector with mean $\mathbf{0}$ and covariance matrix Γ, its characteristic function is given by:

$$\phi_{\mathbf{Z}}(\mathbf{s}) = e^{-\frac{1}{2}\mathbf{s}'\Gamma\mathbf{s}}.$$

Substituting $\mathbf{s} = P\mathbf{t}$ into this expression gives us the characteristic function of $\mathbf{V}$:

$$\phi_{\mathbf{V}}(\mathbf{t}) = e^{-\frac{1}{2}(P\mathbf{t})'\Gamma(P\mathbf{t})} = e^{-\frac{1}{2}\mathbf{t}'P'\Gamma P\mathbf{t}}.$$

As the covariance matrix of $\mathbf{V}$ is $\Sigma_{\mathbf{V}} = P'\Gamma P$, the characteristic function simplifies to:

$$\phi_{\mathbf{V}}(\mathbf{t}) = e^{-\frac{1}{2}\mathbf{t}'\Sigma_{\mathbf{V}}\mathbf{t}}.$$

This is the characteristic function of a multivariate normal distribution with mean $\mathbf{0}$ and covariance matrix $\Sigma_{\mathbf{V}}$.

This completes the proof. □

For a complete understanding of their distribution in Euclidean space, it is essential to define the PDF of a Gaussian vector. The following lemma provides this density for the nondegenerate case, followed by a crucial remark concerning singular covariance matrices.

LEMMA 1.6.– *If $X \sim \mathcal{N}_n(M_X, K_X)$ with a nondegenerate (invertible) covariance matrix K_X, then its PDF is:*

$$f_X(x) = (2\pi)^{-n/2}(\det K_X)^{-1/2} \exp\left(-\frac{1}{2}(x - M_X)' K_X^{-1}(x - M_X)\right)$$

where $\det K_X$ is the determinant of K_X, and K_X^{-1} is its inverse.

REMARK 1.1.– *It is important to note that the covariance matrix K_X of a Gaussian random vector is not always invertible. It can be singular (i.e. have a determinant of zero) under certain conditions. When K_X is singular, it signifies that the Gaussian random vector X is degenerate. This means there exist linear dependencies among its components. Specifically, one or more components X_i can be expressed as an affine (linear plus constant) transformation of the other components. For example, if $X_n = c_1X_1 + \cdots + c_{n-1}X_{n-1} + c_0$ for some constants c_i, then K_X would be singular. The direct consequence of this degeneracy is that the probability distribution of X is not spread out over the entire n-dimensional space $\mathbb{R}^n$. Instead, it is concentrated on a lower dimensional affine subspace of $\mathbb{R}^n$. As the distribution is confined to such a subspace (which has zero Lebesgue measure in $\mathbb{R}^n$), the Gaussian vector does not possess a PDF with respect to the Lebesgue measure on $\mathbb{R}^n$. In these cases, we might refer to its distribution as a degenerate Gaussian or note that its density exists only with respect to a lower dimensional Lebesgue measure on the affine subspace it occupies.*

1.2.1. *Isserlis' theorem for Gaussian moments*

When dealing with expectations of products of Gaussian random variables, especially for orders higher than two, a powerful tool known as *Isserlis' theorem* (also known as Wick's theorem or the Gaussian moment theorem) significantly simplifies calculations. It provides a general formula for the expected value of the product of any number of zero-mean jointly Gaussian random variables.

THEOREM 1.1.– *Let $X_1, X_2, \ldots, X_n$ be jointly Gaussian random variables with zero mean ($\mathbb{E}[X_i] = 0$ for all i). Let $K_{ij} = \mathbb{E}[X_iX_j]$ be their covariances.*

1) If n is an odd integer, then $\mathbb{E}[X_1X_2 \cdots X_n] = 0$.

2) If n is an even integer, then:

$$\mathbb{E}[X_1 X_2 \cdots X_n] = \sum \prod_{pairs} \mathbb{E}[X_a X_b],$$

where the sum is over all distinct ways of partitioning the n variables into $n/2$ pairs. Each product consists of the expected values of these pairs.

Proof. The proof of Isserlis' theorem can be found in Vignat (2012) and the references therein. □

REMARK 1.2 (Isserlis' theorem for Gaussian variables).– *For a zero-mean multivariate Gaussian vector $X = (X_1, \ldots, X_n)$ with covariance matrix $K_X = \mathbb{E}[XX']$:*

1) Third-order moment: for any three components X_i, X_j, X_k, the third-order moment is zero.

$$\mathbb{E}[X_i X_j X_k] = 0.$$

2) Fourth-order moment: for any four components X_i, X_j, X_k, X_l, the fourth-order moment is the sum of all distinct pairings of the covariances.

$$\mathbb{E}[X_i X_j X_k X_l] = \mathbb{E}[X_i X_j]\mathbb{E}[X_k X_l] + \mathbb{E}[X_i X_k]\mathbb{E}[X_j X_l] + \mathbb{E}[X_i X_l]\mathbb{E}[X_j X_k]. \quad [1.1]$$

This corresponds to the three pairings: (X_i, X_j) and (X_k, X_l); (X_i, X_k) and (X_j, X_l); and (X_i, X_l) and (X_j, X_k).

As a direct consequence of Isserlis' theorem, we get the following corollary which provides a fundamental relationship between the covariance of two zero-mean Gaussian random variables of their squares and their original covariance.

COROLLARY 1.1.– *For any two jointly zero-mean Gaussian random variables X and Y, the covariance of their squares is given by:*

$$Cov(X^2, Y^2) = 2(Cov(X, Y))^2.$$

Proof. To prove this, we start with the definition of covariance:

$$\text{Cov}(X^2, Y^2) = \mathbb{E}[X^2 Y^2] - \mathbb{E}[X^2]\mathbb{E}[Y^2].$$

Since X and Y are zero-mean Gaussian random variables, we know that:

$$\mathbb{E}[X^2] = \text{Var}(X) \quad \text{and} \quad \mathbb{E}[Y^2] = \text{Var}(Y).$$

Also, by Theorem 1.1, for zero-mean Gaussian random variables, the fourth-order moment $\mathbb{E}[X_1 X_2 X_3 X_4]$ is the sum of products of pairwise covariances. For $\mathbb{E}(X^2 Y^2)$, we have $X_1 = X, X_2 = X, X_3 = Y, X_4 = Y$. The possible pairings are:

– (X_1, X_2) and (X_3, X_4) $\Longrightarrow$ $\mathbb{E}[X \cdot X]\mathbb{E}[Y \cdot Y] = \text{Var}(X)\text{Var}(Y)$;
– (X_1, X_3) and (X_2, X_4) $\Longrightarrow$ $\mathbb{E}[X \cdot Y]\mathbb{E}[X \cdot Y] = (\text{Cov}(X, Y))^2$;
– (X_1, X_4) and (X_2, X_3) $\Longrightarrow$ $\mathbb{E}[X \cdot Y]\mathbb{E}[X \cdot Y] = (\text{Cov}(X, Y))^2$.

Summing these pairings, we get:

$$\mathbb{E}[X^2Y^2] = \text{Var}(X)\text{Var}(Y) + 2(\text{Cov}(X, Y))^2$$

Now substitute this back into the covariance formula:

$$\begin{aligned}\text{Cov}(X^2, Y^2) &= \text{Var}(X)\text{Var}(Y) + 2(\text{Cov}(X, Y))^2 - \text{Var}(X)\text{Var}(Y)\\ &= 2(\text{Cov}(X, Y))^2,\end{aligned}$$

which completes the proof. □

Moving beyond linear transformations, understanding higher order moments of Gaussian vectors, especially those involving quadratic forms, is crucial. The following lemma provides a powerful and widely used result for the expectation of the product of two quadratic forms from a zero-mean Gaussian vector. Its technical proof relies on Isserlis' theorem, but given its direct utility in this book, we will state the result without derivation.

LEMMA 1.7.– *Let $\mathbf{Y} \sim \mathcal{N}_n(\mathbf{0}, \mathbf{\Sigma})$ be an n-dimensional zero-mean Gaussian vector with covariance matrix $\mathbf{\Sigma}$. For any symmetric $n \times n$ matrices $\mathbf{A}$ and $\mathbf{B}$, the expectation of the product of two quadratic forms $\mathbf{Y}'\mathbf{A}\mathbf{Y}$ and $\mathbf{Y}'\mathbf{B}\mathbf{Y}$ is given by:*

$$\mathbb{E}[(\mathbf{Y}'\mathbf{A}\mathbf{Y})(\mathbf{Y}'\mathbf{B}\mathbf{Y})] = Tr(\mathbf{A}\mathbf{\Sigma})Tr(\mathbf{B}\mathbf{\Sigma}) + 2Tr(\mathbf{A}\mathbf{\Sigma}\mathbf{B}\mathbf{\Sigma}).$$

Here, $Tr(\mathbf{M})$ denotes the trace of a square matrix $\mathbf{M}$, which is defined as the sum of its diagonal elements: $Tr(\mathbf{M}) = \sum_{i=1}^{n} M_{ii}$.

As a direct application of Lemma 1.7, the following corollary derives the expected value of a quadratic form multiplied by a squared linear form.

COROLLARY 1.2.– *Let $\mathbf{Y} \sim \mathcal{N}_n(\mathbf{0}, \mathbf{\Sigma})$ be an n-dimensional zero-mean Gaussian vector with covariance matrix $\mathbf{\Sigma}$. For any symmetric $n \times n$ matrix $\mathbf{A}$ and any constant $n \times 1$ vector $\mathbf{b}$, the expected value of the product of the quadratic form $\mathbf{Y}'\mathbf{A}\mathbf{Y}$ and the squared linear form $(\mathbf{b}'\mathbf{Y})^2$ is given by:*

$$\mathbb{E}[(\mathbf{Y}'\mathbf{A}\mathbf{Y})(\mathbf{b}'\mathbf{Y})^2] = 2Tr(\mathbf{A}\mathbf{\Sigma}\mathbf{b}\mathbf{b}'\mathbf{\Sigma}) + Tr(\mathbf{A}\mathbf{\Sigma})\mathbf{b}'\mathbf{\Sigma}\mathbf{b}.$$

The full proof of Corollary 1.2 is provided as Exercise 1.6, where hints for its derivation can be found.

1.2.2. *Moments of quadratic and squared linear forms*

We now present a particularly useful result for calculating the expectation of the square of a linear form of a centered Gaussian random vector.

LEMMA 1.8.– *Let* $\mathbf{Z} = (Z_1, \ldots, Z_n)'$ *be a centered Gaussian random vector (i.e.* $\mathbb{E}[\mathbf{Z}] = \mathbf{0}$*) with a non-singular covariance matrix* $\mathbf{\Gamma}_Z$*. Let* $\mathbf{t} = (t_1, \ldots, t_n)'$ *be a deterministic vector. Then:*

$$\mathbb{E}[(\mathbf{t}'\mathbf{\Gamma}_Z^{-1}\mathbf{Z})^2] = \mathbf{t}'\mathbf{\Gamma}_Z^{-1}\mathbf{t}.$$

Proof. Let the linear form be $Y = \mathbf{t}'\mathbf{\Gamma}_Z^{-1}\mathbf{Z}$. We want to compute $\mathbb{E}[Y^2]$.

Since Y is a scalar, $Y^2 = YY$. We can write this as:

$$Y^2 = (\mathbf{t}'\mathbf{\Gamma}_Z^{-1}\mathbf{Z})(\mathbf{t}'\mathbf{\Gamma}_Z^{-1}\mathbf{Z}).$$

For any scalar c, $c = c'$. Applied to our scalar linear form, $(\mathbf{t}'\mathbf{\Gamma}_Z^{-1}\mathbf{Z})' = \mathbf{t}'\mathbf{\Gamma}_Z^{-1}\mathbf{Z}$. By the property of transpose of a product $(AB)' = B'A'$, we also have:

$$(\mathbf{t}'\mathbf{\Gamma}_Z^{-1}\mathbf{Z})' = \mathbf{Z}'(\mathbf{\Gamma}_Z^{-1})'(\mathbf{t}')' = \mathbf{Z}'(\mathbf{\Gamma}_Z^{-1})'\mathbf{t}.$$

Given that $\mathbf{\Gamma}_Z$ is a covariance matrix, it is symmetric, that is, $\mathbf{\Gamma}_Z = \mathbf{\Gamma}'_Z$. Consequently, its inverse is also symmetric, $(\mathbf{\Gamma}_Z^{-1})' = \mathbf{\Gamma}_Z^{-1}$. Therefore, we can equate the scalar form with its transpose:

$$\mathbf{t}'\mathbf{\Gamma}_Z^{-1}\mathbf{Z} = \mathbf{Z}'\mathbf{\Gamma}_Z^{-1}\mathbf{t}.$$

Using this, we can rewrite the product within the expectation as:

$$\mathbb{E}[(\mathbf{t}'\mathbf{\Gamma}_Z^{-1}\mathbf{Z})^2] = \mathbb{E}[(\mathbf{t}'\mathbf{\Gamma}_Z^{-1}\mathbf{Z})(\mathbf{Z}'\mathbf{\Gamma}_Z^{-1}\mathbf{t})]. \quad [1.2]$$

The vectors $\mathbf{t}$ and the matrix $\mathbf{\Gamma}_Z^{-1}$ are deterministic constants. By the linearity of the expectation operator, these constant matrices and vectors can be moved outside the expectation:

$$\mathbb{E}[\mathbf{t}'\mathbf{\Gamma}_Z^{-1}\mathbf{Z}\mathbf{Z}'\mathbf{\Gamma}_Z^{-1}\mathbf{t}] = \mathbf{t}'\mathbf{\Gamma}_Z^{-1}\mathbb{E}[\mathbf{Z}\mathbf{Z}']\mathbf{\Gamma}_Z^{-1}\mathbf{t}. \quad [1.3]$$

Since $\mathbf{Z}$ is a centered random vector $\mathbf{Z}$, its covariance matrix is defined as:

$$\text{Cov}(\mathbf{Z}) = \mathbb{E}[(\mathbf{Z} - \mathbb{E}[\mathbf{Z}])(\mathbf{Z} - \mathbb{E}[\mathbf{Z}])'] = \mathbb{E}[\mathbf{Z}\mathbf{Z}'].$$

By definition, the covariance matrix of $\mathbf{Z}$ is $\mathbf{\Gamma}_Z$. Therefore, we have $\mathbb{E}[\mathbf{Z}\mathbf{Z}'] = \mathbf{\Gamma}_Z$.

Substitute this back into the expression [1.3] by [1.2], we get:

$$\mathbb{E}[(\mathbf{t}'\mathbf{\Gamma}_Z^{-1}\mathbf{Z})^2] = \mathbf{t}^T\mathbf{\Gamma}_Z^{-1}\mathbb{E}[\mathbf{Z}\mathbf{Z}']\mathbf{\Gamma}_Z^{-1}\mathbf{t} = \mathbf{t}'\mathbf{\Gamma}_Z^{-1}\mathbf{\Gamma}_Z\mathbf{\Gamma}_Z^{-1}\mathbf{t}.$$

Since $\mathbf{\Gamma}_Z^{-1}\mathbf{\Gamma}_Z = I_n$ (the identity matrix of dimension n), the expression simplifies to:

$$\mathbb{E}[(\mathbf{t}^T\mathbf{\Gamma}_Z^{-1}\mathbf{Z})^2] = \mathbf{t}^T\mathbf{\Gamma}_Z^{-1}\mathbf{t}. \qquad \square$$

To further characterize Gaussian vectors, we next derive the expected value of a quadratic form.

LEMMA 1.9.– *Let $X \sim \mathcal{N}_n(\mathbf{0}, K_X)$ be an n-dimensional zero-mean multivariate Gaussian vector with covariance matrix K_X. Let $\mathbf{A}$ be a constant $n \times n$ matrix. Then, the expected value of the quadratic form $X'\mathbf{A}X$ is given by:*

$$\mathbb{E}[X'\mathbf{A}X] = Tr(\mathbf{A}K_X).$$

Proof. Let $X = (X_1, \dots, X_n)'$ be the Gaussian vector and $\mathbf{A}$ be an $n \times n$ matrix with elements A_{ij}. The quadratic form $X'\mathbf{A}X$ can be written in summation notation as:

$$X'\mathbf{A}X = \sum_{i=1}^{n}\sum_{j=1}^{n} A_{ij}X_iX_j.$$

Taking the expected value of both sides and using the linearity of expectation, we get:

$$\mathbb{E}[X'\mathbf{A}X] = \mathbb{E}\left[\sum_{i=1}^{n}\sum_{j=1}^{n} A_{ij}X_iX_j\right] = \sum_{i=1}^{n}\sum_{j=1}^{n} A_{ij}\mathbb{E}[X_iX_j].$$

Since X is a zero-mean vector, we have $\mathbb{E}[X_iX_j] = \text{Cov}(X_i, X_j)$. By definition, $\text{Cov}(X_i, X_j)$ is the (i, j)th element of the covariance matrix K_X, denoted as $(K_X)_{ij}$. Substituting this into the expectation equation, we have:

$$\mathbb{E}[X'\mathbf{A}X] = \sum_{i=1}^{n}\sum_{j=1}^{n} A_{ij}(K_X)_{ij} \quad (*)$$

Now, let us consider the trace of the product $\mathbf{A}K_X$. Let $\mathbf{C} = \mathbf{A}K_X$. The elements of $\mathbf{C}$ are $C_{ik} = \sum_{p=1}^{n} A_{ip}(K_X)_{pk}$. The trace of $\mathbf{C}$ is the sum of its diagonal elements:

$$\text{Tr}(\mathbf{A}K_X) = \sum_{i=1}^{n} C_{ii} = \sum_{i=1}^{n}\sum_{p=1}^{n} A_{ip}(K_X)_{pi}.$$

Since the covariance matrix K_X is symmetric (i.e. $(K_X)_{pi} = (K_X)_{ip}$), we can substitute this property into the trace sum:

$$\text{Tr}(\mathbf{A}K_X) = \sum_{i=1}^{n}\sum_{p=1}^{n} A_{ip}(K_X)_{ip}.$$

By simply re-indexing the dummy summation variable p to j, we obtain:

$$\text{Tr}(\mathbf{A}K_X) = \sum_{i=1}^{n}\sum_{j=1}^{n} A_{ij}(K_X)_{ij} \quad (**)$$

Comparing equations $(*)$ and $(**)$, we see that:

$$\mathbb{E}[X'\mathbf{A}X] = \text{Tr}(\mathbf{A}K_X).$$

This completes the proof. □

A direct application of the previous lemma yields a particularly insightful result.

COROLLARY 1.3.– *If* $\mathbf{Z} \sim \mathcal{N}_n(\mathbf{0}, \Gamma)$ *is an* n*-dimensional zero-mean Gaussian vector with an invertible covariance matrix* Γ*, then:*

$$\mathbb{E}\left[\mathbf{Z}'\Gamma^{-1}\mathbf{Z}\right] = n.$$

Proof. This result is a direct application of Lemma 1.9. We identify the terms from the corollary with those in the lemma: let $X = \mathbf{Z}$, the covariance matrix $K_X = \mathbf{\Gamma}$, and the constant matrix $\mathbf{A} = \mathbf{\Gamma}^{-1}$. Since $\mathbf{Z}$ is a zero-mean Gaussian vector, we can apply Lemma 1.9:

$$\mathbb{E}[\mathbf{Z}'\mathbf{\Gamma}^{-1}\mathbf{Z}] = \text{Tr}(\mathbf{\Gamma}^{-1}\mathbf{\Gamma}) = \text{Tr}(I_n),$$

where I_n is the $n \times n$ identity matrix. The trace of the identity matrix is the sum of its diagonal elements, which are all 1s, therefore $\text{Tr}(I_n) = n$. Thus, we conclude:

$$\mathbb{E}[\mathbf{Z}'\mathbf{\Gamma}^{-1}\mathbf{Z}] = n.$$

□

A particularly important result in the analysis of Gaussian random vectors, completing their characterization through moments, concerns the variance of their quadratic forms.

LEMMA 1.10.– *Let* $X \sim \mathcal{N}_n(\mathbf{0}, K_X)$ *be an* n*-dimensional zero-mean multivariate Gaussian vector with covariance matrix* K_X*. Let* $\mathbf{A}$ *be a constant* $n \times n$ *matrix. Then, the variance of the quadratic form* $X'\mathbf{A}X$ *is given by:*

$$\text{Var}(X'\mathbf{A}X) = 2\,Tr((\mathbf{A}K_X)^2).$$

Proof. The proof for this result is left as an exercise for the reader and can be found with guiding hints in Exercise 1.5. □

Building on the results for the variance and expectation of quadratic forms, we can derive the expected value of the square of a specific quadratic form.

COROLLARY 1.4.– *If* $\mathbf{Z} \sim \mathcal{N}_n(\mathbf{0}, \Gamma)$ *is a zero-mean Gaussian vector with an invertible covariance matrix* Γ*, then:*

$$\mathbb{E}\left[(\mathbf{Z}'\Gamma^{-1}\mathbf{Z})^2\right] = n^2 + 2n.$$

Proof. Let $Q = \mathbf{Z}'\Gamma^{-1}\mathbf{Z}$. We have $\mathbb{E}[Q^2] = \mathrm{Var}(Q) + (\mathbb{E}[Q])^2$. From Corollary 1.3, we directly have:

$$\mathbb{E}[Q] = \mathbb{E}[\mathbf{Z}'\Gamma^{-1}\mathbf{Z}] = n.$$

From Lemma 1.10,

$$\mathrm{Var}(Q) = \mathrm{Var}(\mathbf{Z}'\Gamma^{-1}\mathbf{Z}) = 2\mathrm{Tr}((\Gamma^{-1}\Gamma)^2) = 2\,\mathrm{Tr}(I_n) = 2n.$$

Therefore,

$$\mathbb{E}[Q^2] = n^2 + 2n. \qquad \square$$

1.3. Gaussian processes

Having established the fundamental properties of Gaussian random variables and vectors, we now extend these concepts to the realm of *stochastic processes*. This extension leads us to Gaussian processes, a cornerstone of statistical modeling with wide-ranging applications.

1.3.1. *Generalities on stochastic processes*

Before diving into Gaussian processes specifically, let us set the stage with some essential definitions for stochastic processes. A stochastic process can be thought of as a collection of random variables indexed by some set, often representing time or space.

DEFINITION 1.4.– *Consider a non-empty set* $\mathbb{T}$*, a probability space* $(\Omega, \mathcal{F}, \mathbb{P})$*, and a function* $X : \mathbb{T}\times\Omega \to \mathbb{R}^d$*, where* $d \geq 1$*. If, for any fixed* $t \in \mathbb{T}$*,* $X(t, \cdot)$ *is an* $\mathbb{R}^d$*-valued random variable on* $(\Omega, \mathcal{F}, \mathbb{P})$*, then* X *is called a stochastic process with values in* $\mathbb{R}^d$*. We commonly denote a stochastic process as* $X = (X_t)_{t\in\mathbb{T}}$*, where* $X_t(\omega) = X(t, \omega)$ *for* $\omega \in \Omega$ *and* $t \in \mathbb{T}$*.*

This fundamental definition comes with key associated concepts:

– For a fixed $\omega \in \Omega$, the function $X(\cdot, \omega)$ is termed a *trajectory* or a *sample path* of the process X. This represents a single realization of the process over its entire index set $\mathbb{T}$.

– The *mean function* of X is defined by $m : t \in \mathbb{T} \longmapsto \mathbb{E}[X_t]$. This function exists if, for all $t \in \mathbb{T}$, every component of X_t is integrable (i.e. $\mathbb{E}[|X_{t,i}|] < \infty$ for all $i = 1, \ldots, d$). This function describes the average behavior of the process at each point $t \in \mathbb{T}$.

– The *covariance function* of X is defined by:

$$C : (s,t) \in \mathbb{T}^2 \longmapsto C(s,t) = \mathrm{Cov}(X_s, X_t).$$

This function exists if, for all $t \in \mathbb{T}$, every component of X_t is square-integrable (i.e. $\mathbb{E}[|X_{t,i}|^2] < \infty$ for all $i = 1, \ldots, d$). $C(s,t)$ is a $d \times d$ matrix that quantifies the linear dependence between the process values at any two points s and t in the index set.

We now introduce the formal concept of the finite-dimensional distributions.

DEFINITION 1.5.– *The finite-dimensional distributions (f.d.d.'s) of a stochastic process $X = (X_t)_{t\in\mathbb{T}}$ are the joint probability distributions of the random vectors $(X_{t_1}, \ldots, X_{t_n})$, for every choice of $n \geq 1$ and any set of indices $(t_1, \ldots, t_n) \in \mathbb{T}^n$. These distributions are crucial as they capture the probabilistic relationships among any finite number of points in the process*[1].

When comparing different stochastic processes, we use specific terms to describe their relationships, focusing on how their sample paths or distributions align.

DEFINITION 1.6 (relationships between stochastic processes).– *Consider two stochastic processes $X = (X_t)_{t\in\mathbb{T}}$ and $Y = (Y_t)_{t\in\mathbb{T}}$.*

– The processes X and Y are said to be indistinguishable if:

$$\mathbb{P}(\{\omega \in \Omega : X_t(\omega) = Y_t(\omega), \forall t \in \mathbb{T}\}) = 1.$$

In simpler terms, X and Y are indistinguishable if they share the exact same sample paths almost surely. This is the strongest form of equivalence.

– The processes X and Y are said to be modifications if for every $t \in \mathbb{T}$:

$$\mathbb{P}(\{\omega \in \Omega : X_t(\omega) = Y_t(\omega)\}) = 1.$$

This means that for any specific time t, the random variables X_t and Y_t are almost surely equal. While indistinguishable processes are always modifications, the reverse is not necessarily true (e.g. if paths can differ on a set of measure zero that varies with t).

---

1 While the finite-dimensional distributions are essential for defining the process's existence, they are generally insufficient to uniquely characterize properties of the sample paths, such as continuity or differentiability.

– The processes X and Y are said to be versions if they have the same finite-dimensional distributions. This implies that for any $n \geq 1$ and any $(t_1, \dots, t_n) \in \mathbb{T}^n$, the random vectors $(X_{t_1}, \dots, X_{t_n})$ and $(Y_{t_1}, \dots, Y_{t_n})$ have the same joint distribution.

– The processes X and Y are said to be independent if for any positive integers $m, n \geq 1$, any finite set of time points $s_1, \dots, s_m \in \mathbb{T}$ for X, and any finite set of time points $t_1, \dots, t_n \in \mathbb{T}$ for Y, the random vector $(X_{s_1}, \dots, X_{s_m})$ is stochastically independent of the random vector $(Y_{t_1}, \dots, Y_{t_n})$.

– The process $Y = (Y_t)_{t\in\mathbb{T}}$ is an independent copy of X if it possesses the same finite-dimensional distributions as X and is stochastically independent of X. This means that for any finite collection of random variables from X (e.g. $X_{s_1}, \dots, X_{s_m}$) and any finite collection of random variables from Y (e.g. $Y_{t_1}, \dots, Y_{t_n}$), these two collections are jointly independent.

To better understand the nuances of these definitions, let us establish the precise relationships between the three types of process equivalences.

PROPOSITION 1.2.– *Let X and Y be two stochastic processes defined on the same probability space $(\Omega, \mathcal{F}, \mathbb{P})$ with index set $\mathbb{T}$. The following implications hold:*

1) If X and Y are indistinguishable, then they are modifications.

2) If X and Y are modifications, then they are versions.

In summary, we have the hierarchy:

$$\textit{Indistinguishable} \implies \textit{Modifications} \implies \textit{Versions}$$

The reverse implications do not hold in general.

Proof. The proof of this proposition directly follows from the definitions and can be found in standard stochastic process literature (see, e.g. Karatzas and Shreve (1998)). The fact that the reverse implications do not hold in general is demonstrated by the counterexamples provided in Exercises 1.8 and 1.9. □

However, a significant special case arises when processes possess continuous sample paths, allowing us to establish a stronger equivalence.

PROPOSITION 1.3.– *Let $X = (X_t)_{t\in\mathbb{T}}$ and $Y = (Y_t)_{t\in\mathbb{T}}$ be two stochastic processes, where $\mathbb{T}$ is an interval of $\mathbb{R}$ (e.g. $[0, \infty)$ or $\mathbb{R}$). If X and Y are modifications of each other and both processes have almost surely continuous sample paths, then they are indistinguishable.*

Proof. Let $X = (X_t)_{t\in\mathbb{T}}$ and $Y = (Y_t)_{t\in\mathbb{T}}$ be two stochastic processes defined on the same probability space $(\Omega, \mathcal{F}, \mathbb{P})$.

Since X and Y are modifications of each other, by definition, for every $t \in \mathbb{T}$, we have $\mathbb{P}(X_t = Y_t) = 1$.

Let $Q = \mathbb{Q} \cap \mathbb{T}$ be the countable set of rational numbers within the index set $\mathbb{T}$. For each $q \in Q$, we know $\mathbb{P}(X_q = Y_q) = 1$. Consider the event:

$$A_q = \{\omega \in \Omega : X_q(\omega) \neq Y_q(\omega)\}.$$

We have $\mathbb{P}(A_q) = 0$ for all $q \in Q$. Now consider the event $A = \bigcup_{q \in Q} A_q$, which represents the event that $X_q \neq Y_q$ for at least one rational q. By the subadditivity of probability for a countable union, we have:

$$\mathbb{P}(A) = \mathbb{P}\left(\bigcup_{q \in Q} A_q\right) \leq \sum_{q \in Q} \mathbb{P}(A_q) = \sum_{q \in Q} 0 = 0.$$

Therefore, $\mathbb{P}(A) = 0$. This implies that:

$$\mathbb{P}(A^c) = \mathbb{P}(\{\omega \in \Omega : X_q(\omega) = Y_q(\omega), \forall q \in Q\}) = 1.$$

Let $\Omega_0 = A^c$.

Next, both processes X and Y have almost surely continuous sample paths. This means there exists a set $\Omega_1 \subseteq \Omega$ with $\mathbb{P}(\Omega_1) = 1$ such that for every $\omega \in \Omega_1$, both $t \mapsto X_t(\omega)$ and $t \mapsto Y_t(\omega)$ are continuous functions on $\mathbb{T}$.

Let $\Omega^* = \Omega_0 \cap \Omega_1$. Since $\mathbb{P}(\Omega_0) = 1$ and $\mathbb{P}(\Omega_1) = 1$, it follows that $\mathbb{P}(\Omega^*) = 1$. Now, consider any $\omega \in \Omega^*$. For this ω: on the one hand, the functions $t \mapsto X_t(\omega)$ and $t \mapsto Y_t(\omega)$ are continuous on $\mathbb{T}$ (because $\omega \in \Omega_1$). On the other hand, $X_q(\omega) = Y_q(\omega)$ for all $q \in Q = \mathbb{Q} \cap \mathbb{T}$ (because $\omega \in \Omega_0$).

Since the set of rational numbers Q is dense in $\mathbb{T}$, and continuous functions that agree on a dense set must agree everywhere, we can conclude that for every $\omega \in \Omega^*$, $X_t(\omega) = Y_t(\omega)$ for all $t \in \mathbb{T}$.

Since $\mathbb{P}(\Omega^*) = 1$, this means that the event $\{\omega \in \Omega : X_t(\omega) = Y_t(\omega), \forall t \in \mathbb{T}\}$ has probability 1. By definition, this means X and Y are indistinguishable. □

Beyond their basic definitions, some stochastic processes exhibit particular temporal dependencies, which are vital for modeling dynamic systems.

DEFINITION 1.7.– *Consider* $X = (X_t)_{t \in \mathbb{T}}$ *a stochastic process, where* $\mathbb{T}$ *denotes a non-empty sub-interval of* $\mathbb{R}$ *(often representing time).*

– The process X has stationary increments if for any $n \geq 1$, any sequence of time points $t_1, \dots, t_n \in \mathbb{T}$, and any time shift $h \in \mathbb{R}$ such that $t_i + h \in \mathbb{T}$ for all $i = 1, \dots, n$, the joint distribution of the increments:

$$(X_{t_2} - X_{t_1}, \dots, X_{t_n} - X_{t_{n-1}})$$

is the same as the joint distribution of the shifted increments:

$$(X_{t_2+h} - X_{t_1+h}, \dots, X_{t_n+h} - X_{t_{n-1}+h}).$$

This means that the joint distribution of the increments is invariant under time shifts.

– The process X has independent increments if for any $n \geq 1$ and any sequence of ordered time points $t_1 < t_2 < \cdots < t_n$ in $\mathbb{T}$, the random variables $X_{t_2} - X_{t_1}$, $X_{t_3} - X_{t_2}, \dots, X_{t_n} - X_{t_{n-1}}$ are mutually independent. This implies that changes over non-overlapping time intervals are independent.

– The process X is said to be H-self-similar (or to possess H-self-similarity) for some constant $H > 0$, if for any $a > 0$, the process $(X_{at})_{t\in\mathbb{T}}$ has the same finite-dimensional distributions as $(a^H X_t)_{t\in\mathbb{T}}$. Formally, this means:

$$(X_{at})_{t\in\mathbb{T}} \stackrel{d}{=} (a^H X_t)_{t\in\mathbb{T}}.$$

This property implies that if we rescale time by a factor a, the process itself scales by a factor of a^H, maintaining its statistical properties.

1.3.2. *Gaussian processes: definition and main properties*

With the general concepts of stochastic processes in place, we can now precisely define Gaussian processes, which are characterized by the Gaussian nature of all their finite-dimensional distributions.

Consider a non-empty set $\mathbb{T}$, which serves as the index set for our process (e.g. time and spatial coordinates).

DEFINITION 1.8.– *A stochastic process $X = (X_t)_{t\in\mathbb{T}}$ is called a Gaussian process if for every $n \in \mathbb{N}$ and any finite collection of indices $(t_1, t_2, \dots, t_n) \in \mathbb{T}^n$, the corresponding random vector $(X_{t_1}, X_{t_2}, \dots, X_{t_n})$ is a Gaussian vector.*

Equivalently, X is a Gaussian process if, for every $n \in \mathbb{N}$, any $(t_1, t_2, \dots, t_n) \in \mathbb{T}^n$, and any coefficients $(a_1, a_2, \dots, a_n) \in \mathbb{R}^n$, the linear combination $a_1 X_{t_1} + a_2 X_{t_2} + \cdots + a_n X_{t_n}$ is a Gaussian random variable. This definition directly leverages our prior understanding of Gaussian vectors and their linear transformations.

REMARK 1.3.– *A standout feature of Gaussian processes, stemming directly from the characteristics of Gaussian vectors (see, e.g., Lemma 1.4), is that their entire probability law is uniquely and completely determined by just two functions:*

– *its mean function* $m : t \in \mathbb{T} \longmapsto \mathbb{E}[X_t]$;

– *its covariance function* $C : (s,t) \in \mathbb{T}^2 \longmapsto \mathrm{Cov}(X_s, X_t)$.

This unique characterization of Gaussian processes by their mean and covariance functions, as highlighted in the preceding remark, profoundly simplifies their analysis and specification. This fundamental property not only facilitates their application in modeling complex phenomena but also implies that crucial stochastic attributes, such as self-similarity and independence, are fully determined by these two functions. The precise conditions governing these properties are formally presented in the following proposition.

PROPOSITION 1.4.– *Let* $X = (X_t)_{t\in\mathbb{T}}$ *and* $Y = (Y_t)_{t\in\mathbb{T}}$ *be two Gaussian processes, with respective mean functions* $\mu_X(t) = \mathbb{E}[X_t]$ *and* $\mu_Y(t) = \mathbb{E}[Y_t]$, *and covariance functions* $C_X(s,t) = Cov(X_s, X_t)$ *and* $C_Y(s,t) = Cov(Y_s, Y_t)$.

1) A Gaussian process X *is H-self-similar (as defined in Definition 1.7) if and only if its mean function satisfies* $\mu_X(at) = a^H \mu_X(t)$ *and its covariance function satisfies:*

$$C_X(as, at) = a^{2H} C_X(s,t)$$

for all $s,t \in \mathbb{T}$ *and* $a > 0$. *In particular, for a centered Gaussian process (i.e.* $\mu_X(t) = 0$ *for all* t*), it is H-self-similar if and only if its covariance function* $C_X(s,t) = \mathbb{E}[X_s X_t]$ *satisfies* $C_X(as, at) = a^{2H} C_X(s,t)$.

2) The Gaussian processes X *and* Y *are independent if and only if their cross-covariance function is identically zero. That is, for all* $s,t \in \mathbb{T}$*:*

$$Cov(X_s, Y_t) = \mathbb{E}[(X_s - \mu_X(s))(Y_t - \mu_Y(t))] = 0.$$

If X *and* Y *are centered Gaussian processes (i.e.* $\mu_X(t) = 0$ *and* $\mu_Y(t) = 0$ *for all* t*), this condition simplifies to* $\mathbb{E}[X_s Y_t] = 0$ *for all* $s,t \in \mathbb{T}$.

A particularly useful property for constructing and manipulating Gaussian processes is their closure under linear combinations.

PROPOSITION 1.5.– *If* $X = (X_t)_{t\in\mathbb{T}}$ *and* $Y = (Y_t)_{t\in\mathbb{T}}$ *are two independent Gaussian processes, then any linear combination* $Z = aX + bY = (aX_t + bY_t)_{t\in\mathbb{T}}$ *(for* $a, b \in \mathbb{R}$*) is also a Gaussian process.*

Proof. To show that Z is a Gaussian process, we need to demonstrate that any finite linear combination of its components is a Gaussian random variable. Consider an arbitrary linear combination $\sum_{i=1}^{n} c_i Z_{t_i}$ for $c_i \in \mathbb{R}$ and $t_i \in \mathbb{T}$. Substituting the definition of Z_{t_i}:

$$\sum_{i=1}^{n} c_i Z_{t_i} = \sum_{i=1}^{n} c_i (aX_{t_i} + bY_{t_i}) = a \sum_{i=1}^{n} c_i X_{t_i} + b \sum_{i=1}^{n} c_i Y_{t_i}.$$

Let $U = a\sum_{i=1}^{n} c_i X_{t_i}$ and $V = b\sum_{i=1}^{n} c_i Y_{t_i}$. Since X is a Gaussian process, the linear combination $\sum_{i=1}^{n} c_i X_{t_i}$ is a Gaussian random variable. Consequently, U, which is a scalar multiple of a Gaussian random variable, is also Gaussian. Similarly, since Y is a Gaussian process, the linear combination $\sum_{i=1}^{n} c_i Y_{t_i}$ is a Gaussian random variable. Therefore, V is also Gaussian. Furthermore, because X and Y are independent processes, the random variables U and V are independent Gaussian random variables. A fundamental property of Gaussian distributions is that the sum of two independent Gaussian random variables is itself a Gaussian random variable (see Lemma 1.1). Thus, $U + V = \sum_{i=1}^{n} c_i Z_{t_i}$ is a Gaussian random variable, which completes the proof. □

1.3.3. *Existence of gaussian processes with prescribed covariance function*

Having defined Gaussian processes as collections of random variables whose finite-dimensional distributions are Gaussian, a natural and critical question arises: *Under what conditions can such a process actually exist* for a given mean and covariance function? This question is answered by a fundamental result rooted in the broader theory of stochastic processes.

First, we need a specific property for functions that can serve as covariance functions.

DEFINITION 1.9.– *A real-valued function $f : \mathbb{T} \times \mathbb{T} \to \mathbb{R}$ is called positive semi-definite if for any $n \in \mathbb{N} \setminus \{0\}$, and for any choice of $t_1, t_2, \ldots, t_n \in \mathbb{T}$ and $a_1, a_2, \ldots, a_n \in \mathbb{R}$, the following inequality holds:*

$$\sum_{j,k=1}^{n} a_j a_k f(t_j, t_k) \geq 0.$$

The existence of a Gaussian process with a specified mean and covariance function is guaranteed by a specialized version of the celebrated *Kolmogorov existence theorem*. While the general Kolmogorov theorem ensures the existence of a stochastic process, given its finite-dimensional distributions, for Gaussian processes, this simplifies considerably. As we have established, the entire probabilistic structure of a Gaussian process is uniquely determined by its mean and covariance functions. This leads to the following powerful result:

THEOREM 1.2 (Kolmogorov existence theorem for Gaussian processes).– *Let $\mathbb{T}$ be any non-empty set. Let $m : \mathbb{T} \to \mathbb{R}$ be any real-valued function (serving as the mean function). Let $C : \mathbb{T} \times \mathbb{T} \to \mathbb{R}$ be a real-valued function (serving as the covariance function) such that: C is symmetric: $C(s,t) = C(t,s)$ for all $s,t \in \mathbb{T}$, and C is positive semi-definite. Then, there exists a Gaussian process $X = (X_t)_{t\in\mathbb{T}}$ with mean function $m(t)$ and covariance function $C(s,t)$.*

Proof. The proof of this fundamental theorem, which relies on constructing the appropriate probability space and random variables, can be found in standard texts on stochastic processes, such as Dudley (2002) or Mishura and Shevchenko (2017). □

To illustrate the concept of a positive semi-definite function and its direct application in defining a Gaussian process, let us consider a classic example:

EXAMPLE 1.1.– *Consider the case where the index set is $\mathbb{T} = \mathbb{R}^+ = [0,\infty)$, and let the candidate covariance function be $f(s,t) = s \wedge t := \min(s,t);\ \ s,t \in \mathbb{R}^+$. We will demonstrate that this function f is indeed positive semi-definite.*

The key insight is to express $s \wedge t$ as an inner product of functions in an L_2 space. For any $s,t \in \mathbb{R}^+$ and a sufficiently large $T \geq s \vee t$ (where $s \vee t := \max(s,t)$), we can write:

$$s \wedge t = \int_0^T \mathbf{1}_{[0,s]}(u)\mathbf{1}_{[0,t]}(u)\,du = \left\langle \mathbf{1}_{[0,s]}, \mathbf{1}_{[0,t]} \right\rangle_{L_2([0,T],\lambda_1)},$$

where λ_1 is the Lebesgue measure on $\mathbb{R}$, and $\langle\cdot,\cdot\rangle_{L_2([0,T],\lambda_1)}$ denotes the standard scalar product in the Hilbert space $L_2([0,T],\lambda_1)$ (the space of square-integrable functions on $[0,T]$). Here, $\mathbf{1}_{[0,x]}$ is the indicator function, which is 1 on $[0,x]$ and 0 otherwise.

Now, for any $n \in \mathbb{N}$, any real coefficients $a_1,\dots,a_n \in \mathbb{R}$, and any ordered time points $0 \leq t_1 \leq \cdots \leq t_n$, choosing $T \geq t_n$, we can examine the sum:

$$\begin{aligned}\sum_{j,k=1}^{n} a_j a_k f(t_j,t_k) &= \sum_{j,k=1}^{n} a_j a_k \left\langle \mathbf{1}_{[0,t_j]}, \mathbf{1}_{[0,t_k]} \right\rangle_{L_2([0,T],\lambda_1)} \\ &= \left\langle \sum_{j=1}^{n} a_j \mathbf{1}_{[0,t_j]}, \sum_{k=1}^{n} a_k \mathbf{1}_{[0,t_k]} \right\rangle_{L_2([0,T],\lambda_1)} \\ &= \left\| \sum_{j=1}^{n} a_j \mathbf{1}_{[0,t_j]} \right\|^2_{L_2([0,T],\lambda_1)} \\ &\geq 0.\end{aligned}$$

The second equality follows from the bilinearity of the inner product, and the third from the definition of the squared norm in an L_2 space. Since the squared norm of *any function is always non-negative, the condition for positive semi-definiteness is satisfied.*

This example provides the foundation for defining one of the most important and widely studied stochastic processes: the Wiener process.

DEFINITION 1.10.– *A stochastic process $(W_t)_{t\geq 0}$ is called a Wiener process (or often, Brownian motion) if it is a Gaussian process with:*

– *a zero mean function: $m(t) := \mathbb{E}[W_t] = 0$ for all $t \geq 0$;*

– *a covariance function given by: $C(s,t) := s \wedge t = \min(s,t)$ for all $s, t \geq 0$.*

REMARK 1.4.– *It follows directly from Example 1.1, where we proved the positive semi-definiteness of $f(s,t) = s \wedge t$, and Theorem 1.2, that Definition 1.10 correctly establishes the existence of such a Gaussian process, ensuring that the Wiener process is a well-defined mathematical object.*

The Wiener process has several crucial properties that make it invaluable in modeling random phenomena.

PROPOSITION 1.6 (properties of the Wiener process).– *A Wiener process $(W_t)_{t\geq 0}$ possesses the following fundamental properties:*

1) $W_0 = 0$ almost surely.

2) For any $0 \leq t_1 < t_2 < \cdots < t_n$, the increments $W_{t_2} - W_{t_1}$, $W_{t_3} - W_{t_2}, \ldots, W_{t_n} - W_{t_{n-1}}$ are mutually independent random variables.

3) For any $0 \leq s < t$, the increment $W_t - W_s$ is a Gaussian random variable with mean 0 and variance $t - s$, that is, $W_t - W_s \sim \mathcal{N}(0, t-s)$.

While the definitions above guarantee the existence of Gaussian processes based on their mean and covariance functions, they do not inherently tell us about the smoothness or continuity of their sample paths. For many applications, particularly in stochastic calculus, the regularity of these paths (trajectories) is paramount. This is where the powerful Kolmogorov continuity theorem comes into play.

THEOREM 1.3 (Kolmogorov continuity theorem).– *Let $X = \{X_t, t \in [0,T]\}$ be a stochastic process defined on a compact interval $[0,T]$.*

1) If there exist strictly positive constants K, α and β such that for all $0 \leq s < t \leq T$:

$$\mathbb{E}|X_t - X_s|^{\alpha} \leq K|t-s|^{1+\beta}, \qquad [1.4]$$

then X possesses a continuous modification. That is, there exists a process $Y = (Y_t)_{t\in[0,T]}$ such that Y is a modification of X (i.e. $P(X_t = Y_t) = 1$ for all t) and almost all of its sample paths $t \mapsto Y_t(\omega)$ are continuous functions.

2) Furthermore, if stochastic process $X = \{X_t, t \in [0,T]\}$ *satisfies condition* [1.4], *then it possesses a modification whose trajectories are Hölder continuous of any order* $\gamma < \beta/\alpha$. *This implies that for almost all sample paths, there exists a constant* $C(\omega)$ *such that* $|Y_t(\omega) - Y_s(\omega)| \leq C(\omega)|t-s|^\gamma$ *for all* $s,t \in [0,T]$.

Building upon the general Kolmogorov continuity theorem, a more specialized and often more direct result exists for Gaussian processes, which are central to many stochastic models. This specialized theorem, often referred to as the Kolmogorov–Chentsov theorem for Gaussian processes, leverages their unique properties to establish sample path regularity based solely on the second moments of their increments.

THEOREM 1.4 (Kolmogorov–Chentsov theorem for Gaussian processes).– *Let* $X = \{X_t\}_{t\in[0,T]}$ *be a centered Gaussian process defined on a compact interval* $[0,T]$. *If there exist a strictly positive constant* C *and an exponent* $\nu > 0$ *such that for all* $0 \leq s < t \leq T$:

$$\mathbb{E}[(X_t - X_s)^2] \leq C|t-s|^\nu. \qquad [1.5]$$

Then, for any $\gamma \in (0, \nu/2)$, *the process* X *possesses a modification whose sample paths are almost surely* γ*-Hölder continuous. That is, for almost all sample paths, there exists a random constant* $C(\omega)$ *such that* $|Y_t(\omega) - Y_s(\omega)| \leq C(\omega)|t-s|^\gamma$ *for all* $s,t \in [0,T]$.

Proof. Let $X = \{X_t\}_{t\in[0,T]}$ be a centered Gaussian process satisfying the condition [1.5]. Since X is a Gaussian process, its increments $Z_{s,t} = X_t - X_s$ are centered Gaussian random variables. Therefore, by Lemma 1.2, for any $p \geq 1$, there exists a positive constant C_p such that:

$$\mathbb{E}[|Z|^p] = C_p(\mathbb{E}[Z^2])^{p/2}.$$

Applying this to the increment $Z_{s,t} = X_t - X_s$, we have:

$$\mathbb{E}[|X_t - X_s|^p] = C_p(\mathbb{E}[(X_t - X_s)^2])^{p/2}.$$

Now, substituting the given variance bound $\mathbb{E}[(X_t - X_s)^2] \leq C|t-s|^\nu$:

$$\mathbb{E}[|X_t - X_s|^p] \leq C_p(C|t-s|^\nu)^{p/2} = (C_p C^{p/2})|t-s|^{p\nu/2}.$$

This inequality holds for any $p \geq 1$.

Now, we compare this with the condition of Kolmogorov continuity Theorem 1.3:

$$\mathbb{E}|X_t - X_s|^\alpha \leq K|t-s|^{1+\beta}.$$

We choose $\alpha = p$. Then, we can identify:

– $K = C_p C^{p/2}$. Since $C > 0$ and $C_p > 0$ for any $p \geq 1$, we have $K > 0$.

– $1 + \beta = p\nu/2$. This implies $\beta = p\nu/2 - 1$.

For Kolmogorov's Theorem 1.3 to guarantee Hölder continuity (Part 2), it requires the constant β to be strictly positive. The condition $\beta > 0$ translates to $p\nu/2 - 1 > 0$, or $p\nu/2 > 1$, which means $p > 2/\nu$. Since $\nu > 0$, we can always choose an integer p sufficiently large to satisfy this condition (e.g. take $p = \lfloor 2/\nu \rfloor + 1$, or simply choose a large integer).

With $\alpha = p$ and $\beta = p\nu/2 - 1$, Kolmogorov's Theorem 1.3 states that X possesses a modification whose trajectories are Hölder continuous of any order $\gamma < \beta/\alpha$. Substituting our values for α and β:

$$\gamma < \frac{p\nu/2 - 1}{p} = \frac{\nu}{2} - \frac{1}{p}.$$

This holds for any p satisfying $p > 2/\nu$. To obtain the strongest possible Hölder exponent, we can let p be arbitrarily large. As $p \to \infty$, the term $1/p$ approaches 0. Therefore, for any $\gamma \in (0, \nu/2)$, we can always choose a sufficiently large integer p such that $\gamma < \frac{\nu}{2} - \frac{1}{p}$, thus satisfying the conditions of the theorem.

Hence, the Gaussian process X possesses a modification whose sample paths are almost surely γ-Hölder continuous for any $\gamma \in (0, \nu/2)$. □

REMARK 1.5.– *The Kolmogorov continuity theorem, as stated above, is commonly presented for processes indexed by a one-dimensional time interval. However, for applications involving stochastic processes on higher dimensional spatial domains, a generalized version is often required. The following theorem extends this fundamental result to processes defined on compact subsets of* $\mathbb{R}^d$.

THEOREM 1.5 (Kolmogorov continuity theorem for multi-dimensional index sets).– *Let* $X = \{X_x, x \in S\}$ *be a stochastic process defined on a compact subset* $S \subset \mathbb{R}^d$.

1) If there exist strictly positive constants K, $p \geq 1$, *and* $\epsilon > 0$ *such that for all* $x, y \in S$:

$$\mathbb{E}[|X_x - X_y|^p] \leq K|x - y|^{d+\epsilon}, \qquad [1.6]$$

then X *possesses a continuous modification. That is, there exists a process* $Y = \{Y_x, x \in S\}$ *such that* Y *is a modification of* X *(i.e.* $P(X_x = Y_x) = 1$ *for all* $x \in S$*) and almost all of its sample paths* $x \mapsto Y_x(\omega)$ *are continuous functions on* S.

2) Furthermore, if the stochastic process $X = \{X_x, x \in S\}$ *satisfies the condition* [1.6], *then it possesses a modification whose trajectories are Hölder continuous of any order* γ *such that* $0 < \gamma < \epsilon/p$. *This implies that for almost all sample paths, there exists a constant* $C(\omega)$ *such that* $|Y_x(\omega) - Y_y(\omega)| \leq C(\omega)|x - y|^\gamma$ *for all* $x, y \in S$.

1.3.4. *Nondegenerate Gaussian processes*

Among the diverse family of Gaussian processes, nondegenerate Gaussian processes form a crucial subclass. These processes are not only fully characterized by their mean and covariance functions but also possess an additional, desirable property: their finite-dimensional distributions are well-behaved, preventing perfect linear dependencies between random variables at distinct time points.

This "nicely behaved" property is formally captured by the invertibility of their covariance matrices.

DEFINITION 1.11.– *A Gaussian process* $\{X_t\}_{t\in\mathbb{T}}$ *is defined as nondegenerate if, for every finite set of distinct time points* $t_1, t_2, \ldots, t_n \in \mathbb{T}$ *(for any* $n \geq 1$*), the corresponding covariance matrix* $\Sigma = (Cov(X_{t_i}, X_{t_j}))_{1\leq i,j\leq n}$ *is invertible.*

This definition implies that for nondegenerate Gaussian processes, all of their finite-dimensional distributions are *non-singular,* meaning their PDFs exist and are well defined. In practical terms, this means there are no perfect linear dependencies among the random variables $X_{t_1}, X_{t_2}, \ldots, X_{t_n}$ for any distinct set of time points.

Such dependencies would indicate that one variable could be perfectly predicted from a linear combination of others, which is often an undesirable characteristic in modeling.

The following lemma establishes an important equivalence, showing that the concept of a nondegenerate Gaussian process is intrinsically linked to the positive definiteness of its covariance matrices.

LEMMA 1.11.– *A Gaussian process* $\{X_t\}_{t\in\mathbb{T}}$ *is nondegenerate if, and only if, for every finite set of distinct time points* $t_1, t_2, \ldots, t_n \in \mathbb{T}$ *(for any* $n \geq 1$*), the corresponding covariance matrix* $\Sigma = (Cov(X_{t_i}, X_{t_j}))_{1\leq i,j\leq n}$ *is strictly positive definite. That is, for any non-zero vector* $\mathbf{a} \in \mathbb{R}^n$*, the quadratic form* $\mathbf{a}'\Sigma\mathbf{a} > 0$.

Proof. Let $\Sigma = (\text{Cov}(X_{t_i}, X_{t_j}))_{1\leq i,j\leq n}$ be the $n \times n$ covariance matrix.

Assume that the process X is non-degenrate; that is the covariance matrix Σ is invertible. By definition of matrix invertibility, this means that Σ has a trivial null space; in other words, if $\Sigma\mathbf{a} = \mathbf{0}$ for some vector $\mathbf{a} \in \mathbb{R}^n$, then it must necessarily be that $\mathbf{a}$ is the zero vector ($\mathbf{a} = \mathbf{0}$).

We also know that by its very definition, Σ is symmetric, that is $\Sigma_{ij} = \Sigma_{ji}$ for all i, j, and positive semi-definite, that is for any vector $\mathbf{a} \in \mathbb{R}^n$, the quadratic form $\mathbf{a}'\Sigma\mathbf{a} \geq 0$. This holds because

$$\mathbf{a}'\Sigma\mathbf{a} = \text{Var}\left(\sum_{i=1}^{n} a_i X_{t_i}\right),$$

and variance is inherently nonnegative.

Now, let us proceed by contradiction. Suppose that Σ is *not* strictly positive definite. This would mean that there exists at least one non-zero vector $\mathbf{a}_0 \in \mathbb{R}^n$ (i.e. $\mathbf{a}_0 \neq \mathbf{0}$) such that the quadratic form is zero:

$$\mathbf{a}_0'\Sigma\mathbf{a}_0 = 0.$$

Since Σ is a positive semi-definite matrix, a fundamental property states that if $\mathbf{x}'M\mathbf{x} = 0$ for a positive semi-definite matrix M, then it must follow that $M\mathbf{x} = \mathbf{0}$ (see Lemma A.1). Applying this property to our covariance matrix Σ (which is positive semi-definite) and the vector $\mathbf{a}_0$, we get that $\Sigma\mathbf{a}_0 = \mathbf{0}$. However, this result, $\Sigma\mathbf{a}_0 = \mathbf{0}$ with $\mathbf{a}_0 \neq \mathbf{0}$, means that $\mathbf{a}_0$ is a non-zero vector in the null space of Σ. This is precisely the definition of a singular (non-invertible) matrix. This directly contradicts our initial assumption that Σ is invertible. Therefore, our supposition that Σ is not strictly positive definite must be false. Consequently, if the Gaussian process is nondegenerate, its covariance matrix must be strictly positive definite.

Now let us assume that Σ is strictly positive definite. This means that for any non-zero vector $\mathbf{a} \neq \mathbf{0}$, $\mathbf{a}'\Sigma\mathbf{a} > 0$. Suppose, for contradiction, that Σ is *not* invertible. This implies that Σ is singular, meaning its null space is non-trivial. Thus, there exists a non-zero vector $\mathbf{a}_0 \neq \mathbf{0}$ such that $\Sigma\mathbf{a}_0 = \mathbf{0}$. Multiplying by $\mathbf{a}_0'$ from the left yields:

$$\mathbf{a}_0'\Sigma\mathbf{a}_0 = \mathbf{a}_0'\mathbf{0} = 0.$$

However, this directly contradicts our assumption that Σ is strictly positive definite (which requires $\mathbf{a}_0^T\Sigma\mathbf{a}_0 > 0$ for any non-zero $\mathbf{a}_0$). Therefore, if Σ is strictly positive definite, it must be invertible. This implies the Gaussian process is non-degenerate. □

EXAMPLE 1.2.– *The Brownian motion on* $(0, +\infty)$ *is a nondegenerate Gaussian process, because for any finite set of distinct time points* $t_1, t_2, \ldots, t_n \in (0, \infty)$*, the covariance matrix* $\Sigma = (Cov(B_{t_i}, B_{t_j}))_{1\leq i,j\leq n} = (\min(t_i, t_j))_{1\leq i,j\leq n}$ *is invertible.*

Indeed, let us assume, without loss of generality, that $0 < t_1 < t_2 < \cdots < t_n$. This gives the matrix:

$$\Sigma = \begin{pmatrix} t_1 & t_1 & t_1 & \ldots & t_1 \\ t_1 & t_2 & t_2 & \ldots & t_2 \\ t_1 & t_2 & t_3 & \ldots & t_3 \\ \vdots & \vdots & \vdots & \ddots & \vdots \\ t_1 & t_2 & t_3 & \ldots & t_n \end{pmatrix}$$

To show that Σ is invertible, we will demonstrate that its determinant is non-zero. Upon applying the sequence of elementary row operations (specifically, $R_i \leftarrow R_i - R_{i-1}$ for $i = n, n-1, \dots, 2$) to the covariance matrix Σ, we come back to the following upper triangular matrix, Σ_m:

$$\Sigma_m = \begin{pmatrix} t_1 & t_1 & t_1 & t_1 & \dots & t_1 \\ 0 & t_2 - t_1 & t_2 - t_1 & t_2 - t_1 & \dots & t_2 - t_1 \\ 0 & 0 & t_3 - t_2 & t_3 - t_2 & \dots & t_3 - t_2 \\ 0 & 0 & 0 & t_4 - t_3 & \dots & t_4 - t_3 \\ \vdots & \vdots & \vdots & \vdots & \ddots & \vdots \\ 0 & 0 & 0 & 0 & \dots & t_n - t_{n-1} \end{pmatrix}.$$

Since these row operations do not change the determinant of the matrix:

$$\det(\Sigma) = \det(\Sigma_m) = t_1 \cdot (t_2 - t_1) \cdot (t_3 - t_2) \cdot \dots \cdot (t_n - t_{n-1}) = t_1 \prod_{k=2}^{n} (t_k - t_{k-1}).$$

Given that the time points are distinct and ordered ($0 < t_1 < t_2 < \dots < t_n$), it implies that $t_1 > 0$ and all differences $(t_k - t_{k-1})$ are strictly positive, ensuring that $\det(\Sigma) > 0$. This confirms the invertibility of the covariance matrix and, consequently, that the Brownian motion on $(0, +\infty)$ is a nondegenerate Gaussian process.

1.4. Gaussian random measures and Wiener integration

1.4.1. *Gaussian random measures*

In this section, we lay the foundational concepts of Gaussian random measures, which are essential for defining various types of noise in stochastic analysis. We then introduce specific instances, including the crucial Brownian measure and its higher dimensional generalization, space–time white noise.

DEFINITION 1.12.– *A Gaussian random measure $\mathcal{M}$ on a measurable space $(S, \mathcal{S})$ is a mapping that assigns a real-valued random variable $\mathcal{M}(A)$ to every set $A \in \mathcal{S}$ such that the following two conditions hold:*

1) For any finite collection of pairwise disjoint sets $A_1, A_2, \dots, A_n \in \mathcal{S}$, the measure is additive:

$$\mathcal{M}\left(\bigcup_{i=1}^{n} A_i\right) = \sum_{i=1}^{n} \mathcal{M}(A_i) \quad \textit{almost surely (a.s.).}$$

2) For any choice of $n \geq 1$ and any collection of sets $A_1, A_2, \dots, A_n \in \mathcal{S}$, the random vector

$$(\mathcal{M}(A_1), \mathcal{M}(A_2), \dots, \mathcal{M}(A_n))$$

is a jointly Gaussian vector.

REMARK 1.6.– *The entire distribution of a Gaussian random measure is completely determined by its mean and covariance structure.*

– We will primarily focus on centered Gaussian random measures, for which the mean is zero: $\mathbb{E}[\mathcal{M}(A)] = 0$ *for all* $A \in \mathcal{S}$.

– For any two sets $A, B \in \mathcal{S}$, *the covariance function* $\Gamma(A, B)$ *completely specifies the joint distribution:*

$$\mathbb{E}[\mathcal{M}(A)\mathcal{M}(B)] = \Gamma(A, B).$$

The function $\Gamma(A, B)$ *must satisfy properties of a covariance function (symmetry and positive semi-definiteness).*

1.4.2. *Brownian measure on* $\mathbb{R}$

The concept of a Gaussian random measure provides a rigorous foundation for understanding various types of "white noise". A primary example is the Brownian measure:

DEFINITION 1.13.– *A Brownian measure* W *on* $\mathbb{R}$ *is a specific type of Gaussian random measure defined on* $(\mathbb{R}, \mathcal{B}_b(\mathbb{R}))$ *(where* $\mathcal{B}_b(\mathbb{R})$ *is the set of bounded Borel sets on* $\mathbb{R}$*) whose covariance function* $\Gamma(A, B)$ *takes the form:*

$$\mathbb{E}[W(A)W(B)] = \lambda(A \cap B),$$

where λ *denotes the Lebesgue measure on* $\mathbb{R}$.

This particular covariance structure implies two fundamental properties for a Brownian measure:

1) The variance of $W(A)$ is simply the Lebesgue measure of the set A:

$$\text{Var}(W(A)) = \mathbb{E}[W(A)^2] = \lambda(A).$$

2) If two sets A and B are disjoint (i.e. $A \cap B = \emptyset$), then $\lambda(A \cap B) = 0$, implying $\mathbb{E}[W(A)W(B)] = 0$. Since $W(A)$ and $W(B)$ are Gaussian, zero covariance implies independence. Thus, the increments of a Brownian measure over disjoint sets are independent.

In integral notation, the expression "$dW(s)$" is a shorthand, referring to integration with respect to this Brownian measure. When we write an integral like $\displaystyle\int f(s)dW(s)$, it represents a stochastic integral of the function f with respect to the Brownian measure W.

1.4.3. *Space–time white noise*

Crucial for stochastic partial differential equations, including the one rigorously studied in this book, is the concept of *space–time white noise*, which generalizes the 1D Brownian measure to a product space.

DEFINITION 1.14.– *A space–time white noise, often denoted by $W(ds, dy)$, is a specific type of Gaussian random measure defined on the product space $(\mathbb{R} \times \mathbb{R}^d, \mathcal{B}_b(\mathbb{R} \times \mathbb{R}^d))$. Its covariance function for any two bounded Borel sets $A, B \in \mathcal{B}_b(\mathbb{R} \times \mathbb{R}^d)$ is given by the Lebesgue measure of their intersection:*

$$\mathbb{E}(W(A)W(B)) = \lambda_{1+d}(A \cap B),$$

where λ_{1+d} is the $(1 + d)$-dimensional Lebesgue measure on $\mathbb{R} \times \mathbb{R}^d$.

For product sets $A = I_A \times B_A$ and $B = I_B \times B_B$, where $I_A, I_B \in \mathcal{B}_b(\mathbb{R})$ (often representing time intervals) and $B_A, B_B \in \mathcal{B}_b(\mathbb{R}^d)$ (representing spatial regions), the covariance can equivalently be written as:

$$\mathbb{E}(W(I_A \times B_A)W(I_B \times B_B)) = \lambda_1(I_A \cap I_B)\lambda_d(B_A \cap B_B),$$

where λ_1 and λ_d are the Lebesgue measures on $\mathbb{R}$ and $\mathbb{R}^d$, respectively. This detailed form highlights its "whiteness" in both temporal and spatial dimensions.

Similar to the one-dimensional Brownian measure, this implies that the increments of a space–time white noise over disjoint space–time regions are independent. This characteristic makes space–time white noise a standard model for driving noise in stochastic partial differential equations.

1.4.4. *Gaussian random fields*

Building upon the concept of Gaussian processes, where the random variables are indexed by a one-dimensional parameter (typically, time $t \in \mathbb{R}$ or $t \in [0, \infty)$), we now extend this notion to higher dimensions. A Gaussian random field is a collection of random variables indexed by a multi-dimensional parameter space.

DEFINITION 1.15.– *A Gaussian random field $X = \{X_{\mathbf{x}}, \mathbf{x} \in \mathcal{D}\}$ is a collection of random variables indexed by a set $\mathcal{D} \subset \mathbb{R}^d$ for some integer $d \geq 1$ (or a more general topological space), such that every finite collection of these random variables, $\{X_{\mathbf{x}_1}, X_{\mathbf{x}_2}, \dots, X_{\mathbf{x}_n}\}$, has a multivariate normal (Gaussian) distribution.*

Similar to Gaussian processes, a Gaussian random field is completely characterized by its *mean function* $m(\mathbf{x}) = \mathbb{E}[X_{\mathbf{x}}]$ and its *covariance function*

$$C(\mathbf{x}, \mathbf{y}) = \mathbb{E}[(X_{\mathbf{x}} - m(\mathbf{x}))(X_{\mathbf{y}} - m(\mathbf{y}))]$$

for all $\mathbf{x}, \mathbf{y} \in \mathcal{D}$. If $m(\mathbf{x}) = 0$ for all $\mathbf{x}$, the field is said to be *centered.*

Examples:

– Consider the temperature at various points on a geographical map. Here, $\mathbf{x}$ could represent the (latitude, longitude) coordinates of a location, and $X_{\mathbf{x}}$ would be the temperature at that specific location. The entire collection of temperatures across the map forms a Gaussian random field if its statistical properties are Gaussian.

– Another example could be the intensity of pixels in a grayscale image, where $\mathbf{x}$ represents the (x, y) coordinates of a pixel, and $X_{\mathbf{x}}$ is the pixel's intensity value.

1.4.5. *Wiener integration: definitions and isometries*

This section provides the formal definitions of the Wiener integral with respect to both 1D Wiener processes and space–time white noise, along with their crucial Itô isometry properties. These concepts are fundamental for building stochastic models and performing calculations in infinite-dimensional stochastic calculus.

1.4.5.1. *Wiener integral with respect to a 1D Wiener process* $W(y)$

Here, we recall the definition of the Itô (or Wiener) integral with respect to a standard one-dimensional Wiener process, also called Brownian motion (see Definition 1.10).

Let $W = \{W(y); y \in \mathbb{R}\}$ be a standard 1D Wiener process. The Wiener integral of a deterministic function $g : \mathbb{R} \to \mathbb{R}$ with respect to $W(y)$ is defined in stages:

1) For a simple function $g(y) = \sum_{j=1}^{n} c_j \mathbf{1}_{[a_j, b_j)}(y)$, where $c_j \in \mathbb{R}$ and $[a_j, b_j)$ are disjoint intervals, the integral is defined as:

$$\int_{\mathbb{R}} g(y) dW(y) := \sum_{j=1}^{n} c_j (W(b_j) - W(a_j)).$$

This definition ensures linearity and consistency with the properties of Brownian motion increments.

2) For any function $g \in L^2(\mathbb{R})$ (i.e. $\int_{\mathbb{R}} |g(y)|^2 dy < \infty$), there exists a sequence of simple functions g_n such that $g_n \to g$ in $L^2(\mathbb{R})$. The Wiener integral of g is then defined as the $L^2(\Omega)$ limit of the integrals of the simple functions:

$$\int_{\mathbb{R}} g(y) dW(y) := \lim_{n \to \infty} \int_{\mathbb{R}} g_n(y) dW(y).$$

This limit exists and is independent of the choice of approximating sequence g_n. The resulting integral is a centered Gaussian random variable.

1.4.5.2. *Itô isometry for 1D Wiener process*

THEOREM 1.6.– *The Wiener integral is an L^2-isometry. For any two deterministic functions $g_1, g_2 \in L^2(\mathbb{R})$, the following holds:*

$$\mathbb{E}\left[\left(\int_{\mathbb{R}} g_1(y)dW(y)\right)\left(\int_{\mathbb{R}} g_2(y)dW(y)\right)\right] = \int_{\mathbb{R}} g_1(y)g_2(y)dy.$$

In particular, for $g_1 = g_2$, this yields the L^2-norm isometry:

$$\mathbb{E}\left[\left(\int_{\mathbb{R}} g(y)dW(y)\right)^2\right] = \int_{\mathbb{R}} g(y)^2dy.$$

This isometry is a cornerstone of stochastic calculus, allowing the computation of variances and covariances of Wiener integrals by simple integration in the L^2 space of the integrands.

1.4.5.3. *Wiener integral with respect to space–time white noise*

Consider a standard space–time white noise $W = \{W(I \times A); I \subset \mathbb{R}, A \subset \mathbb{R}^d,$ measurable$\}$.

When considering integrals of the form $\displaystyle\int f(s, y)W(ds, dy)$, or more simply $\displaystyle\int f(s, y)dW(s, y)$, the white noise $W(s, y)$ is understood as the underlying random measure for the measure space $(\mathbb{R} \times \mathbb{R}^d, \mathcal{B}(\mathbb{R} \times \mathbb{R}^d), dsdy)$.

Let $S = \mathbb{R} \times \mathbb{R}^d$ and μ be the Lebesgue measure $dsdy$ on S. The Wiener integral of a deterministic function $F : S \to \mathbb{R}$ with respect to the standard space–time white noise W is constructed in stages:

1) For a simple function $F(s, y) = \displaystyle\sum_{j=1}^{n} c_j \mathbf{1}_{I_j \times A_j}(s, y)$, where $I_j \times A_j$ are disjoint measurable rectangles, the integral is defined as:

$$\int_S F(s, y)dW(s, y) := \sum_{j=1}^{n} c_j W(I_j \times A_j).$$

2) For any function $F \in L^2(S, \mu)$ (i.e. $\displaystyle\int_S |F(s, y)|^2 dsdy < \infty$), there exists a sequence of simple functions F_n such that $F_n \to F$ in $L^2(S, \mu)$. The Wiener integral of F is then defined as the $L^2(\Omega)$ limit of the integrals of the simple functions:

$$\int_S F(s, y)dW(s, y) := \lim_{n\to\infty} \int_S F_n(s, y)dW(s, y).$$

This limit exists and is independent of the choice of approximating sequence F_n. The resulting integral is a centered Gaussian random variable.

1.4.5.4. *Itô isometry for space–time white noise*

THEOREM 1.7.– *The Wiener integral is an L^2-isometry. For any two deterministic functions $F, G \in L^2(\mathbb{R} \times \mathbb{R}^d, dsdy)$, the following holds:*

$$\mathbb{E}\left[\left(\int_{\mathbb{R}}\int_{\mathbb{R}^d} F(s,y)dW(s,y)\right)\left(\int_{\mathbb{R}}\int_{\mathbb{R}^d} G(s,y)dW(s,y)\right)\right] \\ = \int_{\mathbb{R}}\int_{\mathbb{R}^d} F(s,y)G(s,y)dsdy.$$

In particular, for $F = G$, this yields the L^2-norm isometry:

$$\mathbb{E}\left[\left(\int_{\mathbb{R}}\int_{\mathbb{R}^d} F(s,y)dW(s,y)\right)^2\right] = \int_{\mathbb{R}}\int_{\mathbb{R}^d} F(s,y)^2 dsdy.$$

This isometry is fundamental for computing covariances and variances of Wiener integrals.

1.5. Exercises

EXERCISE 1.1.– *Show that if X is a Gaussian vector, then $aX + b$ is also a Gaussian vector for any scalar $a \in \mathbb{R}$ and vector $b \in \mathbb{R}^n$.*

EXERCISE 1.2.– *Provide a rigorous proof for Lemma 1.2, which details the central moments of a Gaussian variable $X \sim \mathcal{N}(m, \sigma^2)$. Ensure your proof covers both the odd and even cases for n, and clearly demonstrates how the double factorial arises.*

Hint: consider first the case of a standard Gaussian variable $Z \sim \mathcal{N}(0,1)$. The characteristic function or integration by parts can be helpful.

EXERCISE 1.3.– *For a Gaussian random variable $X \sim \mathcal{N}(m, \sigma^2)$, derive its first few raw moments:*

– $\mathbb{E}[X] = m$
– $\mathbb{E}[X^2] = m^2 + \sigma^2$
– $\mathbb{E}[X^3] = m^3 + 3m\sigma^2$
– $\mathbb{E}[X^4] = m^4 + 6m^2\sigma^2 + 3\sigma^4$

Hint: these can be derived using the characteristic function or by expanding $(m + \sigma Z)^n$ where $Z \sim \mathcal{N}(0,1)$ and utilizing the known moments of a standard Gaussian variable.

EXERCISE 1.4.– *Consider a two-dimensional random vector* $X = (X_1, X_2)'$. *Let* $X_1 \sim \mathcal{N}(0,1)$ *be a standard Gaussian variable, and let* $X_2 = X_1$.

1) Show that X *is indeed a Gaussian vector.*

2) Determine the expectation vector $M_X = \mathbb{E}[X]$.

3) Compute the covariance matrix $K_X = \mathrm{Cov}(X)$.

4) Show that K_X *is singular by calculating its determinant.*

EXERCISE 1.5.– *Let* $X \sim \mathcal{N}_n(\mathbf{0}, K_X)$ *be an* n*-dimensional zero-mean multivariate Gaussian vector with covariance matrix* K_X*, and let* $\mathbf{A}$ *be a constant* $n \times n$ *matrix.*

1) Show that $\mathbb{E}[(X'\mathbf{A}X)^2] = \sum_{i,j,k,l=1}^{n} A_{ij}A_{kl}\mathbb{E}[X_iX_jX_kX_l]$.

2) Applying Isserlis' theorem and manipulating the summations, show that $\mathbb{E}[(X'\mathbf{A}X)^2]$ *can be expressed as:*

$$\mathbb{E}[(X'\mathbf{A}X)^2] = (Tr(\mathbf{A}K_X))^2 + 2Tr((\mathbf{A}K_X)^2).$$

3) Using the expression obtained in the previous part and Lemma 1.9, prove that the variance of the quadratic form $X'\mathbf{A}X$ *is given by:*

$$\mathrm{Var}(X'\mathbf{A}X) = 2\,\mathrm{Tr}((\mathbf{A}K_X)^2).$$

Hint: remember that K_X *is a symmetric matrix* ($K_{uv} = K_{vu}$) *and that trace properties include* $Tr(\mathbf{MN}) = Tr(\mathbf{NM})$.

EXERCISE 1.6.– *This exercise guides you through the proof of Corollary 1.2, building on fundamental matrix identities and properties of Gaussian vectors.*

1) Prove the following matrix trace identity: for any $n \times 1$ *column vector* $\mathbf{b}$ *and any* $n \times n$ *matrix* $\mathbf{\Sigma}$*, show that:*

$$Tr((\mathbf{bb}')\mathbf{\Sigma}) = \mathbf{b}'\mathbf{\Sigma}\mathbf{b}.$$

Hint: you may find the cyclic property of the trace, $Tr(\mathbf{CD}) = Tr(\mathbf{DC})$*, particularly useful. Also, remember that the trace of a* 1×1 *matrix is simply its single element.*

2) Let $\mathbf{Y} \sim \mathcal{N}_n(\mathbf{0}, \mathbf{\Sigma})$ *be an* n*-dimensional zero-mean Gaussian vector with covariance matrix* $\mathbf{\Sigma}$*. For any symmetric* $n \times n$ *matrix* $\mathbf{A}$ *and any constant* $n \times 1$ *vector* $\mathbf{b}$*, show that the expected value of the product of the quadratic form* $\mathbf{Y}'\mathbf{AY}$ *and the squared linear form* $(\mathbf{b}'\mathbf{Y})^2$ *is given by:*

$$\mathbb{E}[(\mathbf{Y}'\mathbf{AY})(\mathbf{b}'\mathbf{Y})^2] = 2Tr(\mathbf{A\Sigma bb}'\mathbf{\Sigma}) + Tr(\mathbf{A\Sigma})\mathbf{b}'\mathbf{\Sigma b}.$$

Hint: begin by rewriting the squared linear form $(\mathbf{b}'\mathbf{Y})^2$ *as a quadratic form involving a symmetric matrix* $\mathbf{B}$*. Then, apply Lemma 1.7 with appropriate matrices and utilize the trace identity proven in part (a) to simplify the resulting expression.*

EXERCISE 1.7.– *Consider the function $C : \mathbb{R}^+ \times \mathbb{R}^+ \to \mathbb{R}$ defined by $C(s,t) = s \cdot t$ for $s, t \in \mathbb{R}^+ = [0, \infty)$. Can a zero-mean Gaussian process exist with $C(s,t) = s \cdot t$ as its covariance function? Justify your answer.*

EXERCISE 1.8.– *Consider a probability space $(\Omega, \mathcal{F}, \mathbb{P})$ and let U be a random variable uniformly distributed on the interval $[0,1]$. Let $\mathbb{T} = [0,1]$ be our index set. Define two stochastic processes, $X = (X_t)_{t\in\mathbb{T}}$ and $Y = (Y_t)_{t\in\mathbb{T}}$, as follows:*

– $X_t(\omega) = 0$ *for all $t \in \mathbb{T}$ and all $\omega \in \Omega$. (This is the zero process)*

– $Y_t(\omega) = \begin{cases} 1 & \text{if } t = U(\omega) \\ 0 & \text{if } t \neq U(\omega) \end{cases}$ *for all $t \in \mathbb{T}$ and all $\omega \in \Omega$.*

Show that X and Y are modifications of each other, but they are not indistinguishable.

EXERCISE 1.9.– *Consider two independent standard Brownian motions, $X = (B_t^{(1)})_{t\geq 0}$ and $Y = (B_t^{(2)})_{t\geq 0}$, defined on the same probability space $(\Omega, \mathcal{F}, \mathbb{P})$. Show that X and Y are versions of each other, but they are not modifications.*

EXERCISE 1.10.– *Two-sided Brownian motion.*

Let $(W_t)_{t\geq 0}$ be a standard Brownian motion on $[0,\infty)$ with $W_0 = 0$. We define a process $B = (B_t)_{t\in\mathbb{R}}$ on the entire real line as follows:

$$B_t = \begin{cases} W_t & \text{for } t \geq 0 \\ \tilde{W}_{-t} & \text{for } t < 0 \end{cases}$$

where $\tilde{W} = (\tilde{W}_t)_{t\geq 0}$ is an independent copy of the standard Brownian motion W.

1) Demonstrate that the process $B = (B_t)_{t\in\mathbb{R}}$ is a centered Gaussian process.

2) Show that the covariance function of B is given by:

$$\mathbb{E}[B_s B_t] = \frac{1}{2}\left(|t| + |s| - |t-s|\right)$$

for all $s, t \in \mathbb{R}$. Explicitly verify this formula for all necessary cases: ($s,t \geq 0$; $s,t < 0$; and $s \geq 0, t < 0$).

3) Verify that the increments of B are Gaussian: for any $s < t$, $B_t - B_s$ is a Gaussian random variable with mean 0 and variance $|t-s|$.

4) Discuss the properties of stationary and independent increments for this two-sided Brownian motion, explaining why these hold based on its construction and the properties of standard Brownian motion.

2

Key Fractional Gaussian Processes: Building Blocks for the Generalized Fractional Brownian Motion

In this chapter, we will explore key fractional Gaussian processes, including fractional Brownian motion (fBm), its odd and even variants, and sub-fractional Brownian motion (sfBm). These discussions will provide the essential tools and insights needed to understand the generalized fractional Brownian motion, the main subject of investigation in this book.

2.1. Fractional Brownian motion

Fractional Brownian motion (fBm), a fundamental fractional Gaussian process, is a continuous-time Gaussian process distinguished by its mean function and a unique covariance function that inherently captures long-range dependence. Widely recognized for its ability to model phenomena exhibiting self-similarity and long-memory, fBm finds significant applications in diverse fields such as finance, telecommunications and hydrology. It is about the process defined by:

DEFINITION 2.1.– *Any centered Gaussian process $B^H = (B_t^H)_{t\in\mathbb{R}}$ is called a two-sided fractional Brownian motion (tsfBm) with Hurst parameter $H \in (0,1)$ if its covariance function is given by:*

$$\mathbb{E}[B_s^H B_t^H] = \frac{1}{2}\left(|t|^{2H} + |s|^{2H} - |t-s|^{2H}\right), \quad \textit{for all } s,t \in \mathbb{R}.$$

Similarly, if the process is defined for $t \in [0,+\infty)$, it is called a fractional Brownian motion.

To rigorously establish the mathematical existence of such a process and thus fully characterize it, we must first prove that its proposed covariance function is positive semi-definite. Although this result has been rigorously proven elsewhere (see Proposition 1.6 in Nourdin (2012)), for the benefit and completeness of this book, we will present the detailed proof. The following lemma provides this essential preliminary step.

LEMMA 2.1.– *The function f defined for all $s, t \in \mathbb{R}$ by the equality:*

$$f(s,t) = \frac{1}{2}\left(|t|^{2H} + |s|^{2H} - |t-s|^{2H}\right), \qquad [2.1]$$

is positive semi-definite.

Proof. Consider $n \in \mathbb{N}$, $t_1, t_2, ..., t_n \in \mathbb{R}$ and $a_1, a_2, ..., a_n \in \mathbb{R}$. To obtain the announced result, we will prove that:

$$\sum_{j,k=1}^{n} a_j a_k f(t_j, t_k) \geq 0.$$

We first note that, if we denote by C the finite constant $C = \int_0^\infty s^{-1-2H}(1-e^{-s^2})ds$, then for all real t we have:

$$|t|^{2H} = C^{-1}\int_0^\infty s^{-1-2H}(1-e^{-s^2t^2})ds,$$

and consequently, for all $j, k \in \{1, 2, ..., n\}$:

$$\begin{aligned}
&|t_j|^{2H} + |t_k|^{2H} - |t_j - t_k|^{2H} \\
&= C^{-1}\left[\int_0^\infty s^{-1-2H}(1-e^{-s^2t_j^2})(1-e^{-s^2t_k^2})ds \right.\\
&\qquad \left. + \int_0^\infty s^{-1-2H}e^{-s^2(t_j^2+t_k^2)}(e^{2s^2t_jt_k}-1)ds\right] \\
&= C^{-1}\left[\int_0^\infty s^{-1-2H}(1-e^{-s^2t_j^2})(1-e^{-s^2t_k^2})ds \right.\\
&\qquad \left. + \int_0^\infty s^{-1-2H}e^{-s^2(t_j^2+t_k^2)}\sum_{p=1}^{\infty}\frac{(2s^2t_jt_k)^p}{p!}ds\right]. \\
&= C^{-1}\left[\int_0^\infty s^{-1-2H}(1-e^{-s^2t_j^2})(1-e^{-s^2t_k^2})ds \right.\\
&\qquad \left. + \sum_{p=1}^{\infty}\frac{2^p}{p!}\int_0^\infty s^{-1+2p-2H}t_j^pe^{-s^2t_j^2}t_k^pe^{-s^2t_k^2}ds\right].
\end{aligned}$$

Thus,

$$\sum_{j,k}^{n} a_j a_k f(t_j, t_k) = (2C)^{-1}\left[\int_0^{\infty} s^{-1-2H}\left(\sum_{k=1}^{n} a_k(1-e^{-s^2t_k^2})\right)^2 ds + \sum_{p=1}^{\infty}\frac{2^p}{p!}\int_0^{\infty} s^{-1+2p-2H}\left(\sum_{j=1}^{n} a_j t_j^p e^{-s^2t_j^2}\right)^2 ds\right].$$

Since $C > 0$, $s^{-1-2H} > 0$ for $s > 0$, $s^{-1+2p-2H} > 0$ for $s > 0$, and each integrand is a square, the integrals are nonnegative. Furthermore, each term in the sum over p is nonnegative. Therefore, the entire expression is greater than or equal to zero. This concludes the proof that $f(s,t)$ is positive semi-definite. □

REMARK 2.1.– *The symmetry of the covariance function, $f(s,t) = f(t,s)$ for all $s,t \in \mathbb{R}$, combined with the proven positive semi-definiteness, directly invokes Kolmogorov's existence theorem (Theorem 1.2). Consequently, we are assured of the mathematical existence of a centered Gaussian process B^H possessing precisely this covariance structure.*

REMARK 2.2.– *A particularly significant insight arises when the Hurst parameter is $H = 1/2$. In this specific instance, the tsfBm $B^{1/2}$ reduces precisely to the two-sided Brownian motion (see Exercise 1.10). When restricted to $t \in [0, +\infty)$, $B^{1/2}$ becomes the standard Brownian motion (see Definition 1.10). This remarkable connection underscores that fBm serves as a powerful and generalized extension of the well-established Brownian motion, broadening its applicability to phenomena exhibiting long-range dependence.*

In the following proposition, we consolidate several fundamental properties of fBm, including one of its crucial integral representations. Proofs for these characteristics are extensively detailed in various seminal works, such as Nourdin (2012); Nualart (2006); Samorodnitsky and Taqqu (1994), and the references therein.

PROPOSITION 2.1.– *Let $B^H = \left(B_t^H\right)_{t\in\mathbb{R}}$ be an fBm with Hurst parameter H and covariance function defined by equation [2.1]. Then, B^H exhibits the following key characteristics:*

1) B^H is self-similar with index H. This implies that for any $a \in (0,\infty)$, the processes $\left(B_{at}^H\right)_{t\geq 0}$ and $\left(a^H B_t^H\right)_{t\geq 0}$ are versions (i.e. they have the same finite-dimensional distributions).

2) The increments of B^H are stationary. This property means that the statistical properties of its increments depend only on the time difference, not on the absolute *time. For B^H, being a centered Gaussian process, this stationarity is precisely quantified by the explicit form of the covariance function of its increments, given by:*

$$\mathrm{Cov}(B_t^H - B_s^H, B_u^H - B_v^H) = \frac{1}{2}\left(|t-v|^{2H} - |t-u|^{2H} - |s-v|^{2H} + |s-u|^{2H}\right). \quad [2.2]$$

This expression clearly depends only on time differences and not absolute time. As a consequence, the variance of a single increment is given by:

$$\mathbb{E}[(B_t^H - B_s^H)^2] = \mathrm{Cov}(B_t^H - B_s^H, B_t^H - B_s^H) = |t-s|^{2H}. \quad [2.3]$$

3) The sample paths of B^H are almost surely Hölder continuous with any exponent $\alpha < H$. This property quantifies the local regularity of the paths.

4) For all $H \in (0,1)$ and $s, t \in \mathbb{R}$, its covariance function can be expressed as:

$$\mathbb{E}\left[B^H(t)B^H(s)\right] = c_H^2 \int_0^\infty \frac{\sin(xs)\sin(xt) + (1-\cos(xs))(1-\cos(xt))}{x^{1+2H}} dx, \quad [2.4]$$

where c_H^2 is a constant defined by:

$$c_H^2 = \frac{\Gamma(1+2H)\sin(\pi H)}{\pi}. \quad [2.5]$$

2.2. Odd fractional Brownian motion

Having outlined the foundational properties of fBm, we now introduce a specific type of centered Gaussian process characterized by a unique covariance structure. This process is a crucial building block for understanding more generalized fractional Gaussian processes and exhibits distinct properties compared to the standard fBm.

DEFINITION 2.2.– *Any centered Gaussian process $B^{H,o} = \left(B_t^{H,o}\right)_{t\in\mathbb{R}}$ is called a two-sided odd fractional Brownian motion with Hurst parameter $H \in (0,1)$ if its covariance function is given by:*

$$\mathbb{E}[B_s^{H,o} B_t^{H,o}] = \frac{1}{4}\left(|t+s|^{2H} - |t-s|^{2H}\right), \quad \textit{for all } s, t \in \mathbb{R}. \quad [2.6]$$

Similarly, if the process is defined for $t \in [0, +\infty)$, it is called an odd fractional Brownian motion (ofBm).

REMARK 2.3.– *The ofBm was first introduced in Dzhaparidze and Zanten (2004), where the authors utilized it as a tool to obtain a series expansion of fBm. However, they did not formally name the process or provide a detailed characterization of its properties. This process is formally named ofBm in Mishura and Zili (2025) and also in the present monograph, where some of its main properties are demonstrated.*

To rigorously establish the mathematical existence of such a process and thus fully characterize it, we must first prove that its proposed covariance function is positive semi-definite and symmetric. The following lemma provides this essential preliminary step.

LEMMA 2.2.– *The function f_o defined for all $s, t \in \mathbb{R}$ by the equality*

$$f_o(s,t) = \frac{1}{4}\left(|t+s|^{2H} - |t-s|^{2H}\right), \qquad [2.7]$$

is positive semi-definite and symmetric.

Proof. The function f_o clearly satisfy the symmetry property. Let $n \in \mathbb{N} \setminus \{0\}$, $t_1, t_2, ..., t_n \in \mathbb{R}$, and $a_1, a_2, ..., a_n \in \mathbb{R}$. To demonstrate positive semi-definiteness, we must show that

$$\sum_{j,k=1}^{n} a_j a_k f_o(t_j, t_k) \geq 0.$$

First note that for any real t and $H \in (0,1)$, the following integral representation holds: $|t|^{2H} = C^{-1}\int_0^\infty x^{-1-2H}(1 - e^{-x^2t^2})dx$, where C is the finite constant given by: $C = \int_0^\infty x^{-1-2H}(1-e^{-x^2})dx$. Utilizing this representation, for all $j, k \in \{1, 2, ..., n\}$, we can write:

$$\begin{aligned}
&|t_j + t_k|^{2H} - |t_j - t_k|^{2H} \\
&= C^{-1}\int_0^\infty x^{-1-2H}\left((1 - e^{-x^2(t_j+t_k)^2}) - (1 - e^{-x^2(t_j-t_k)^2})\right)dx \\
&= C^{-1}\int_0^\infty x^{-1-2H}\left(e^{-x^2(t_j-t_k)^2} - e^{-x^2(t_j+t_k)^2}\right)dx \\
&= C^{-1}\int_0^\infty x^{-1-2H}e^{-x^2t_j^2}e^{-x^2t_k^2}\left(e^{2x^2t_kt_j} - e^{-2x^2t_kt_j}\right)dx \\
&= C^{-1}\int_0^\infty x^{-1-2H}e^{-x^2t_j^2}e^{-x^2t_k^2}\left(\sum_{p=0}^{\infty}\frac{(2x^2t_jt_k)^p}{p!} - \sum_{p=0}^{\infty}\frac{(-2x^2t_jt_k)^p}{p!}\right)dx \\
&= 2C^{-1}\int_0^\infty x^{-1-2H}e^{-x^2t_j^2}e^{-x^2t_k^2}\sum_{p=0}^{\infty}\frac{(2x^2t_jt_k)^{2p+1}}{(2p+1)!}dx.
\end{aligned}$$

Therefore, the quadratic form becomes:

$$\begin{aligned}
&\sum_{j,k=1}^{n} a_j a_k f_o(t_j, t_k) \\
&= \frac{1}{4} \sum_{j,k=1}^{n} a_j a_k \left(|t_j + t_k|^{2H} - |t_j - t_k|^{2H} \right) \\
&= \frac{1}{4} \cdot 2C^{-1} \sum_{j,k=1}^{n} a_j a_k \int_0^{\infty} x^{-1-2H} e^{-x^2 t_j^2} e^{-x^2 t_k^2} \sum_{p=0}^{\infty} \frac{(2x^2 t_j t_k)^{2p+1}}{(2p+1)!} dx \\
&= \frac{1}{2C} \int_0^{\infty} x^{-1-2H} \left(\sum_{j,k=1}^{n} a_j a_k e^{-x^2 t_j^2} e^{-x^2 t_k^2} \sum_{p=0}^{\infty} \frac{(2x^2 t_j t_k)^{2p+1}}{(2p+1)!} \right) dx \\
&= \frac{1}{2C} \sum_{p=0}^{\infty} \frac{2^{2p+1}}{(2p+1)!} \int_0^{\infty} x^{1-2H+4p} \sum_{j,k=1}^{n} \left(a_k e^{-x^2 t_k^2} t_k^{2p+1} \right) \left(a_j e^{-x^2 t_j^2} t_j^{2p+1} \right) dx \\
&= \frac{1}{2C} \sum_{p=0}^{\infty} \frac{2^{2p+1}}{(2p+1)!} \int_0^{\infty} x^{1-2H+4p} \left(\sum_{j=1}^{n} a_j e^{-x^2 t_j^2} t_j^{2p+1} \right)^2 dx.
\end{aligned}$$

The interchange of the infinite sum and the integral is justified by Tonelli's theorem, given that all terms in the sum are nonnegative. Since $C > 0$, $x^{1-2H+4p} > 0$ for $x > 0$, and the sum $\left(\sum_{j=1}^{n} a_j e^{-x^2 t_j^2} t_j^{2p+1}\right)^2$ is a square (hence nonnegative), the entire expression is clearly nonnegative, thus proving the positive semi-definiteness of f_o. □

REMARK 2.4.– *From the symmetry and positive semi-definiteness of f_o defined by equation [2.6], Kolmogorov's existence theorem (Theorem 1.2) guarantees the mathematical existence of a unique centered Gaussian process $B^{H,o}$ characterized by f_o as its covariance function.*

The inherent structure of the ofBm leads to a particularly well-behaved process, as evidenced by its nondegenerate finite-dimensional distributions.

COROLLARY 2.1.– *The ofBm $B^{H,o}$ is a nondegenerate Gaussian process on $(0, +\infty)$. Specifically, for any set of distinct time points $0 < t_1 < t_2 < \cdots < t_n$, the covariance matrix Σ_O of the random vector $(B_{t_1}^{H,o}, \ldots, B_{t_n}^{H,o})$ is invertible.*

Proof. Let Σ be the $n \times n$ covariance matrix corresponding to the chosen distinct time points, with entries given by:

$$\Sigma_{jk} = \mathrm{Cov}(B_{t_j}^{H,o}, B_{t_k}^{H,o}) = f_o(t_j, t_k),$$

where $f_o(s,t)$ is the covariance function defined in equation [2.7]. As established by Lemma 1.11, for the ofBm to be nondegenerate on $(0,+\infty)$, we must demonstrate that Σ is strictly positive definite.

To achieve this, we need to show that for any nonzero vector $a = (a_1,\ldots,a_n)' \in \mathbb{R}^n$, the quadratic form $Q(a) = \sum_{j,k=1}^{n} a_j a_k f_o(t_j,t_k)$ is strictly positive.

Recall the derived expression for $Q(a)$ from the proof of Lemma 2.2,

$$Q(a) = \frac{1}{2C}\sum_{p=0}^{\infty}\frac{2^{2p+1}}{(2p+1)!}\int_0^{\infty} x^{1-2H+4p}\left(\sum_{j=1}^{n} a_j e^{-x^2 t_j^2} t_j^{2p+1}\right)^2 dx.$$

As previously shown, $Q(a) \geq 0$. Now, let us assume, for contradiction, that $Q(a) = 0$ for some nonzero vector a.

Since $C > 0$, $\frac{2^{2p+1}}{(2p+1)!} > 0$, $x^{1-2H+4p} > 0$ for $x > 0$, and the integrand is a square (hence nonnegative), for the integral to be zero, the term inside the square must be zero almost everywhere with respect to x for each term in the sum over p:

$$\sum_{j=1}^{n} a_j e^{-x^2 t_j^2} t_j^{2p+1} = 0 \quad \text{for all } p \geq 0 \text{ and for almost all } x \in (0,\infty).$$

Let $y = x^2$. Since $x \in (0,\infty)$, $y \in (0,\infty)$. The distinct positive time points $0 < t_1 < t_2 < \cdots < t_n$ imply that $\lambda_j = t_j^2$ are distinct positive constants, that is, $0 < \lambda_1 < \lambda_2 < \cdots < \lambda_n$. The equation then transforms to:

$$\sum_{j=1}^{n}(a_j t_j^{2p+1}) e^{-y\lambda_j} = 0 \quad \text{for all } p \geq 0 \text{ and for almost all } y \in (0,\infty).$$

The set of exponential functions $\{e^{-y\lambda_j}\}_{j=1}^{n}$, with distinct positive exponents λ_j, forms a linearly independent set over any interval $(0,\infty)$. Therefore, for their linear combination to be zero almost everywhere, all of the coefficients must necessarily be zero:

$$a_j t_j^{2p+1} = 0 \quad \text{for all } j = 1,\ldots,n \text{ and for each } p \geq 0.$$

Since $t_j > 0$ for all $j = 1,\ldots,n$, it follows that t_j^{2p+1} is never zero. Consequently, we must have $a_j = 0$ for all $j = 1,\ldots,n$.

This demonstrates that $Q(a) = 0$ if and only if a is the zero vector, which by definition means the covariance matrix Σ is strictly positive definite. This completes the proof of the corollary. □

2.3. Even fractional Brownian motion

Following our exploration of fractional and ofBms, we now introduce another key fractional Gaussian process: even fractional Brownian motion (efBm). This process, also a centered Gaussian process, is defined by a unique covariance structure that leads to distinct path properties and specific applications. Our primary aim in this section is to rigorously establish the existence of this process by demonstrating the validity of its defining covariance function, independent of its relation to other fractional processes.

DEFINITION 2.3.– *Any centered Gaussian process* $B^{H,e}$ *with Hurst parameter* $H \in (0,1)$ *is called an even fractional Brownian motion if its covariance function is given by:*

$$\mathbb{E}[B_s^{H,e} B_t^{H,e}] = \frac{1}{2}\left(|t|^{2H} + |s|^{2H}\right) - \frac{1}{4}\left(|t+s|^{2H} + |t-s|^{2H}\right). \qquad [2.8]$$

More specifically:

– If $B^{H,e} = \left(B_t^{H,e}\right)_{t\in\mathbb{R}}$, *it is called a two-sided even fractional Brownian motion, and the covariance holds for all* $s, t \in \mathbb{R}$.

– If $B^{H,e} = \left(B_t^{H,e}\right)_{t\in[0,+\infty)}$, *it is called an even fractional Brownian motion, and the covariance holds for all* $s, t \in [0, +\infty)$.

To rigorously establish the mathematical existence of such a process, we must first demonstrate that its proposed covariance function, $f_e(s,t)$, is valid. This requires proving its symmetry and positive semi-definiteness. The following theorem and its lemma provide this essential step by offering an integral representation that naturally demonstrates the positive semi-definite property.

THEOREM 2.1.– *The function* f_e *defined for every* $s, t \in \mathbb{R}$ *by*

$$f_e(s,t) = \frac{1}{2}\left(s^{2H} + t^{2H}\right) - \frac{1}{4}\left((s+t)^{2H} + |t-s|^{2H}\right) \qquad [2.9]$$

is positive semi-definite.

To prove Theorem 2.1, we first establish the following preliminary lemma, which provides a crucial integral representation for the covariance function f_e.

LEMMA 2.3.– *For a fixed Hurst parameter* $H \in (0,1)$, *let*

$$\Gamma_H(t,r) = \mid t-r \mid^{H-\frac{1}{2}} sign(t-r) + 2 \mid r \mid^{H-\frac{1}{2}} sign(r) - \mid t+r \mid^{H-\frac{1}{2}} sign(t+r)$$

for $t, r \in \mathbb{R}$. *Then, for any* $(s,t) \in \mathbb{R}^2$, $\Gamma_H(t,.) \in L_2(\mathbb{R}, dr)$ *and*

$$< \Gamma_H(t,.), \Gamma_H(s,.) >= 2C_H\left(s^{2H} + t^{2H} - \frac{1}{2}((s+t)^{2H} + \mid s-t \mid^{2H})\right),$$

where C_H is a positive constant and $< .,. >$ denotes the inner product on $L_2(\mathbb{R}, dr)$ defined by $< f, g >= \int_{\mathbb{R}} f(r)g(r)dr$.

Proof. We begin by showing that $\Gamma_H(t,.) \in L_2(\mathbb{R}, dr)$. We note that $\Gamma_H(0,r) = 0$ and

$$\left(\Gamma_H(t,r)\right)^2 = \mid t \mid^{2H-1} \left(\Gamma_H(1, \frac{r}{t})\right)^2$$

for any $r \in \mathbb{R}$ and $t \in \mathbb{R} \setminus \{0\}$.

The function $u \longmapsto \Gamma_H^2(1,u)$ is locally integrable on $\mathbb{R} \setminus \{-1, 0, 1\}$. Since $H \in (0,1)$, the functions $u \longmapsto |1-u|^{H-\frac{1}{2}}\text{sign}(1-u)$,

$$u \longmapsto |u|^{H-\frac{1}{2}}\text{sign}(u) \quad \text{and} \quad u \longmapsto |1+u|^{H-\frac{1}{2}}\text{sign}(1+u)$$

are locally square integrable in the neighborhoods of $u = -1, u = 0$ and $u = 1$. Thus,

$$\int_{-2}^{2} \Gamma_H^2(1,u)du < \infty.$$

For the integrability over infinite intervals, consider $u \geq 2$:

$$\begin{aligned}
&\int_2^{+\infty} \Gamma_H^2(1,u)du \\
&= \int_2^{+\infty} \left(\left(u^{H-\frac{1}{2}} - (u-1)^{H-\frac{1}{2}}\right) - \left((u+1)^{H-\frac{1}{2}} - u^{H-\frac{1}{2}}\right)\right)^2 du \\
&\leq 2\left(\int_2^{+\infty} \left(u^{H-\frac{1}{2}} - (u-1)^{H-\frac{1}{2}}\right)^2 du + \int_2^{+\infty} \left((u+1)^{H-\frac{1}{2}} - u^{H-\frac{1}{2}}\right)^2 du\right) \\
&\leq 2(H-\frac{1}{2})^2\left(\int_2^{+\infty}\int_{u-1}^{u} y^{2H-3}dydu + \int_2^{+\infty}\int_u^{u+1} y^{2H-3}dydu\right) \\
&\leq 2(H-\frac{1}{2})^2\left(\int_2^{+\infty} (u-1)^{2H-3}du + \int_2^{+\infty} u^{2H-3}du\right) < \infty
\end{aligned}$$

because $2H - 3 < -1$. A similar argument holds for $u \leq -2$.

Therefore, $\Gamma_H(t,.) \in L_2(\mathbb{R}, dr)$ for any $t \in \mathbb{R}$.

Now, in order to evaluate $< \Gamma_H(t,.), \Gamma_H(s,.) >$, we first note that:

$$\Gamma_H(t,r) := \gamma_H(t,r) + \gamma_H(-t,r), \quad \text{for } t, r \in \mathbb{R}, \qquad [2.10]$$

where $\gamma_H(t,r) = \mid t - r \mid^{H-\frac{1}{2}} \text{sign}(t-r) + \mid r \mid^{H-\frac{1}{2}} \text{sign}(r)$, for any $(t,r) \in \mathbb{R}^2$.

It is straightforward to see that $\gamma_H(0,r) = 0$ and

$$\left(\gamma_H(t,r)\right)^2 =| t |^{2H-1} \left(\gamma_H(1,\frac{r}{t})\right)^2$$

for any $r \in \mathbb{R}$ and $t \in \mathbb{R} \setminus \{0\}$. As we have shown for Γ_H, it is easy to check that $\gamma_H(t,.) \in L_2(\mathbb{R}, dr)$ for any t. Thus,

$$||\gamma_H(t,.)||_2^2 = \int_{\mathbb{R}} | t |^{2H-1} \gamma_H^2(1,\frac{r}{t})dr =| t |^{2H} \int_{\mathbb{R}} \gamma_H^2(1,u)du =| t |^{2H} C_H, \quad [2.11]$$

with $C_H = ||\gamma_H^2(1,.)||_2^2 > 0$.

By the linearity of $L_2(\mathbb{R}, dr)$, we deduce that $\gamma_H(t,.) - \gamma_H(s,.) \in L_2(\mathbb{R}, dr)$ and by a change of variable ($u = r - s$ or $u = r - t$), we can show that:

$$||\gamma_H(t,.) - \gamma_H(s,.)||_2^2 = C_H \mid t - s \mid^{2H} . \quad [2.12]$$

Thus, by the bilinearity of the inner product and the polarization identity, combining equations [2.11] and [2.12], we get:

$$\begin{aligned} &< \gamma_H(t,.), \gamma_H(s,.) > \\ &= \frac{1}{2}\left(||\gamma_H(t,.)||_2^2 + ||\gamma_H(s,.)||_2^2 - ||\gamma_H(t,.) - \gamma_H(s,.)||_2^2 \right) \\ &= \frac{C_H}{2}\left(\mid t \mid^{2H} + \mid s \mid^{2H} - \mid t - s \mid^{2H} \right). \end{aligned} \quad [2.13]$$

Now, from eqauations [2.10] and [2.13], and using the fact that $< \gamma_H(-t,.), \gamma_H(-s,.) >$), we calculate the main inner product:

$$\begin{aligned} &< \Gamma_H(t,.), \Gamma_H(s,.) > \\ &= < \gamma_H(t,.) + \gamma_H(-t,.), \gamma_H(s,.) + \gamma_H(-s,.) > \\ &= < \gamma_H(t,.), \gamma_H(s,.) > + < \gamma_H(t,.), \gamma_H(-s,.) > \\ &\quad + < \gamma_H(-t,.), \gamma_H(s,.) > + < \gamma_H(-t,.), \gamma_H(-s,.) > \\ &= \frac{C_H}{2}\Bigg(\left(|t|^{2H} + |s|^{2H} - |t-s|^{2H}\right) \quad \text{(from } < \gamma_H(t,.), \gamma_H(s,.) > \text{)} \\ &\quad + \left(|t|^{2H} + |-s|^{2H} - |t-(-s)|^{2H}\right) \quad \text{(from } < \gamma_H(t,.), \gamma_H(-s,.) > \text{)} \\ &\quad + \left(|-t|^{2H} + |s|^{2H} - |-t-s|^{2H}\right) \quad \text{(from } < \gamma_H(-t,.), \gamma_H(s,.) > \text{)} \\ &\quad + \left(|-t|^{2H} + |-s|^{2H} - |-t-(-s)|^{2H}\right) \Bigg) \end{aligned}$$

$$= \frac{C_H}{2}\left(2|t|^{2H} + 2|s|^{2H} - 2|t-s|^{2H} + 2|t|^{2H} + 2|s|^{2H} - 2|t+s|^{2H}\right)$$

$$= 2C_H\left(|t|^{2H} + |s|^{2H} - \frac{1}{2}\left(|t-s|^{2H} + |t+s|^{2H}\right)\right).$$

Thus, we have shown that $< \Gamma_H(t,.), \Gamma_H(s,.) >= 2C_H f_e(s,t)$, which completes the proof of the lemma. □

Proof of Theorem 2.1. Let $n \in \mathbb{N}$, $t_1, t_2, ..., t_n \in \mathbb{R}$ and $a_1, a_2, ..., a_n \in \mathbb{R}$. To prove that f_e is positive semi-definite, we must show that $\sum_{j,k=1}^{n} a_j a_k f_e(t_j, t_k) \geq 0$. By Lemma 2.3, we have:

$$\begin{aligned}
\sum_{j,k=1}^{n} a_j a_k f_e(t_j, t_k) &= \sum_{j,k=1}^{n} a_j a_k \frac{1}{2C_H} < \Gamma_H(t_j,.), \Gamma_H(t_k,.) > \\
&= \frac{1}{2C_H}\left\langle \sum_{j=1}^{n} a_j \Gamma_H(t_j,.), \sum_{k=1}^{n} a_k \Gamma_H(t_k,.)\right\rangle \\
&= \frac{1}{2C_H}\left\| \sum_{j=1}^{n} a_j \Gamma_H(t_j,.)\right\|_2^2 \\
&\geq 0.
\end{aligned}$$

Since C_H is a positive constant, and the square of an L_2 norm is always nonnegative, the quadratic form is indeed non-negative, which proves the positive semi-definiteness of f_e. □

REMARK 2.5.– *The proven symmetry, $f_e(s,t) = f_e(t,s)$, and positive semi-definiteness of the function f_e, defined by equation [2.8], ensure its validity as a covariance function. This, coupled with its real-valued nature, directly invokes Kolmogorov's existence theorem (Theorem 1.2). Consequently, we are assured of the mathematical existence of a unique centered Gaussian process $B^{H,e}$ with f_e as its covariance function.*

By an analogous argument to the proof of Corollary 2.1, we establish the following:

COROLLARY 2.2.– *The efBm ($B^{H,e}$) is a nondegenerate Gaussian process on $(0,+\infty)$. Specifically, for any set of distinct time points $0 < t_1 < t_2 < \cdots < t_n$, the covariance matrix Σ_E of the random vector $(B^{H,e}_{t_1}, \ldots, B^{H,e}_{t_n})$ is symmetric and positive definite, and thus invertible.*

2.4. Relationships between fractional, odd and even fractional Brownian motions

Having rigorously defined fBm, ofBm and efBM independently, it is insightful to explore the inherent connections between these processes. In this section, we demonstrate that both ofBm and efBM can be uniquely characterized as the odd and even parts, respectively, of a two-sided fractional Brownian motion. These relationships not only provide alternative constructive definitions but also highlight the fundamental symmetries underlying these Gaussian processes.

PROPOSITION 2.2.– *A centered Gaussian process $B^{H,o} = (B_t^{H,o})_{t\geq 0}$ is an ofBm if and only if there exists a tsfBm $B^H = (B_t^H)_{t\in\mathbb{R}}$ such that for all $t \geq 0$:*

$$B_t^{H,o} = \frac{1}{2}(B_t^H - B_{-t}^H). \qquad [2.14]$$

Proof. (⇒) Assume $B^{H,o} = (B_t^{H,o})_{t\geq 0}$ is an ofBm. Consider, on the same probability space, an independent efBm $B^{H,e} = (B_t^{H,e})_{t\in\mathbb{R}}$, which is a centered Gaussian process with covariance function defined by equation [2.8]. Now, define a new process $B^H = (B_t^H)_{t\in\mathbb{R}}$ as:

$$B_t^H = B_{|t|}^{H,e} + \text{sign}(t) B_{|t|}^{H,o}. \qquad [2.15]$$

Since $B^{H,e}$ and $B^{H,o}$ are independent Gaussian processes, and B_t^H is a linear combination of their components (specifically, $B_{|t|}^{H,e}$ and $B_{|t|}^{H,o}$ for any given t), it follows directly from Proposition 1.5 that B^H is also a centered Gaussian process. Now, through a straightforward computation, which we leave as an exercise for the reader (see Exercise 2.3), we can verify that for all $s, t \in \mathbb{R}$:

$$\mathbb{E}[B_s^H B_t^H] = \frac{1}{2}\left(|t|^{2H} + |s|^{2H} - |t-s|^{2H}\right).$$

This is the defining covariance function of an fBm. Thus, B^H is an fBm on $\mathbb{R}$.

From [2.15], we have $B_t^H = B_t^{H,e} + B_t^{H,o}$ for $t \geq 0$, and $B_{-t}^H = B_t^{H,e} - B_t^{H,o}$ for $t \geq 0$. Subtracting these two equations for $t \geq 0$:

$$B_t^H - B_{-t}^H = (B_t^{H,e} + B_t^{H,o}) - (B_t^{H,e} - B_t^{H,o}) = 2B_t^{H,o}.$$

Hence, relation [2.14] holds.

(⇐) Conversely, assume there exists an fBm $B^H = (B_t^H)_{t\in\mathbb{R}}$ such that $B_t^{H,o} = \frac{1}{2}(B_t^H - B_{-t}^H)$ for $t \geq 0$. We need to show that $B^{H,o}$ is an ofBm. Since B^H is a centered Gaussian process, $B^{H,o}$ defined as a linear combination of its components is also a centered Gaussian process. Now we compute its covariance function for $s, t \geq 0$:

$$\begin{aligned}\mathbb{E}[B_s^{H,o} B_t^{H,o}] &= \mathbb{E}\left[\frac{1}{2}(B_s^H - B_{-s}^H)\frac{1}{2}(B_t^H - B_{-t}^H)\right] \\ &= \frac{1}{4}\mathbb{E}[B_s^H B_t^H - B_s^H B_{-t}^H - B_{-s}^H B_t^H + B_{-s}^H B_{-t}^H]\end{aligned}$$

Using the covariance function of B^H, $\mathbb{E}[B_u^H B_v^H] = \frac{1}{2}\left(|u|^{2H} + |v|^{2H} - |u-v|^{2H}\right)$:

$$\mathbb{E}[B_s^H B_t^H] = \frac{1}{2}(s^{2H} + t^{2H} - |s-t|^{2H})$$

$$\mathbb{E}[B_s^H B_{-t}^H] = \frac{1}{2}(s^{2H} + (-t)^{2H} - |s-(-t)|^{2H}) = \frac{1}{2}(s^{2H} + t^{2H} - |s+t|^{2H})$$

$$\mathbb{E}[B_{-s}^H B_t^H] = \frac{1}{2}((-s)^{2H} + t^{2H} - |-s-t|^{2H}) = \frac{1}{2}(s^{2H} + t^{2H} - |s+t|^{2H})$$

$$\begin{aligned}\mathbb{E}[B_{-s}^H B_{-t}^H] &= \frac{1}{2}((-s)^{2H} + (-t)^{2H} - |-s-(-t)|^{2H}) \\ &= \frac{1}{2}(s^{2H} + t^{2H} - |s-t|^{2H})\end{aligned}$$

Substituting these back into the expression for $\mathbb{E}[B_s^{H,o} B_t^{H,o}]$:

$$\begin{aligned}\mathbb{E}[B_s^{H,o} B_t^{H,o}] &= \frac{1}{4}\left(\frac{1}{2}(s^{2H} + t^{2H} - |s-t|^{2H}) - \frac{1}{2}(s^{2H} + t^{2H} - |s+t|^{2H})\right. \\ &\quad \left. - \frac{1}{2}(s^{2H} + t^{2H} - |s+t|^{2H}) + \frac{1}{2}(s^{2H} + t^{2H} - |s-t|^{2H})\right) \\ &= \frac{1}{4}\left(|s+t|^{2H} - |s-t|^{2H}\right).\end{aligned}$$

This is precisely the covariance function of an ofBm as defined in equation [2.6]. Therefore, $B^{H,o}$ is an ofBM. □

The relationship between a tsfBm and its odd part has been established. Analogously, we can characterize efBm in terms of a tsfBm.

PROPOSITION 2.3.– *A centered Gaussian process $B^{H,e} = (B_t^{H,e})_{t \geq 0}$ is an efBm if and only if there exists a tsfBm $B^H = (B_t^H)_{t \in \mathbb{R}}$ such that for all $t \geq 0$:*

$$B_t^{H,e} = \frac{1}{2}(B_t^H + B_{-t}^H). \qquad [2.16]$$

Proof. This proposition can be proved by following a similar approach to the proof of Proposition 2.1. The details are left as an exercise for the reader (see Exercise 2.4). □

REMARK 2.6.– *These propositions demonstrate that any tsfBm B^H can be uniquely decomposed into an odd part $B_t^{H,o} = \frac{1}{2}(B_t^H - B_{-t}^H)$ and an even part $B_t^{H,e} = \frac{1}{2}(B_t^H + B_{-t}^H)$. Conversely, by taking independent ofBm and efBm, we can construct a standard fBm. This decomposition highlights the underlying symmetry properties and provides a fundamental link between these seemingly distinct processes.*

REMARK 2.7.– *A particularly interesting case arises when the Hurst parameter $H = 1/2$. In this specific instance, both the ofBm $B^{1/2,o}$ (for $t \geq 0$) and the efBm $B^{1/2,e}$ (for $t \geq 0$) simplify significantly. Let us examine their covariance functions with $H = 1/2$: For $B^{1/2,o}$, with $s, t \geq 0$:*

$$\mathbb{E}[B_s^{1/2,o} B_t^{1/2,o}] = \frac{1}{4}\left(|s+t|^1 - |s-t|^1\right) = \frac{1}{4}\left((s+t) - |s-t|\right)$$
$$= \frac{1}{4}(2\min(s,t)) = \frac{1}{2}\min(s,t).$$

Similarly, for $B^{1/2,e}$, with $s, t \geq 0$:

$$\mathbb{E}[B_s^{1/2,e} B_t^{1/2,e}] = \frac{1}{2}\left(|t|^1 + |s|^1\right) - \frac{1}{4}\left(|t+s|^1 + |t-s|^1\right)$$
$$= \frac{1}{2}(t+s) - \frac{1}{4}((t+s) + |t-s|)$$
$$= \frac{1}{4}(t+s) - \frac{1}{4}|t-s| = \frac{1}{4}(2\min(s,t)) = \frac{1}{2}\min(s,t).$$

This demonstrates that for $H = 1/2$, both $B^{1/2,o}$ (on $[0,\infty)$) and $B^{1/2,e}$ (on $[0,\infty)$) are centered Gaussian processes with covariance function $\frac{1}{2}\min(s,t)$. This implies they are equivalent to a standard Brownian motion scaled by $1/\sqrt{2}$, thus possessing stationary increments.

2.5. Key stochastic characteristics of odd and even fractional Brownian motions

Following their construction from a tsfBm, ofBm $B^{H,o}$ and efBm $B^{H,e}$ possess several defining stochastic characteristics. These properties are crucial for understanding their behavior and distinguishing them within the family of self-similar processes. In the following theorem, we rigorously state these fundamental attributes.

THEOREM 2.2.– *Let $B^{H,o} = (B_t^{H,o})_{t\in\mathbb{R}}$ be an ofBm and $B^{H,e} = (B_t^{H,e})_{t\in\mathbb{R}}$ be an efBm, both with Hurst parameter $H \in (0,1)$.*

1) Both processes are H-self-similar. That is, for any $a > 0$:

$$(B_{at}^{H,o})_{t\in\mathbb{R}} \stackrel{d}{=} (a^H B_t^{H,o})_{t\in\mathbb{R}} \text{ and } (B_{at}^{H,e})_{t\in\mathbb{R}} \stackrel{d}{=} (a^H B_t^{H,e})_{t\in\mathbb{R}}.$$

2) *The ofBm $B^{H,o}$ and the efBm $B^{H,e}$ are independent processes.*

Proof. From Proposition 1.4, for a centered Gaussian process, H-self-similarity is equivalent to its covariance function satisfying

$$\mathbb{E}[X_{at}X_{as}] = a^{2H}\mathbb{E}[X_t X_s].$$

For $B^{H,o}$: using the covariance function from [2.6]:

$$\begin{aligned}\mathbb{E}\left[B^{H,o}(at)B^{H,o}(as)\right] &= \frac{1}{4}\left(|at+as|^{2H} - |at-as|^{2H}\right)\\ &= a^{2H}\left[\frac{1}{4}\left(|t+s|^{2H} - |t-s|^{2H}\right)\right]\\ &= a^{2H}\mathbb{E}\left[B^{H,o}(t)B^{H,o}(s)\right].\end{aligned}$$

This confirms $B^{H,o}$ is H-self-similar.

Similarly, for the efBm, we use its covariance function from [2.8]:

$$\begin{aligned}&\mathbb{E}\left[B^{H,e}(at)B^{H,e}(as)\right]\\ &= 2^{2H-2}\left(|at|^{2H} + |as|^{2H}\right) - \frac{1}{4}\left(|at+as|^{2H} + |at-as|^{2H}\right)\\ &= a^{2H}\mathbb{E}\left[B^{H,e}(t)B^{H,e}(s)\right].\end{aligned}$$

Therefore, $B^{H,e}$ is also H-self-similar.

Again from Proposition 1.4, to establish independence of $B^{H,o}$ and $B^{H,e}$, for Gaussian processes it suffices to show that their cross-covariance is zero. We compute $\mathbb{E}[B_t^{H,o}B_s^{H,e}]$ for any $s, t \in \mathbb{R}$:

$$\begin{aligned}\mathbb{E}[B_t^{H,o}B_s^{H,e}] &= \mathbb{E}\left[\left(\frac{B_t^H - B_{-t}^H}{2}\right)\left(\frac{B_s^H + B_{-s}^H}{2}\right)\right]\\ &= \frac{1}{4}\mathbb{E}[(B_t^H - B_{-t}^H)(B_s^H + B_{-s}^H)]\\ &= \frac{1}{4}\left(\mathbb{E}[B_t^H B_s^H] + \mathbb{E}[B_t^H B_{-s}^H] - \mathbb{E}[B_{-t}^H B_s^H] - \mathbb{E}[B_{-t}^H B_{-s}^H]\right).\end{aligned}$$

Expanding this expression and using the explicit form of the covariance function of B^H in equation [2.1], we find that the terms cancel out, leading to:

$$\mathbb{E}[B_t^{H,o}B_s^{H,e}] = 0.$$

Since their cross-covariance is zero, and they are Gaussian processes, $B^{H,o}$ and $B^{H,e}$ are independent. □

We now turn our attention to the core stochastic characteristics of ofBm and efBm. A particularly insightful property to investigate is the variance of their increments, as it directly reveals information about their inherent path regularity and scaling properties.

THEOREM 2.3.– *Let* $B^{H,o} = (B_t^{H,o})_{t\geq 0}$ *be an ofBm and* $B^{H,e} = (B_t^{H,e})_{t\in\mathbb{R}}$ *be an efBm, both with Hurst parameter* $H \in (0,1)$.

1) The variance of increments for the ofBm is given by:

$$\mathbb{E}[(B_t^{H,o} - B_s^{H,o})^2] = 2^{2H-2}\left(|t|^{2H} + |s|^{2H}\right) - \frac{1}{2}\left(|t+s|^{2H} - |t-s|^{2H}\right), \quad [2.17]$$

for all $s, t \geq 0$.

2) The variance of increments for the efBm is given by:

$$\mathbb{E}[(B_t^{H,e} - B_s^{H,e})^2] = \frac{1}{2}\left(|t+s|^{2H} + |t-s|^{2H}\right) - 2^{2H-2}\left(|t|^{2H} + |s|^{2H}\right) \; hbox, \quad [2.18]$$

for all $s, t \in \mathbb{R}$.

3) Neither ofBm nor efBm possess stationary increments, except when $H = 1/2$*, where their increments are indeed stationary. However, for* $H \neq 1/2$*, they exhibit a characteristic power-law scaling. For any* $s, t \geq 0$*, there exist positive constants* C *and* D *(which may depend on* H *and the domain of* s, t *but not on* s *and* t *individually) such that:*

$$C|t-s|^{2H} \leq \mathbb{E}[(B_t^{H,o} - B_s^{H,o})^2] \leq D|t-s|^{2H},$$

$$C'|t-s|^{2H} \leq \mathbb{E}[(B_t^{H,e} - B_s^{H,e})^2] \leq D'|t-s|^{2H}.$$

This property is sometimes referred to as a "quasi-helix", "quasi-Hurst" or "local scaling" property, indicating their path regularity akin to fBm.

Proof. The explicit expressions for the variance of increments, as presented in equations [2.17] and [2.18], are obtained by direct computation, which is left as an exercise for the reader (see Exercise 2.5).

Regarding the nonstationarity and power-law scaling (point 3): As shown in Remark 2.7, a notable exception occurs when $H = 1/2$. In this specific case, both $B^{1/2,o}$ (on $[0,\infty)$) and $B^{1/2,e}$ (on $[0,\infty)$) are equivalent to a standard Brownian motion scaled by $1/\sqrt{2}$, and thus their increments are indeed stationary.

For $H \neq 1/2$, without loss of generality, let us assume $t \geq s \geq 0$. This simplifies the absolute values: $|t| = t$, $|s| = s$, $|t+s| = t+s$ and $|s-t| = t-s$. The expression [2.17] becomes:

$$\mathbb{E}[(B_t^{H,o} - B_s^{H,o})^2] = 2^{2H-2}(t^{2H} + s^{2H}) - \frac{1}{2}((t+s)^{2H} - (t-s)^{2H}).$$

To establish the desired bounds, we can factor out t^{2H} and introduce the ratio $x = s/t$. Since $t \geq s$, x will be in the interval $[0, 1]$.

$$\mathbb{E}[(B_t^{H,o} - B_s^{H,o})^2] = t^{2H} \left[2^{2H-2} \left(1 + \left(\frac{s}{t}\right)^{2H}\right) - \frac{1}{2}\left(\left(1 + \frac{s}{t}\right)^{2H} - \left(1 - \frac{s}{t}\right)^{2H}\right)\right]$$
$$= t^{2H}\left[2^{2H-2}(1 + x^{2H}) - \frac{1}{2}((1 + x)^{2H} - (1 - x)^{2H})\right].$$

Let us define the function:

$$\psi_H(x) = 2^{2H-2}(1 + x^{2H}) - \frac{1}{2}((1 + x)^{2H} - (1 - x)^{2H}).$$

So, the variance can be written as $\mathbb{E}[(B_t^{H,o} - B_s^{H,o})^2] = t^{2H}\psi_H(x)$.

Our goal is to show that $C(t-s)^{2H} \leq t^{2H}\psi_H(x) \leq D(t-s)^{2H}$ with two positive constants C and D. Here, we recall that we have taken $x = s/t$.

Dividing by t^{2H} (which is positive):

$$C\left(\frac{t-s}{t}\right)^{2H} \leq \psi_H(x) \leq D\left(\frac{t-s}{t}\right)^{2H},$$

or

$$C(1 - x)^{2H} \leq \psi_H(x) \leq D(1 - x)^{2H}.$$

This means we need to prove that the ratio function $R_H(x) = \dfrac{\psi_H(x)}{(1-x)^{2H}}$ is bounded above and below by positive constants D and C for $x \in [0, 1]$.

The function $\psi_H(x)$ is a sum and difference of terms involving x^{2H}, which is continuous for $x \geq 0$. The denominator $(1 - x)^{2H}$ is continuous and nonzero for $x \in [0, 1)$. Therefore, $R_H(x)$ is continuous for all $x \in [0, 1)$.

Let us evaluate $R_H(x)$ as $x \to 0$:

$$\psi_H(0) = 2^{2H-2}(1 + 0^{2H}) - \frac{1}{2}((1 + 0)^{2H} - (1 - 0)^{2H}) = 2^{2H-2}.$$

So, $R_H(0) = \dfrac{2^{2H-2}}{(1-0)^{2H}} = 2^{2H-2}$. Since $H \in (0, 1)$, $2H - 2 \in (-2, 0)$, meaning 2^{2H-2} is a positive, finite constant.

This is the most crucial part. Let $x = 1 - \epsilon$ for a small $\epsilon > 0$. As $x \to 1^-$, $\epsilon \to 0^+$. The denominator becomes $(1 - (1 - \epsilon))^{2H} = \epsilon^{2H}$.

Now, let us analyze the numerator, $\psi_H(1 - \epsilon)$:

$$\psi_H(1-\epsilon) = 2^{2H-2}(1 + (1-\epsilon)^{2H}) - \frac{1}{2}((1 + (1-\epsilon))^{2H} - (1 - (1-\epsilon))^{2H})$$

$$= 2^{2H-2}(1 + (1-\epsilon)^{2H}) - \tfrac{1}{2}((2-\epsilon)^{2H} - \epsilon^{2H}).$$

We use the first-order Taylor expansion for $(a - b)^\alpha$:

$$(a-b)^\alpha = a^\alpha - \alpha a^{\alpha-1}b + O(b^2)$$

For small b/a, we use the approximation $(a - b)^\alpha \approx a^\alpha - \alpha a^{\alpha-1}b$, where $\approx$ denotes approximation up to the first order. Specifically, we have:

$$(1-\epsilon)^{2H} \approx 1 - 2H\epsilon$$
$$(2-\epsilon)^{2H} = 2^{2H}(1-\epsilon/2)^{2H} \approx 2^{2H}(1 - 2H(\epsilon/2)) = 2^{2H}(1 - H\epsilon)$$

Substitute these approximations back into the expression for $\psi_H(1 - \epsilon)$:

$$\begin{aligned}\psi_H(1-\epsilon) &\approx 2^{2H-2}(1 + (1-2H\epsilon)) - \frac{1}{2}(2^{2H}(1-H\epsilon) - \epsilon^{2H}) \\ &= 2^{2H-2}(2 - 2H\epsilon) - \frac{1}{2}2^{2H}(1-H\epsilon) + \frac{1}{2}\epsilon^{2H} \\ &= 2^{2H-1} - 2^{2H-1}H\epsilon - 2^{2H-1} + 2^{2H-1}H\epsilon + \frac{1}{2}\epsilon^{2H} \\ &= \frac{1}{2}\epsilon^{2H}.\end{aligned}$$

All of this allows us to get:

$$\lim_{x\to 1^-} R_H(x) = \lim_{\epsilon\to 0+} \frac{\frac{1}{2}\epsilon^{2H}}{\epsilon^{2H}} = \frac{1}{2}.$$

This is also a positive, finite constant.

Since the function $R_H(x) = \dfrac{\psi_H(x)}{(1-x)^{2H}}$ is continuous on the interval $[0, 1)$ and has finite, positive limits at both $x = 0$ and as $x \to 1^-$, it is therefore bounded above and below by positive constants on the closed interval $[0, 1]$. Let:

$$C = \min_{x\in[0,1]} R_H(x) \text{ and } D = \max_{x\in[0,1]} R_H(x).$$

These constants are guaranteed to be positive. Therefore, for all $x \in [0, 1]$:

$$C \le \frac{\psi_H(x)}{(1-x)^{2H}} \le D,$$

which implies

$$C(t-s)^{2H} \leq \mathbb{E}[(B_t^{H,o} - B_s^{H,o})^2] \leq D(t-s)^{2H}.$$

Since we initially assumed $t \geq s$, we have $|t-s| = t-s$. If $s > t$, the symmetry of the variance formula ensures that the inequality still holds with $|t-s|$. Thus, the quasi-Hurst inequalities are directly proven for the ofBm. □

COROLLARY 2.3.– *Let $H \in (0,1)$. Both ofBm $B^{H,o}$ and efBm $B^{H,e}$ admit versions whose sample paths are almost surely Hölder continuous of order strictly less than H.*

Proof. We will prove the result for ofBm $B^{H,o}$. The proof for efBm $B^{H,e}$ is analogous. First, recall that $B^{H,o}$ is a centered Gaussian process, and that from Theorem 2.3, Part 3, we have the following upper bound for the variance of the increments of $B^{H,o}$ for $s, t \geq 0$:

$$\mathbb{E}[(B_t^{H,o} - B_s^{H,o})^2] \leq D|t-s|^{2H}.$$

Therefore, by the Kolmogorov–Chentsov theorem for Gaussian processes [1.4], $B^{H,o}$ possesses a modification whose sample paths are almost surely γ-Hölder continuous for any order $\gamma \in (0, \nu/2)$ with $\nu = 2H$, which simplifies to $\gamma \in (0, H)$.

Thus, ofBm admits a version whose sample paths are almost surely Hölder continuous of order strictly less than H. □

Building on the established path continuity for ofBm and efBm, we now introduce a significant connection between efBm and a related process, the sfBm. This connection highlights the foundational nature of efBm within stochastic analysis.

REMARK 2.8.– *A particularly significant connection emerges when considering the sfBm, a well-known process in the literature extensively studied for its distinct properties. The sfBm, often denoted as $\xi^H = (\xi_t^H)_{t \geq 0}$, is a centered Gaussian process defined by its covariance function:*

$$g(s,t) = s^{2H} + t^{2H} - \frac{1}{2}\left((s+t)^{2H} + |t-s|^{2H}\right),$$

or equivalently, through the construction:

$$\xi_t^H = \frac{B_t^H + B_{-t}^H}{\sqrt{2}},$$

where $B^H = (B_t^H)_{t \in \mathbb{R}}$ is a tsfBm. For more on sfBm, see Mishura and Zili (2018); Bojdecki et al. (2004a); Tudor (2007).

Upon comparing the covariance function of sfBm with that of the efBm $f_e(s,t)$ *defined in equation [2.8], we observe that* $g(s,t) = 2 \cdot f_e(s,t)$. *This implies that the sfBm* ξ^H *is directly proportional to the efBm* $B^{H,e}$ *by a scaling factor of* $\sqrt{2}$. *Specifically, since both* ξ^H *and* $\sqrt{2}B^{H,e}$ *are centered Gaussian processes with identical covariance functions, they are equivalent (i.e. they have the same finite-dimensional distributions). Furthermore, as both sfBm and efBm have almost surely continuous sample paths, their equivalence, by Proposition 1.3, implies they are indistinguishable. This equivalence highlights that our rigorously defined efBm is not just a theoretical construct, but a fundamental building block closely related to established processes in stochastic analysis.*

This connection also offers an alternative perspective on the quasi-Hurst inequalities for ofBm $B^{H,o}$. *Since standard fBm* B^H *exhibits direct Hurst scaling (Proposition 2.1, point 2), and efBm* $B^{H,e}$ *inherits quasi-Hurst inequalities from its equivalence to a scaled sfBm (as just shown in this remark, and also in Mishura and Zili (2018); Bojdecki et al. (2004a) and Tudor (2007)), it can be deduced that* $B^{H,o}$ *also satisfies these inequalities due to its linear relationship with* B^H *and* $B^{H,e}$.

2.5.1. *Integral representations of covariance functions*

Beyond the variance of increments, integral representations provide a powerful tool for characterizing these processes. They often simplify proofs and reveal deeper mathematical structures.

PROPOSITION 2.4.– *Let* $B^{H,o}$ *be an ofBm and* $B^{H,e}$ *be an efBm, both with Hurst parameter* $H \in (0,1)$.

1) The covariance function of the ofBm can be expressed as:

$$\mathbb{E}[B^{H,o}(t)B^{H,o}(s)] = c_H^2 \int_0^\infty x^{-1-2H} \sin(sx)\sin(tx)dx, \quad [2.19]$$

for all $s,t \in \mathbb{R}$, *where* c_H^2 *is the constant defined in equation [2.5].*

2) The covariance function of the efBm can be expressed as:

$$\mathbb{E}[B^{H,e}(t)B^{H,e}(s)] = c_H^2 \int_0^\infty x^{-1-2H}(1-\cos(sx))(1-\cos(tx))dx, \quad [2.20]$$

for all $s,t \in \mathbb{R}$, *where* c_H^2 *is the constant defined in equation [2.5].*

Proof. We start by proving the integral representation for the ofBm. Consider the following integral identity (e.g. Erdelyi et al. 1954, section 1.4, equation 11):

$$\int_0^\infty x^{\nu-1}\sin(ax)\sin(xy)dx = \pi\frac{|y-a|^{-\nu} - |y+a|^{-\nu}}{4\Gamma(1-\nu)\sin(\nu\pi/2)}$$

This identity holds for every $y > 0$, $a > 0$ and $\nu \in \mathbb{C} \setminus \{0\}$ with $-2 < \text{Re}(\nu) < 1$.

In our context, we set $a = s$, $y = t$ and $\nu = -2H$. Since $H \in (0, 1)$, we have $-2 < -2H < 0$, which satisfies the condition on ν. For $s > 0$ and $t > 0$, substituting these values yields:

$$\int_0^\infty x^{-2H-1} \sin(sx) \sin(xt) dx = \pi \frac{|t-s|^{2H} - |t+s|^{2H}}{4\Gamma(1+2H) \sin(-H\pi)}$$

Using the property $\sin(-x) = -\sin(x)$, we can rewrite $\sin(-H\pi) = -\sin(H\pi)$. This leads to:

$$\int_0^\infty x^{-2H-1} \sin(sx) \sin(xt) dx = -\pi \frac{|t-s|^{2H} - |t+s|^{2H}}{4\Gamma(1+2H) \sin(H\pi)}$$

Rearranging the terms to match the covariance function of odd fBm:

$$\frac{|t+s|^{2H} - |t-s|^{2H}}{4} = \frac{\Gamma(1+2H) \sin(H\pi)}{\pi} \int_0^\infty x^{-2H-1} \sin(sx) \sin(xt) dx$$

Recalling the definition of $c_H^2 = \frac{\Gamma(1+2H) \sin(\pi H)}{\pi}$ from equation [2.5], we have:

$$\mathbb{E}[B^{H,o}(t) B^{H,o}(s)] = c_H^2 \int_0^\infty x^{-1-2H} \sin(sx) \sin(tx) dx$$

This establishes equation [2.19] for $s > 0, t > 0$.

Now, we extend this to all $s, t \in \mathbb{R}$. If $s > 0$ and $t < 0$, let $t' = -t > 0$. Since the sine function is odd ($\sin(-x) = -\sin(x)$), we have:

$$\begin{aligned} I &= \int_0^\infty x^{-2H-1} \sin(sx) \sin(xt) dx \\ &= \int_0^\infty x^{-2H-1} \sin(sx)(-\sin(xt')) dx \\ &= -\int_0^\infty x^{-2H-1} \sin(sx) \sin(xt') dx. \end{aligned}$$

Using the result for positive arguments:

$$\begin{aligned} I &= -\left(\frac{\pi}{\Gamma(1+2H) \sin(H\pi)} \frac{|t'-s|^{2H} - |t'+s|^{2H}}{4} \right) \\ &= \frac{\pi}{\Gamma(1+2H) \sin(H\pi)} \frac{|t'+s|^{2H} - |t'-s|^{2H}}{4} \end{aligned}$$

Substituting $t' = -t$:

$$I = \frac{\pi}{\Gamma(1+2H)\sin(H\pi)} \frac{|-t+s|^{2H} - |-t-s|^{2H}}{4}$$

$$= \frac{\pi}{\Gamma(1+2H)\sin(H\pi)} \frac{|t-s|^{2H} - |t+s|^{2H}}{4}.$$

This matches the form of the covariance function of odd fBm with $s > 0, t < 0$. A similar argument holds for $s < 0, t > 0$ and $s < 0, t < 0$. Thus, equation [2.19] is valid for all $s, t \in \mathbb{R}$.

Now, the integral representation for the efBm (equation [2.20]) can be derived by leveraging the decomposition of the fBm and the integral representations for both standard fBm and odd fBm.

We know that a tsfBm B^H can be decomposed into its odd and even parts (as shown in the previous section):

$$\mathbb{E}[B^H(t)B^H(s)] = \mathbb{E}[B^{H,o}(t)B^{H,o}(s)] + \mathbb{E}[B^{H,e}(t)B^{H,e}(s)].$$

We also have the integral representation [2.4] for the standard fBm:

$$\mathbb{E}[B^H(t)B^H(s)] = c_H^2 \int_0^\infty \frac{\sin(xs)\sin(xt) + (1-\cos(xs))(1-\cos(xt))}{x^{1+2H}} dx$$

and we have just proved the integral representation for the ofBm (equation [2.19]):

$$\mathbb{E}[B^{H,o}(t)B^{H,o}(s)] = c_H^2 \int_0^\infty x^{-1-2H} \sin(sx)\sin(tx) dx$$

By rearranging the decomposition equation, we get:

$$\mathbb{E}[B^{H,e}(t)B^{H,e}(s)] = \mathbb{E}[B^H(t)B^H(s)] - \mathbb{E}[B^{H,o}(t)B^{H,o}(s)].$$

Substituting the integral representations:

$$\begin{aligned}\mathbb{E}[B^{H,e}(t)B^{H,e}(s)] &= c_H^2 \int_0^\infty \frac{\sin(xs)\sin(xt) + (1-\cos(xs))(1-\cos(xt))}{x^{1+2H}} dx \\ &\quad - c_H^2 \int_0^\infty x^{-1-2H} \sin(sx)\sin(tx) dx \\ &= c_H^2 \int_0^\infty x^{-1-2H} (1-\cos(sx))(1-\cos(tx)) dx.\end{aligned}$$

This completes the proof of equation [2.20]. □

2.6. Series expansions for odd, even and standard fractional Brownian motions (Dzhaparidze–van Zanten's approach)

In sections 2.2 and 2.3, we defined the ofBm and efBm, $B^{H,o}$ and $B^{H,e}$, respectively, through their distinct covariance structures, and subsequently demonstrated their relationship to the standard fBm.

A powerful alternative perspective, and indeed the original context in which these components were formally introduced, is through the work of Dzhaparidze and van Zanten (Dzhaparidze and Zanten 2004). They defined $B^{H,o}$ and $B^{H,e}$ explicitly as the odd and even parts of the fBm B^H, leveraging this decomposition primarily to derive novel series expansions for B^H itself. Their method provides profound insights into the underlying structural properties of fBm.

In this section, we recall the main results obtained by Dzhaparidze and van Zanten and their subsequent series representations. These representations are often expressed in terms of Bessel functions, utilizing their zeros to form an orthogonal basis for the relevant function spaces, offering a constructive approach to the fBm.

We begin by recalling essential definitions related to Bessel functions, which are foundational to the series expansions presented hereafter. For $\nu \neq -1, -2, \dots$, the Bessel function J_ν of the first kind of order ν is defined on the region $\{z \in \mathbb{C}; |\arg(z)| < \pi\}$ as the absolutely convergent sum:

$$J_\nu(z) = \sum_{n=0}^{\infty} \frac{(-1)^n (z/2)^{\nu+2n}}{\Gamma(n+1)\Gamma(\nu+n+1)}.$$

It is well known that for $\nu > -1$, the function J_ν possesses a countable number of real, positive and simple zeros (see Watson 1944, Chapter 15). These zeros, denoted $x_1 < x_2 < \dots$, can be arranged in ascending order of magnitude and become arbitrarily large. The following lemma summarizes crucial asymptotic properties of the Bessel function and its positive zeros (see Watson 1944).

LEMMA 2.4 (Dzhaparidze and Zanten (2004) and Watson (1944)).– *For $\nu > -1$, let J_ν be the Bessel function of the first kind of order ν and let $z_1 < z_2 < \dots$ be its positive zeros.*

Then,

1) $z_n \sim n\pi$, *and* $J^2_{1+\nu}(z_n) \sim 2/(n\pi^2)$ *as* $n \to \infty$;

2) $J_{\nu-1}(z_n) + J_{\nu+1}(z_n) = 0$, *for every* $n = 1, 2, \dots$;

3) $\forall p > 0, \; \sum_{n=1}^{\infty} \frac{1}{z_n^{p+2} J^2_{1+\nu}(z_n)} < \infty.$

Let $H \in (0,1)$ be arbitrary. The constant c_H^2 used in the subsequent theorems is given by:

$$c_H^2 = \frac{\Gamma(1+2H)\sin \pi H}{\pi} \qquad [2.21]$$

where Γ is Euler's gamma function.

The following theorem presents the precise series representations for the covariance functions of the ofBm and efBm components, which are foundational for the full fBm expansion.

THEOREM 2.4 (Theorems 4.2 and 4.3 in Dzhaparidze and Zanten (2004)).– *Let $H \in (0,1)$ be arbitrary. The constant c_H^2 is given by equation [2.21].*

1) Let $x_1 < x_2 < \cdots$ be the positive, real zeros of J_{-H}. For $n \in \mathbb{N}$, we define:

$$\sigma_n^2 = \frac{2c_H^2}{x_n^{2H} J_{1-H}^2(x_n)}. \qquad [2.22]$$

Then for all $s,t \in [0,1]$, the covariance function of the ofBm $B^{H,o}$ is given by:

$$\mathbb{E}[B_s^{H,o} B_t^{H,o}] = \sum_{n=1}^{\infty} \frac{\sin(x_n s)\sin(x_n t)}{x_n^2}\sigma_n^2,$$

where the series converges absolutely and uniformly in $(s,t) \in [0,1] \times [0,1]$.

2) Let $y_1 < y_2 < \cdots$ be the positive, real zeros of J_{1-H}. For $n \in \mathbb{N}$, we define:

$$\tau_n^2 = \frac{2c_H^2}{y_n^{2H} J_{-H}^2(y_n)}. \qquad [2.23]$$

Then for all $s,t \in [0,1]$, the covariance function of the efBm $B^{H,e}$ is given by:

$$\mathbb{E}[B_s^{H,e} B_t^{H,e}] = \sum_{n=1}^{\infty} \frac{(1-\cos(y_n s))(1-\cos(y_n t))}{y_n^2}\tau_n^2,$$

where the series converges absolutely and uniformly in $(s,t) \in [0,1] \times [0,1]$.

By combining the covariance series representations for the ofBm and efBm provided in Theorem 2.4, we obtain the full series expansion for the standard fBm covariance, as stated in the following corollary.

COROLLARY 2.4.– *Let $H \in (0,1)$ be arbitrary. The covariance function of the fBm B^H for $s,t \in [0,1]$ is given by:*

$$\mathbb{E}[B_s^H B_t^H] = \sum_{n=1}^{\infty} \frac{\sin(x_n s)\sin(x_n t)}{x_n^2}\sigma_n^2 + \sum_{n=1}^{\infty} \frac{(1-\cos(y_n s))(1-\cos(y_n t))}{y_n^2}\tau_n^2, \qquad [2.24]$$

where x_n are the positive zeros of J_{-H}, y_n are the positive zeros of J_{1-H} and σ_n^2, τ_n^2 are defined by equations [2.22] and [2.23], respectively.

Proof. Since the tow processes $B^{H,o}$ and $B^{H,e}$ are independent, we have:

$$\mathbb{E}[B_s^H B_t^H] = \mathbb{E}[B_s^{H,o} B_t^{H,o}] + \mathbb{E}[B_s^{H,e} B_t^{H,e}], \quad \forall s,t \in [0,1].$$

Therefore, combining the series representations from Theorem 2.4, we get the states expansion [2.24]. □

The following theorem presents the key result of Dzhaparidze and van Zanten's work (2004): a complete, constructive series expansion that defines the fBm B^H itself in terms of the Bessel function zeros.

THEOREM 2.5.– *Let $H \in (0,1)$ be arbitrary. Define σ_n^2 and τ_n^2 using the positive zeros of the Bessel functions, x_n (for J_{-H}) and y_n (for J_{1-H}), as given in equations [2.22] and [2.23].*

Let $X_1, X_2, \ldots$ and $Y_1, Y_2, \ldots$ be two independent sequences of independent, centered Gaussian random variables on a common probability space, with Var$(X_n) = \sigma_n^2$ and Var$(Y_n) = \tau_n^2$.

The random process defined by the first series, $B_t^{H,o} = \sum_{n=1}^{\infty} \frac{\sin(x_n t)}{x_n} X_n$, is the ofBm, and the random process defined by the second series, $B_t^{H,e} = \sum_{n=1}^{\infty} \frac{1-\cos(y_n t)}{y_n} Y_n$, is the efBm.

Then, the fBm $B^H = (B_t^H)_{t\in[0,1]}$ is given by the sum:

$$B_t^H = \sum_{n=1}^{\infty} \frac{\sin(x_n t)}{x_n} X_n + \sum_{n=1}^{\infty} \frac{1-\cos(y_n t)}{y_n} Y_n.$$

With probability 1, both series converge absolutely and uniformly in $t \in [0,1]$, and B^H is an fBm with Hurst index H.

For a detailed proof for Theorem 2.5, an interested reader is referred to the paper by Dzhaparidze and van Zanten (2004), specifically Theorems 4.4 and 4.5.

REMARK 2.9.– *The series expansion presented in Theorem 2.5 is particularly valuable for the numerical simulation of fBm sample paths. By truncating the infinite series to a finite number of terms, we can generate approximate sample paths of B^H. The Gaussian nature and independence of the random variables X_n and Y_n, coupled with the explicit coefficients σ_n^2 and τ_n^2 and the well-defined Bessel function zeros, provide a concrete and efficient method for generating these paths in computational environments. This approach offers a direct alternative to other simulation methods that may rely on Cholesky decomposition of large covariance matrices or circulant embedding techniques. The uniform convergence properties of the series further guarantee that increasing the number of terms leads to a progressively more accurate representation of the true fBm path.*

2.7. Exercises

EXERCISE 2.1.– *Prove that the ofBm $B^{H,o}$ is a nondegenerate Gaussian process on the interval $(-\infty, 0)$. You may follow the same steps used in the proof for its nondegeneracy on $(0, +\infty)$.*

EXERCISE 2.2.–

1) Let $n = 2$ and consider the distinct time points $t_1 = -1$ and $t_2 = 1$, both in $\mathbb{R} \setminus \{0\}$. Compute the 2×2 covariance matrix Σ of the random vector $(B_{t_1}^{H,o}, B_{t_2}^{H,o})$. Recall that $Cov(B_s^{H,o}, B_t^{H,o}) = \frac{1}{4}\left(|t+s|^{2H} - |t-s|^{2H}\right)$.

2) Is the covariance matrix Σ invertible? Justify your answer by calculating its determinant.

3) Deduce whether the ofBm is a nondegenerate Gaussian process on $\mathbb{R} \setminus \{0\}$. Explain your reasoning.

EXERCISE 2.3.– *Verify that the covariance function of $B_t^H = B_{|t|}^{H,e} + sign(t) B_{|t|}^{H,o}$ is indeed $\mathbb{E}[B_s^H B_t^H] = \frac{1}{2}\left(|t|^{2H} + |s|^{2H} - |t-s|^{2H}\right)$ for all $s, t \in \mathbb{R}$, by discussing the different cases based on the signs of s and t. Recall that $B^{H,e}$ and $B^{H,o}$ are independent, so their cross-covariance is zero.*

EXERCISE 2.4.–

1) Let $B^{H,e}$ be an efBm. Let $B^{H,o}$ be an independent ofBm. Consider the process $B^H = (B_t^H)_{t \in \mathbb{R}}$ defined as:

$$B_t^H = B_{|t|}^{H,e} + sign(t) B_{|t|}^{H,o}. \quad [2.25]$$

Prove that this process B^H is indeed an fBm on $\mathbb{R}$. Once this is established, conclude that $B_t^{H,e} = \dfrac{1}{2}(B_t^H + B_{-t}^H)$ for $t \geq 0$.

2) *Assume that there exists an fBm* $B^H = (B_t^H)_{t\in\mathbb{R}}$ *such that* $B_t^{H,e} = \frac{1}{2}(B_t^H + B_{-t}^H)$ *for all* $t \geq 0$. *Prove that* $B^{H,e}$ *is an efBm.*

EXERCISE 2.5.– *Prove the formulas for the variance of increments as stated in Theorem 2.3, that is, show that:*

1) $\mathbb{E}[(B_t^{H,o} - B_s^{H,o})^2] = 2^{2H-2}\left(|t|^{2H} + |s|^{2H}\right) - \frac{1}{2}\left(|t+s|^{2H} - |t-s|^{2H}\right)$.

2) $\mathbb{E}[(B_t^{H,e} - B_s^{H,e})^2] = \frac{1}{2}\left(|t+s|^{2H} + |t-s|^{2H}\right) - 2^{2H-2}\left(|t|^{2H} + |s|^{2H}\right)$.

3

The Zili Generalized Fractional Brownian Motion: Definition and Core Properties

This chapter introduces the generalized fractional Brownian motion, the central subject of this book. This novel and comprehensive Gaussian process is designed as a versatile model that simultaneously generalizes both fractional Brownian motion (fBm) and sub-fractional Brownian motion (sfBm), effectively bridging a critical gap in stochastic modeling. By integrating arbitrary linear combinations of the two-sided fBm and its time-reversed counterpart, our generalized process offers a flexible framework capable of capturing a wider spectrum of behaviors, including both stationary and non-stationary increment structures. To avoid any confusion with recent work by Pang and Taqqu (2019), who also termed their extension of fBm "generalized fractional Brownian motion", our specific generalization of fBm, introduced herein, will be referred to as the Zili generalized fractional Brownian motion (ZgfBm) throughout this book.

3.1. Definition and main stochastic properties

We begin by rigorously defining the ZgfBm, which Zili initially introduced in 2017 (Zili 2017). Let $H \in (0,1)$ be the Hurst parameter, and let a, b be two real constants such that $(a,b) \neq (0,0)$.

DEFINITION 3.1.– *The ZgfBm of parameters* a, b *and* H, *is a process* $Z^H = \{Z_t^H(a,b); t \geq 0\}$, *defined on the probability space* $(\Omega, \mathcal{F}, \mathbb{P})$ *by:*

$$\forall t \in \mathbb{R}_+, \quad Z_t^H = Z_t^H(a,b) = aB_t^H + bB_{-t}^H, \qquad [3.1]$$

where $(B_t^H)_{t\in\mathbb{R}}$ *is a two-sided fractional Brownian motion with Hurst parameter* H.

This definition reveals the ZgfBm's capacity for generalization:

– If $a = 1$ and $b = 0$, then $Z_t^H = B_t^H$ for $t \geq 0$, which is precisely a standard fractional Brownian motion.

– If $a = b = \dfrac{1}{\sqrt{2}}$, then $Z_t^H = \dfrac{1}{\sqrt{2}}(B_t^H + B_{-t}^H)$, which corresponds to a sfBm, as demonstrated in Remark 2.8.

Furthermore, the ZgfBm readily reduces to a standard Brownian motion. This occurs if $H = \dfrac{1}{2}$ when $a = b = \frac{1}{\sqrt{2}}$, or more directly, when $a = 1$ and $b = 0$ with $H = \dfrac{1}{2}$. This inherent flexibility, enabling the ZgfBm to encompass fractional Brownian motion, sfBm, and standard Brownian motion, serves as a significant motivation for its introduction. It thus provides a powerful and versatile tool for modeling a broad spectrum of natural phenomena, particularly those exhibiting diverse increment structures, both stationary and non-stationary.

The following lemma summarizes some fundamental properties of the generalized fractional Brownian motion, derived directly from its definition and the known characteristics of two-sided fractional Brownian motion.

LEMMA 3.1.– *The ZgfBm $(Z_t^H(a,b))_{t\in\mathbb{R}_+}$ satisfies the following properties:*

1) Z^H is a centered Gaussian process.

2) $\forall s \in \mathbb{R}_+, \forall t \in \mathbb{R}_+,$

$$Cov\Big(Z_t^H(a,b), Z_s^H(a,b)\Big)$$
$$= \frac{1}{2}(a+b)^2\left(s^{2H} + t^{2H}\right) - ab(t+s)^{2H} - \frac{a^2+b^2}{2} \mid t - s \mid^{2H}.$$

3) $\forall t \in \mathbb{R}_+, \quad E\left[Z_t^H(a,b)^2\right] = \left(a^2 + b^2 - (2^{2H} - 2)ab\right) t^{2H}.$

4) For every $t > 0$, the variance $E\left[Z_t^H(a,b)^2\right]$ is strictly positive for all $(a,b) \in \mathbb{R}^2 \setminus \{(0,0)\}$.

Proof.

1) Since B_t^H is a centered Gaussian process and linear combinations of jointly Gaussian processes are also Gaussian, $Z_t^H = aB_t^H + bB_{-t}^H$ is immediately a Gaussian process. Being centered follows directly from $\mathbb{E}[B_t^H] = 0$ for all t, which implies:

$$\mathbb{E}[Z_t^H] = a\mathbb{E}[B_t^H] + b\mathbb{E}[B_{-t}^H] = a \cdot 0 + b \cdot 0 = 0.$$

2) We use the definition $Z_t^H = aB_t^H + bB_{-t}^H$ and $Z_s^H = aB_s^H + bB_{-s}^H$.

$$\begin{aligned}\mathrm{Cov}(Z_t^H, Z_s^H) &= \mathbb{E}\left[(aB_t^H + bB_{-t}^H)(aB_s^H + bB_{-s}^H)\right] \\ &= a^2\mathbb{E}[B_t^H B_s^H] + ab\mathbb{E}[B_t^H B_{-s}^H] + ba\mathbb{E}[B_{-t}^H B_s^H] + b^2\mathbb{E}[B_{-t}^H B_{-s}^H]\end{aligned}$$

Now, substitute the covariance function of a two-sided fBm given in [2.1]:

$$\mathbb{E}[B_t^H B_s^H] = \mathbb{E}[B_{-t}^H B_{-s}^H] = \frac{1}{2}(|t|^{2H} + |s|^{2H} - |t-s|^{2H}),$$

$$\mathbb{E}[B_t^H B_{-s}^H] = \mathbb{E}[B_{-t}^H B_s^H] = \frac{1}{2}(|t|^{2H} + |s|^{2H} - |t+s|^{2H}).$$

Substitute these back into the covariance expression:

$$\begin{aligned}
&\mathrm{Cov}(Z_t^H, Z_s^H) \\
&= (a^2+b^2)\frac{1}{2}(|t|^{2H} + |s|^{2H} - |t-s|^{2H}) + 2ab\frac{1}{2}(|t|^{2H} + |s|^{2H} - |t+s|^{2H}) \\
&= \frac{1}{2}(a+b)^2(|t|^{2H} + |s|^{2H}) - ab|t+s|^{2H} - \frac{a^2+b^2}{2}|t-s|^{2H}.
\end{aligned}$$

Since $t, s \in \mathbb{R}_+$, $|t| = t$, $|s| = s$, $|t+s| = t+s$. This matches the given formula.

3) This is a special case of the covariance function when $s = t$.

$$\begin{aligned}
\mathbb{E}\left[Z_t^H(a,b)^2\right] &= \mathrm{Cov}(Z_t^H(a,b), Z_t^H(a,b)) \\
&= \frac{1}{2}(a+b)^2(t^{2H} + t^{2H}) - ab(t+t)^{2H} - \frac{a^2+b^2}{2}|t-t|^{2H} \\
&= (a^2+b^2-(2^{2H}-2)ab)t^{2H}.
\end{aligned}$$

This matches the given formula.

4) We need to show that the coefficient $\left(a^2+b^2-(2^{2H}-2)ab\right)$ is strictly positive for all $(a,b) \in \mathbb{R}^2 \setminus \{(0,0)\}$ and $H \in (0,1)$. Let

$$C(a,b,H) = a^2 + b^2 - (2^{2H}-2)ab.$$

– *Case 1*: $a = 0$. In this case, $C(0,b,H) = 0^2 + b^2 - (2^{2H}-2)\cdot 0 \cdot b = b^2$. Since $(a,b) \neq (0,0)$, b must be non-zero, so $b^2 > 0$.

– *Case 2*: $b = 0$. Similarly, $C(a,0,H) = a^2$. Since $(a,b) \neq (0,0)$, a must be non-zero, so $a^2 > 0$.

– *Case 3*: $a \neq 0$ and $b \neq 0$. We can analyze the quadratic form $C(a,b,H)$ by considering it as a quadratic in the ratio $x = a/b$. Dividing by b^2 (which is positive), we get $x^2 - (2^{2H}-2)x + 1$. To determine the sign of this quadratic, we examine its discriminant:

$$\Delta = (2^{2H}-2)^2 - 4.$$

Let us analyze the term $2^{2H} - 2$ for $H \in (0,1)$:

– If $H = 1/2$, then $2^{2H} - 2 = 0$. In this specific case, $\Delta = -4 < 0$. So

$$C(a,b,1/2) = a^2 + b^2,$$

which is strictly positive for $(a,b) \neq (0,0)$.

– If $H \neq 1/2$: The function $f(H) = 2^{2H} - 2$ is strictly increasing on $(0, 1)$.

- As $H \to 0^+$, $f(H) \to 2^0 - 2 = -1$.

- As $H \to 1^-$, $f(H) \to 2^2 - 2 = 2$.

Therefore, for $H \in (0,1) \setminus \{1/2\}$, $2^{2H} - 2$ takes values in $(-1, 2) \setminus \{0\}$. Squaring this range, $(2^{2H} - 2)^2$ will be in $[0, 4) \setminus \{0\}$.

Consequently,

$$\Delta = (2^{2H} - 2)^2 - 4 < 0$$

for all $H \in (0, 1) \setminus \{1/2\}$.

Since the quadratic $x^2 - (2^{2H} - 2)x + 1$ has a positive leading coefficient (1) and a negative discriminant ($\Delta < 0$), it means the quadratic is always strictly positive for all real values of x.

Therefore,

$$C(a, b, H) = b^2((a/b)^2 - (2^{2H} - 2)(a/b) + 1)$$

is strictly positive for all $(a, b) \neq (0, 0)$.

Combining all cases, we conclude that $\left(a^2 + b^2 - (2^{2H} - 2)ab\right) > 0$ for all $(a, b) \in \mathbb{R}^2 \setminus \{(0, 0)\}$.

Thus,

$$\mathbb{E}\left[Z_t^H(a, b)^2\right] > 0$$

for all $t > 0$, whenever $(a, b) \neq (0, 0)$. □

The following proposition establishes the crucial property of self-similarity for the generalized fractional Brownian motion.

PROPOSITION 3.1.– *The generalized fractional Brownian motion* $Z^H = (Z_t^H(a, b))_{t \geq 0}$ *is self-similar with index* H. *That is, for any* $c > 0$, *the process* $(Z_{ct}^H(a, b))_{t \geq 0}$ *has the same finite-dimensional distributions as the process* $(c^H Z_t^H(a, b))_{t \geq 0}$.

Proof. To prove that Z^H is self-similar with index H, we need to show that for any $c > 0$, the processes $X_t = Z_{ct}^H(a, b)$ and $Y_t = c^H Z_t^H(a, b)$ are versions of each other. Since Z^H is a centered Gaussian process, X_t and Y_t are also centered Gaussian processes (as they are linear transformations and scalings of Z^H).

Therefore, it suffices to prove that they have the same covariance function.

Let us compute the covariance function for X_t: for any $s, t \geq 0$,

$$\begin{aligned}
\text{Cov}(X_s, X_t) &= \text{Cov}(Z_{cs}^H(a,b), Z_{ct}^H(a,b)) \\
&= \frac{1}{2}(a+b)^2 \left[(cs)^{2H} + (ct)^{2H}\right] - ab(cs+ct)^{2H} - \frac{a^2+b^2}{2}|cs-ct|^{2H} \\
&= c^{2H}\left[\frac{1}{2}(a+b)^2(s^{2H}+t^{2H}) - ab(s+t)^{2H} - \frac{a^2+b^2}{2}|s-t|^{2H}\right] \\
&= c^{2H}\text{Cov}(Z_s^H(a,b), Z_t^H(a,b)).
\end{aligned}$$

Now, let us compute the covariance function for Y_t: for any $s, t \geq 0$,

$$\begin{aligned}
\text{Cov}(Y_s, Y_t) &= \text{Cov}(c^H Z_s^H(a,b), c^H Z_t^H(a,b)) \\
&= c^{2H}\text{Cov}(Z_s^H(a,b), Z_t^H(a,b)).
\end{aligned}$$

Since $\text{Cov}(X_s, X_t) = \text{Cov}(Y_s, Y_t)$, and both processes are centered Gaussian, their finite-dimensional distributions are identical. Thus, the ZgfBm is self-similar with index H. □

We now investigate a crucial characteristic of the ZgfBm: its non-Markovian nature, which distinguishes it from classical Brownian motion and highlights its long-memory properties for certain parameter configurations.

THEOREM 3.1.– *For all $H \in (0,1) \setminus \{1/2\}$ and $(a,b) \in \mathbb{R}^2 \setminus \{(0,0)\}$, the process $(Z_t^H(a,b))_{t\in\mathbb{R}_+}$ is not a Markovian process.*

Proof. The proof proceeds by contradiction. We will assume that Z^H is a Markovian process for some $H \in (0,1) \setminus \{1/2\}$ and $(a,b) \in \mathbb{R}^2 \setminus \{(0,0)\}$, and show that this leads to a contradiction.

First, recall that Z^H is a centered Gaussian process. For Z^H to be a non-identically zero stochastic process, its variance $\text{Var}(Z_t^H)$ must be strictly positive for $t > 0$. From Lemma 3.1 (specifically, its fourth item and corresponding proof), we know that the variance:

$$\text{Var}(Z_t^H) = \mathbb{E}[(Z_t^H(a,b))^2] = \left(a^2+b^2-(2^{2H}-2)ab\right)t^{2H}$$

is indeed strictly positive for all $t > 0$, when $(a,b) \neq (0,0)$. This confirms that Z^H is a truly stochastic (non-trivial) process under the given conditions.

According to a known characterization for centered Gaussian Markov processes (see Revuz and Yor (1999)), if Z^H were a Markovian process, then for any ordered time points $s < t < u$, the following condition on its covariance function must hold:

$$\text{Cov}(Z_s^H, Z_u^H)\text{Cov}(Z_t^H, Z_t^H) = \text{Cov}(Z_s^H, Z_t^H)\text{Cov}(Z_t^H, Z_u^H). \qquad [3.2]$$

We will test this condition by choosing specific time points and examining the asymptotic behavior of the terms as $t \to \infty$. Let $s = t^{1/2}$, $t = t$ and $u = t^2$, with $t > 1$. This ensures $s < t < u$ for sufficiently large t.

Let $A = \frac{1}{2}(a+b)^2$, $B = ab$, and $D = \frac{a^2+b^2}{2}$. We note that $A - B - D = 0$.

Let $K_{xy} = \mathrm{Cov}(Z_x^H(a,b), Z_y^H(a,b))$. The condition becomes:

$$K_{s,u}K_{t,t} = K_{s,t}K_{t,u}. \qquad [3.3]$$

Using the covariance formula from Lemma 3.1:

$$\mathrm{Cov}(Z_x^H, Z_y^H) = A(x^{2H} + y^{2H}) - B(x+y)^{2H} - D|x-y|^{2H}.$$

We normalize each covariance term by t^{2H} to analyze their asymptotic behavior as $t \to \infty$. Let

$$C_0 = a^2 + b^2 - (2^{2H} - 2)ab = \frac{K_{t,t}}{t^{2H}}.$$

The normalized Markovian condition [3.3], after dividing both sides by $(t^{2H})^2$, becomes:

$$\left(\frac{K_{s,u}}{t^{2H}}\right) C_0 = \left(\frac{K_{s,t}}{t^{2H}}\right)\left(\frac{K_{t,u}}{t^{2H}}\right). \qquad [3.4]$$

We use the Taylor expansion $(1+x)^\gamma = 1 + \gamma x + \frac{\gamma(\gamma-1)}{2}x^2 + O(x^3)$ as $x \to 0$.

We derive the asymptotic expansions for each term as $t \to \infty$:

$$\begin{aligned}
\frac{K_{s,u}}{t^{2H}} &= \frac{\mathrm{Cov}(Z_{t^{1/2}}^H, Z_{t^2}^H)}{t^{2H}} \\
&= A(t^{-H} + t^{2H}) - B(1 + t^{-3/2})^{2H} - Dt^{2H}(1 - t^{-3/2})^{2H} \\
&= (A-D)t^{2H} + 2H(D-B)t^{2H-3/2} + At^{-H} - B + O(t^{2H-3}) \quad \text{as } t \to \infty.
\end{aligned}$$

$$\begin{aligned}
\frac{K_{s,t}}{t^{2H}} &= \frac{\mathrm{Cov}(Z_{t^{1/2}}^H, Z_t^H)}{t^{2H}} \\
&= A(t^{-H} + 1) - B(1 + t^{-1/2})^{2H} - D(1 - t^{-1/2})^{2H} \\
&= (A - B - D) + At^{-H} - 2H(B+D)t^{-1/2} + O(t^{-1}) \\
&= At^{-H} - 2H(B+D)t^{-1/2} + O(t^{-1}) \quad \text{as } t \to \infty \quad (\text{since } A - B - D = 0).
\end{aligned}$$

$$\frac{K_{t,u}}{t^{2H}} = \frac{\mathrm{Cov}(Z_t^H, Z_{t^2}^H)}{t^{2H}}$$
$$= A(1+t^{2H}) - Bt^{2H}(1+t^{-1})^{2H} - Dt^{2H}(1-t^{-1})^{2H}$$
$$= (A-B-D)t^{2H} + 2H(D-B)t^{2H-1} + A + O(t^{2H-2}) \quad \text{as } t \to \infty$$
$$= 2H(D-B)t^{2H-1} + A + O(t^{2H-2}) \quad (\text{since } A-B-D=0).$$

We now analyze two main cases for H.

Case 1: $1/2 < H < 1$

– The leading term of $\frac{K_{s,u}}{t^{2H}}$ is $(A-D)t^{2H}$, since $2H > 2H-3/2$ and $2H > -H$.

– The leading term of $\frac{K_{s,t}}{t^{2H}}$ is $-2H(B+D)t^{-1/2}$, since $-1/2 > -H$.

Note: $D - B = \frac{(a-b)^2}{2} \geq 0$, $B + D = \frac{a^2+b^2+2ab}{2} = A \geq 0$.

So, $D-B$ can be zero if $a=b$, but $B+D$ is generally not zero unless $a=b=0$.

– The leading term of $\frac{K_{t,u}}{t^{2H}}$ is $2H(D-B)t^{2H-1}$, since $2H-1>0$ for $H>1/2$.

Substituting these into the normalized Markov condition [3.4]: the LHS is dominated by: $C_0(A-D)t^{2H}$ as $t \to \infty$.

The RHS is dominated by the product of the leading terms of $\frac{K_{s,t}}{t^{2H}}$ and $\frac{K_{t,u}}{t^{2H}}$:

$$\left(2H(D-B)t^{-1/2}\right)\left(2H(D-B)t^{2H-1}\right) = 4H^2(D-B)^2 t^{2H-3/2} \quad \text{as } t \to \infty.$$

Comparing the leading powers of t: the LHS has a term of order t^{2H}, while the RHS has a highest term of order $t^{2H-3/2}$.

Since $2H > 2H - 3/2$, for the equality to hold asymptotically as $t \to \infty$, the coefficient of the highest power of t on the LHS must be zero. Therefore, we must have $C_0(A-D) = 0$. Since $A - D = ab$, this implies $C_0 \cdot ab = 0$. This requires either $ab = 0$ or $C_0 = 0$. We analyze these two sub-cases:

Case 1.1: $ab = 0$

Since $(a,b) \neq (0,0)$, this implies either $a=0$ (and $b \neq 0$) or $b=0$ (and $a \neq 0$). In either situation, we have:

$$A = \frac{1}{2}(a^2+b^2), \; B = 0, \; \text{and } D = \frac{1}{2}(a^2+b^2).$$

Thus, $A = D$ and $B = 0$. This makes $A - D = 0$, so the t^{2H} term on the LHS vanishes as required. Now, we compare the next dominant terms of LHS and RHS. The next dominant term on the LHS is from $C_0 \cdot 2H(D - B)t^{2H-3/2}$.

Substituting $B = 0, D = A = \frac{1}{2}(a^2 + b^2)$, and $C_0 = a^2 + b^2$ (since $ab = 0$): LHS leading term:

$$(a^2 + b^2) \cdot 2H(\frac{1}{2}(a^2 + b^2))t^{2H-3/2} = H(a^2 + b^2)^2 t^{2H-3/2}.$$

The RHS leading term remains $4H^2(D - B)^2 t^{2H-3/2}$.

Substituting $B = 0, D = A = \frac{1}{2}(a^2 + b^2)$: RHS leading term:

$$4H^2(\frac{1}{2}(a^2 + b^2))^2 t^{2H-3/2} = H^2(a^2 + b^2)^2 t^{2H-3/2}.$$

Equating the coefficients of $t^{2H-3/2}$:

$$H(a^2 + b^2)^2 = H^2(a^2 + b^2)^2.$$

Since $(a, b) \neq (0, 0)$, $(a^2 + b^2)^2 \neq 0$. Also, $H \neq 0$. Thus, $1 = H$. This result contradicts our assumption for this case, which is $1/2 < H < 1$. Therefore, the Markov condition cannot hold if $ab = 0$ for $1/2 < H < 1$.

Case 1.2: $C_0 = 0$

The condition $C_0 = 0$ implies $a^2 + b^2 - (2^{2H} - 2)ab = 0$. As the previously analyzed $ab = 0$ scenario covers cases where $b = 0$, we can assume $b \neq 0$ here. Dividing by b^2, we obtain a quadratic equation in (a/b):

$$\left(\frac{a}{b}\right)^2 - (2^{2H} - 2)\left(\frac{a}{b}\right) + 1 = 0.$$

The discriminant is $\Delta = (2^{2H} - 2)^2 - 4$.

Since $1/2 < H < 1$, we have $2H \in (1, 2)$. This implies $2^{2H} \in (2, 4)$. Therefore, $2^{2H} - 2 \in (0, 2)$. Consequently, $(2^{2H} - 2)^2 \in (0, 4)$. This means:

$$\Delta = (2^{2H} - 2)^2 - 4 < 0.$$

A quadratic equation with real coefficients and a negative discriminant has no real solutions for a/b. This contradicts the assumption that a and b are real constants (not both zero) that would make $C_0 = 0$. Thus, the Markov condition cannot hold for $C_0 = 0$ when $1/2 < H < 1$.

Case 2: $0 < H < 1/2$

– The leading term of $\frac{K_{s,u}}{t^{2H}}$ is $(A - D)t^{2H}$, since $2H > 0$ and $2H > -H$.

– The leading term of $\frac{K_{s,t}}{t^{2H}}$ is At^{-H}, since $-H > -1/2$.

– The leading term of $\frac{K_{t,u}}{t^{2H}}$ is A, since $2H - 1 < 0$.

Substituting these into the normalized Markov condition [3.4]:

– The LHS is dominated by: $C_0(A - D)t^{2H}$ as $t \to \infty$.

– The RHS is dominated by the product of the leading terms of $\frac{K_{s,t}}{t^{2H}}$ and $\frac{K_{t,u}}{t^{2H}}$:

$$(At^{-H})(A) = A^2t^{-H} \quad \text{as } t \to \infty.$$

Comparing the leading powers of t:

– The LHS has a term of order t^{2H}, while the RHS has a term of order t^{-H}.

Since $2H > -H$ for $H > 0$, for the equality to hold asymptotically as $t \to \infty$, the coefficient of the highest power of t on the LHS must be zero. Therefore, we must have $C_0(A - D) = 0$. Since $A - D = ab$, this implies $C_0 \cdot ab = 0$. This requires either $ab = 0$ or $C_0 = 0$. We analyze these two sub-cases:

Case 2.1: $ab = 0$

As in Case 1.1, this implies either $a = 0$ ($b \neq 0$) or $b = 0$ ($a \neq 0$). In both scenarios, we have $A = D = \frac{1}{2}(a^2 + b^2)$ and $B = 0$. This makes $A - D = 0$, so the t^{2H} term on the LHS vanishes as required.

Now, we need to compare the next dominant terms. The full expansion for $\frac{K_{s,u}}{t^{2H}}$ with $A = D, B = 0$ is:

$$\frac{K_{s,u}}{t^{2H}} = At^{-H} + 2HAt^{2H-3/2} - H(2H - 1)At^{2H-3} + O(t^{2H-9/2}).$$

Since $0 < H < 1/2$, we have $-H > 2H - 3/2$. Thus, the leading term on the LHS is $C_0 \cdot At^{-H}$. Substituting $C_0 = a^2 + b^2$ and $A = \frac{1}{2}(a^2 + b^2)$: LHS leading term:

$$(a^2 + b^2) \cdot \frac{1}{2}(a^2 + b^2)t^{-H} = \frac{1}{2}(a^2 + b^2)^2t^{-H}.$$

The RHS leading term is A^2t^{-H}. Substituting $A = \frac{1}{2}(a^2 + b^2)$: RHS leading term:

$$(\frac{1}{2}(a^2 + b^2))^2t^{-H} = \frac{1}{4}(a^2 + b^2)^2t^{-H}.$$

Equating the coefficients of t^{-H}:

$$\frac{1}{2}(a^2+b^2)^2 = \frac{1}{4}(a^2+b^2)^2.$$

Since, $(a,b) \neq (0,0)$, $(a^2+b^2)^2 \neq 0$. This leads to the clear contradiction $\frac{1}{2} = \frac{1}{4}$. Therefore, the Markov condition cannot hold if $ab = 0$ for $0 < H < 1/2$.

Case 2.2: $C_0 = 0$

The condition $C_0 = 0$ implies $a^2 + b^2 - (2^{2H} - 2)ab = 0$. As before, assuming $b \neq 0$, we have:

$$\left(\frac{a}{b}\right)^2 - (2^{2H} - 2)\left(\frac{a}{b}\right) + 1 = 0.$$

The discriminant of this quadratic equation is $\Delta = (2^{2H} - 2)^2 - 4$.

Since $0 < H < 1/2$, we have $2H \in (0,1)$. This implies $2^{2H} \in (1,2)$. Therefore, $2^{2H} - 2 \in (-1,0)$. Consequently, $(2^{2H} - 2)^2 \in (0,1)$.

This means the discriminant $\Delta = (2^{2H} - 2)^2 - 4$ is negative (specifically, $\Delta \in (-4,-3)$). A quadratic equation with real coefficients and a negative discriminant has no real solutions for a/b. This contradicts the assumption that a and b are real constants (not both zero) that would make $C_0 = 0$.

Thus, the Markov condition cannot hold for $C_0 = 0$ when $0 < H < 1/2$.

All possible scenarios for $H \in (0,1) \setminus \{1/2\}$ and $(a,b) \in \mathbb{R}^2 \setminus \{(0,0)\}$ have led to a contradiction, thus proving that Z^H is not a Markovian process under these conditions. □

3.2. The increments study and some applications

Let us start with the following lemma, in which we will characterize the second moment increments of the ZgfBm.

LEMMA 3.2.– *For all* $(s,t) \in \mathbb{R}_+^2; s \leq t$:

$$\mathbb{E}\left[\left(Z_t^H(a,b) - Z_s^H(a,b)\right)^2\right]$$
$$= (a^2+b^2) \mid t-s \mid^{2H} - 2^{2H}ab(\mid t \mid^{2H} + \mid s \mid^{2H}) + 2ab \mid t+s \mid^{2H}.$$

Proof. We begin by deriving the second moment of the increments of the ZgfBm. By definition, the variance of the increment can be expanded as:

$$\mathbb{E}\left[(Z_t^H(a,b) - Z_s^H(a,b))^2\right]$$
$$= \mathbb{E}[(Z_t^H(a,b))^2] + \mathbb{E}[(Z_s^H(a,b))^2] - 2\text{Cov}(Z_t^H(a,b), Z_s^H(a,b)).$$

Substituting the expressions for the variance and the covariance directly from Lemma 3.1, we get the stated formula for the second moment of the increments, completing the first assertion. □

COROLLARY 3.1.– *For all* $(s,t) \in \mathbb{R}_+^2; s \leq t$:

$$\gamma(a,b,H)(t-s)^{2H} \leq \mathbb{E}\left[\left(Z_t^H(a,b) - Z_s^H(a,b)\right)^2\right] \leq \nu(a,b,H)(t-s)^{2H} \qquad [3.5]$$

with

$$\gamma(a,b,H) = \left(a^2+b^2-2ab(2^{2H-1}-1)\right)\mathbf{1}_{\mathcal{C}}(a,b,H)+\left(a^2+b^2\right)\mathbf{1}_{\mathcal{D}}(a,b,H)$$

and

$$\nu(a,b,H) = \left(a^2+b^2\right)\mathbf{1}_{\mathcal{C}}(a,b,H) + \left(a^2+b^2-2ab(2^{2H-1}-1)\right)\mathbf{1}_{\mathcal{D}}(a,b,H),$$

$$\mathcal{C} = \left\{(a,b,H) \in \mathbb{R}^2 \setminus \{(0,0)\}\times]0,1[; (H > \frac{1}{2}, ab \geq 0) \text{ or } (H < \frac{1}{2}, ab \leq 0)\right\},$$

$$\mathcal{D} = \left\{(a,b,H) \in \mathbb{R}^2 \setminus \{(0,0)\}\times]0,1[; (H > \frac{1}{2}, ab \leq 0) \text{ or } (H < \frac{1}{2}, ab \geq 0)\right\}.$$

Moreover, the constants in the second assertion inequalities are the best possible.

Proof. We will establish the inequalities for the second moment of the increments in the case where $H > 1/2$ and $ab \geq 0$. In this specific case, according to the corollary, we expect to show:

$$\gamma(a,b,H)(t-s)^{2H} \leq \mathbb{E}\left[(Z_t^H(a,b) - Z_s^H(a,b))^2\right] \leq \nu(a,b,H)(t-s)^{2H}$$

where $\gamma(a,b,H) = \left(a^2+b^2-2ab(2^{2H-1}-1)\right)$ and $\nu(a,b,H) = \left(a^2+b^2\right)$. The proof for other cases proceeds analogously by considering the signs of relevant terms.

From Lemma 3.2, we have:

$$\mathbb{E}\left[(Z_t^H(a,b) - Z_s^H(a,b))^2\right] = (a^2+b^2)(t-s)^{2H} + 2ab\left[(t+s)^{2H} - 2^{2H-1}(t^{2H}+s^{2H})\right].$$

Let

$$Q(s,t) = (t+s)^{2H} - 2^{2H-1}(t^{2H}+s^{2H}).$$

Since $H > 1/2$, the function $x \mapsto x^{2H}$ is convex on $\mathbb{R}_+$. A property of convex functions is that for $s,t \geq 0$:

$$(t+s)^{2H} \leq 2^{2H-1}(t^{2H}+s^{2H}).$$

This implies $Q(s,t) \leq 0$. Since we are in the case where $ab \geq 0$, multiplying $Q(s,t) \leq 0$ by $2ab \geq 0$ preserves the inequality direction: $2abQ(s,t) \leq 0$.

Substituting this back into the expression for $\mathbb{E}[(Z_t^H - Z_s^H)^2]$:

$$\mathbb{E}[(Z_t^H - Z_s^H)^2] \leq (a^2 + b^2)(t-s)^{2H}.$$

This confirms the upper bound $\nu(a,b,H) = (a^2 + b^2)$ for this specific case.

To find the lower bound, we will show that

$$Q(s,t) \geq (1 - 2^{2H-1})(t-s)^{2H}.$$

Let $x = t - s$ (the length of the interval) and treat s as a variable for the starting point of the interval. Consider the function:

$$f_\lambda(x) = 2^{2H-1}((s+x)^{2H} + s^{2H}) - (2s+x)^{2H} - \lambda x^{2H}$$

for $x > 0$. We first check $f_\lambda(0)$:

$$f_\lambda(0) = 2^{2H-1}(s^{2H} + s^{2H}) - (2s)^{2H} - \lambda(0)^{2H} = 2^{2H-1}(2s^{2H}) - 2^{2H}s^{2H} = 0.$$

The derivative of $f_\lambda(x)$ with respect to x is:

$$f'_\lambda(x) = 2H\left[2^{2H-1}(s+x)^{2H-1} - (2s+x)^{2H-1} - \lambda x^{2H-1}\right].$$

To ensure $f_\lambda(x)$ is a decreasing function for all $x > 0$, we require $f'_\lambda(x) \leq 0$. This is equivalent to finding λ such that:

$$2^{2H-1}(s+x)^{2H-1} - (2s+x)^{2H-1} \leq \lambda x^{2H-1}.$$

For $x > 0$, we can divide by x^{2H-1}:

$$\lambda \geq \frac{2^{2H-1}(s+x)^{2H-1} - (2s+x)^{2H-1}}{x^{2H-1}}.$$

Let

$$g(x) = \frac{2^{2H-1}(s+x)^{2H-1} - (2s+x)^{2H-1}}{x^{2H-1}}.$$

We can rewrite $g(x)$ by setting $X = s/x$. Note that as $x \to 0^+$, $X \to \infty$; and as $x \to \infty$, $X \to 0^+$.

$$g(x) = 2^{2H-1}\left(\frac{s}{x} + 1\right)^{2H-1} - \left(2\frac{s}{x} + 1\right)^{2H-1} = h(X)$$

where

$$h(X) = 2^{2H-1}(X+1)^{2H-1} - (2X+1)^{2H-1}$$

for $X \geq 0$. The function $h(X)$ is continuous and can be shown to be strictly decreasing on $[0, \infty)$. Let us evaluate its limits:

$$h(0) = 2^{2H-1}(0+1)^{2H-1} - (2 \cdot 0 + 1)^{2H-1} = 2^{2H-1} - 1.$$

$$\lim_{X\to\infty} h(X) = \lim_{X\to\infty} \left[2^{2H-1} X^{2H-1} \left(1 + \frac{1}{X}\right)^{2H-1} - (2X)^{2H-1} \left(1 + \frac{1}{2X}\right)^{2H-1} \right]$$

$$= \lim_{X\to\infty} X^{2H-1} 2^{2H-1} \left[\left(1 + \frac{1}{X}\right)^{2H-1} - \left(1 + \frac{1}{2X}\right)^{2H-1} \right].$$

Using the generalized binomial theorem $(1+u)^{\alpha} = 1 + \alpha u + O(u^2)$ for small u:

$$\begin{aligned}
&\lim_{X\to\infty} h(X) \\
&= \lim_{X\to\infty} X^{2H-1} 2^{2H-1} \left[\frac{2H-1}{X} - \frac{2H-1}{2X} + O(1/X^2) \right] \\
&= \lim_{X\to\infty} X^{2H-1} 2^{2H-1} \frac{2H-1}{2X} \\
&= \lim_{X\to\infty} 2^{2H-2}(2H-1) X^{2H-2}.
\end{aligned}$$

Since $H < 1$, $2H - 2 < 0$, so $X^{2H-2} \to 0$ as $X \to \infty$. Therefore,

$$\lim_{X\to\infty} h(X) = 0.$$

Thus, the range of $h(X)$ (and $g(x)$) is $(\lim_{X\to\infty} h(X), h(0)] = (0, 2^{2H-1} - 1]$. So,

$$\sup_{x>0} g(x) = 2^{2H-1} - 1.$$

To ensure $f'_\lambda(x) \leq 0$ for all $x > 0$ (making $f_\lambda(x)$ decreasing), we must choose:

$$\lambda \geq \sup_{x>0} g(x).$$

The smallest such that λ is $\lambda^* = 2^{2H-1} - 1$.

Since $H > 1/2$, $\lambda^* > 0$. Since $f_{\lambda^*}(0) = 0$ and $f_{\lambda^*}(x)$ is decreasing for $x > 0$, it follows that $f_{\lambda^*}(x) \leq 0$ for all $x > 0$.

Substituting back the definitions of $f_{\lambda^*}(x)$, using $x = t - s$ and $2s + x = t + s$:

$$2^{2H-1}((s+x)^{2H} + s^{2H}) - (2s+x)^{2H} - (2^{2H-1} - 1)(t-s)^{2H} \leq 0.$$

The first two terms are equivalent to $-Q(s,t)$:

$$-Q(s,t) - (2^{2H-1} - 1)(t-s)^{2H} \leq 0.$$

Rearranging this inequality, we get the lower bound for $Q(s,t)$:

$$Q(s,t) \geq -(2^{2H-1}-1)(t-s)^{2H} = (1-2^{2H-1})(t-s)^{2H}.$$

So, for $H > 1/2$, we have established that:

$$(1-2^{2H-1})(t-s)^{2H} \leq Q(s,t).$$

Since we are in the case where $ab \geq 0$, multiplying by $2ab \geq 0$ preserves the inequality direction:

$$2ab(1-2^{2H-1})(t-s)^{2H} \leq 2abQ(s,t).$$

Adding $(a^2+b^2)(t-s)^{2H}$ to both parts of the above inequality:

$$(a^2+b^2)(t-s)^{2H} + 2ab(1-2^{2H-1})(t-s)^{2H} \leq \mathbb{E}[(Z_t^H - Z_s^H)^2].$$

Factoring out $(t-s)^{2H}$:

$$\left(a^2+b^2-2ab(2^{2H-1}-1)\right)(t-s)^{2H} \leq \mathbb{E}[(Z_t^H - Z_s^H)^2].$$

This confirms the lower bound $\gamma(a,b,H) = \left(a^2+b^2-2ab(2^{2H-1}-1)\right)$ for this specific case ($H > 1/2, ab \geq 0$).

Now, to show that the constants $\gamma(a,b,H)$ and $\nu(a,b,H)$ are the best possible, we demonstrate that they are achieved asymptotically.

Let $x = s/t$ with $0 \leq x \leq 1$. We analyze the limit of $\dfrac{\mathbb{E}[(Z_t^H - Z_s^H)^2]}{(t-s)^{2H}}$ as x approaches its bounds.

$$\frac{\mathbb{E}[(Z_t^H - Z_s^H)^2]}{(t-s)^{2H}} = (a^2+b^2) + 2ab\frac{(1+x)^{2H} - 2^{2H-1}(1+x^{2H})}{(1-x)^{2H}}.$$

Let

$$f(x) = \frac{(1+x)^{2H} - 2^{2H-1}(1+x^{2H})}{(1-x)^{2H}}.$$

Upper bound achievement ($s \to t$ or $x \to 1^-$): as $x \to 1^-$, $f(x)$ is an indeterminate form $0/0$. Applying L'Hôpital's rule twice yields:

$$\lim_{x\to 1^-} f(x) = 0.$$

Therefore,

$$\lim_{s\to t^-} \frac{\mathbb{E}[(Z_t^H - Z_s^H)^2]}{(t-s)^{2H}} = (a^2+b^2) + 2ab\cdot 0 = a^2+b^2 = \nu(a,b,H).$$

This shows that the upper bound is achieved as $s \to t$.

Lower bound achievement ($s \to 0^+$ or $x \to 0^+$): to show that the lower bound $\gamma(a, b, H)$ is the best possible, we examine the limit of the scaled variance as $s \to 0^+$.

As $s \to 0^+$ (for fixed $t > 0$), we have $x = s/t \to 0^+$. Evaluating $f(x)$ directly as $x \to 0^+$, we find:

$$\lim_{x \to 0^+} f(x) = \frac{(1+0)^{2H} - 2^{2H-1}(1+0^{2H})}{(1-0)^{2H}} = \frac{1-2^{2H-1}}{1} = 1 - 2^{2H-1}.$$

Substituting this limit back into the expression for the scaled variance of the increments:

$$\begin{aligned}\lim_{s \to 0^+} \frac{\mathbb{E}[(Z_t^H(a,b) - Z_s^H(a,b))^2]}{(t-s)^{2H}} &= \lim_{s \to 0^+} \left[(a^2+b^2) + 2ab\frac{Q(s,t)}{(t-s)^{2H}}\right] \\ &= (a^2+b^2) + 2ab(1-2^{2H-1}) \\ &= a^2 + b^2 - 2ab(2^{2H-1} - 1) \\ &= \gamma(a,b,H).\end{aligned}$$

This demonstrates that the lower bound $\gamma(a, b, H)$ is achieved as $s \to 0^+$, confirming it is the best possible constant. □

The following result is a first important deduction from Corollary 3.1. It deals with the continuity of the ZgfBm sample paths.

COROLLARY 3.2.– *Let $H \in (0, 1)$ and $(a, b) \neq (0, 0)$. The ZgfBm admits a version whose sample paths are almost surely Hölder continuous of order strictly less than H.*

Proof. The ZgfBm is by a centered Gaussian process, and from Corollary 3.1, we have the following upper bound for the variance of the ZgfBm's increments for any s, t:

$$\mathbb{E}[(Z_t^H(a,b) - Z_s^H(a,b))^2] \leq \nu(a,b,H)|t-s|^{2H}.$$

Thus, by Theorem 1.4, $Z^H(a, b)$ admits a version whose sample paths are almost surely γ-Hölder continuous for any order $\gamma \in (0, \nu/2)$ with $\nu = 2H$, which simplifies to $\gamma \in (0, H)$.

This concludes that the ZgfBm admits a version whose sample paths are almost surely Hölder continuous of order strictly less than H. □

Another key insight from Corollary 3.1 is the non-differentiability of the ZgfBm sample paths.

COROLLARY 3.3.– *Let $H \in (0,1)$ and $(a,b) \neq (0,0)$. For any $t_0 \in \mathbb{R}$:*

$$\lim_{\epsilon \to 0^+} \sup_{t \in [t_0-\epsilon, t_0+\epsilon]} \left| \frac{Z_t^H(a,b) - Z_{t_0}^H(a,b)}{t - t_0} \right| = +\infty, \qquad [3.6]$$

with probability one.

Proof. Our goal is to prove that the increment quotient almost surely blows up as the interval shrinks, implying that the ZgfBm is almost surely nowhere differentiable at t_0. This is equivalent to showing that the probability of the complementary event, where the limit superior is finite, is zero. Let

$$L_{t_0}(\omega) = \lim_{\epsilon \to 0^+} \sup_{t \in [t_0-\epsilon, t_0+\epsilon]} \left| \frac{Z_t^H(\omega,a,b) - Z_{t_0}^H(\omega,a,b)}{t - t_0} \right|.$$

We aim to show $\mathbb{P}(L_{t_0} = +\infty) = 1$, which is equivalent to $\mathbb{P}(L_{t_0} < +\infty) = 0$.

Let M be an arbitrary positive constant. Consider the event F_M, where L_{t_0} is bounded by M:

$$F_M = \{\omega \in \Omega; L_{t_0}(\omega) \leq M\}.$$

If $L_{t_0}(\omega) < +\infty$, then $L_{t_0}(\omega) \leq M$ for some sufficiently large integer M. Thus,

$$\mathbb{P}(L_{t_0} < +\infty) = \mathbb{P}\left(\bigcup_{M=1}^{\infty} F_M \right).$$

If we can show $\mathbb{P}(F_M) = 0$ for every M, then by countable subadditivity:

$$\mathbb{P}\left(\bigcup_{M=1}^{\infty} F_M \right) \leq \sum_{M=1}^{\infty} \mathbb{P}(F_M) = 0.$$

Now, let us analyze the event F_M. Consider a sequence of shrinking intervals defined by $\epsilon_n = 1/n$ for $n \in \mathbb{N}$. Let $S_n(\omega)$ be the supremum of the increment quotient over the interval $[t_0 - 1/n, t_0 + 1/n]$ for a given outcome ω:

$$S_n(\omega) = \sup_{t \in \left[t_0 - \frac{1}{n}, t_0 + \frac{1}{n}\right]} \left| \frac{Z_t^H(\omega,a,b) - Z_{t_0}^H(\omega,a,b)}{t - t_0} \right|.$$

As n increases, the interval $\left[t_0 - \frac{1}{n}, t_0 + \frac{1}{n}\right]$ shrinks. Specifically,

$$\left[t_0 - \frac{1}{n+1}, t_0 + \frac{1}{n+1}\right] \subseteq \left[t_0 - \frac{1}{n}, t_0 + \frac{1}{n}\right].$$

Since the supremum of a function over a smaller set cannot be larger than its supremum over a larger set containing it, it follows that $S_{n+1}(\omega) \leq S_n(\omega)$ for every $\omega \in \Omega$.

Therefore, $S_n(\omega)$ is a non-increasing sequence of random variables.

The limit $L_{t_0}(\omega)$ is precisely $\lim_{n\to+\infty} S_n(\omega)$.

Let $E_n^{(M)}$ be the event that S_n is bounded by M:

$$E_n^{(M)} = \{\omega \in \Omega; S_n(\omega) \leq M\}.$$

Since $S_n(\omega)$ is a non-increasing sequence, if $S_n(\omega) \leq M$ for a given outcome ω, then:

$$S_{n+1}(\omega) \leq S_n(\omega) \leq M.$$

This implies that if $\omega \in E_n^{(M)}$, then $\omega \in E_{n+1}^{(M)}$. Therefore, the sequence of events $\left(E_n^{(M)}\right)_{n\in\mathbb{N}}$ is increasing:

$$E_1^{(M)} \subseteq E_2^{(M)} \subseteq E_3^{(M)} \subseteq \ldots.$$

The event $F_M = \{L_{t_0} \leq M\}$ is the event that the limit of S_n is bounded by M. Since S_n is a non-increasing sequence, $L_{t_0}(\omega) \leq M$ if and only if $S_n(\omega) \leq M$ for all sufficiently large n. This means:

$$F_M = \bigcup_{n=1}^{\infty} E_n^{(M)}.$$

By the continuity property of probability measures for an increasing sequence of events, we have:

$$\mathbb{P}(F_M) = \mathbb{P}\Big(\bigcup_{n=1}^{+\infty} E_n^{(M)}\Big) = \lim_{n\to+\infty} \mathbb{P}\Big(E_n^{(M)}\Big).$$

To prove $\mathbb{P}(F_M) = 0$, it is therefore sufficient to show that $\lim_{n\to+\infty} \mathbb{P}(E_n^{(M)}) = 0$.

Now, consider a specific point $t = t_0 + 1/n$. This point lies within the interval $[t_0 - 1/n, t_0 + 1/n]$ for $n \geq 1$. If the supremum of the increment quotient over the entire interval $[t_0 - 1/n, t_0 + 1/n]$ is bounded by M (i.e. $\omega \in E_n^{(M)}$), then it necessarily follows that the increment quotient at this specific point $t_0 + 1/n$ must also be bounded by M. This provides a useful upper bound for $\mathbb{P}(E_n^{(M)})$ by considering the probability of a simpler event:

$$\mathbb{P}\Big(E_n^{(M)}\Big) \leq \mathbb{P}\left(\left|\frac{Z^H_{t_0+\frac{1}{n}}(a,b) - Z^H_{t_0}(a,b)}{1/n}\right| \leq M\right).$$

Let $\Delta Z_n = Z^H_{t_0+\frac{1}{n}}(a,b) - Z^H_{t_0}(a,b)$. Since $Z^H(a,b)$ is a Gaussian process centered at t_0, ΔZ_n is a centered Gaussian random variable. Its variance is $\sigma_n^2 = \mathbb{E}\Big(\Delta Z_n\Big)^2$.

From Corollary 3.1, for $s = t_0$ and $t = t_0 + 1/n$, we have the lower bound on the variance of increments:

$$\sigma_n^2 \geq \gamma(a,b,H)\left|\left(t_0 + \frac{1}{n}\right) - t_0\right|^{2H} = \gamma(a,b,H)\left(\frac{1}{n}\right)^{2H}.$$

Let $C_L = \gamma(a,b,H)$. Since $(a,b) \neq (0,0)$, $C_L > 0$. Thus, $\sigma_n \geq \sqrt{C_L} n^{-H}$.

For a centered Gaussian random variable X with variance σ^2, we have:

$$\mathbb{P}(|X| \leq \delta) = \frac{1}{\sigma\sqrt{2\pi}} \int_{-\delta}^{\delta} \exp\left(-\frac{x^2}{2\sigma^2}\right) dx.$$

This integral is bounded by the width of the integration interval multiplied by the maximum value of the integrand (which is $1/(\sigma\sqrt{2\pi})$ at $x = 0$). So,

$$\mathbb{P}(|X| \leq \delta) \leq \frac{2\delta}{\sigma\sqrt{2\pi}}.$$

Applying this to ΔZ_n with $\delta = M/n$:

$$\mathbb{P}\left(\Big|\Delta Z_n\Big| \leq \frac{M}{n}\right) \leq \frac{2(M/n)}{\sigma_n\sqrt{2\pi}}.$$

Substituting the lower bound for σ_n:

$$\mathbb{P}\left(\Big|\Delta Z_n\Big| \leq \frac{M}{n}\right) \leq \frac{2M/n}{\sqrt{C_L} n^{-H}\sqrt{2\pi}} = \frac{2M}{\sqrt{2\pi C_L}} n^{H-1}.$$

Since $H \in (0,1)$, we have $H - 1 < 0$. Therefore, as $n \to +\infty$, $n^{H-1} \to 0$. Thus, for any fixed M:

$$\lim_{n\to+\infty} \mathbb{P}\left(\left|Z^H_{t_0+\frac{1}{n}}(a,b) - Z^H_{t_0}(a,b)\right| \leq \frac{M}{n}\right) = 0.$$

Since $\mathbb{P}(E_n^{(M)}) \le \mathbb{P}\left(\left| Z^H_{t_0+\frac{1}{n}}(a,b) - Z^H_{t_0}(a,b) \right| \le \frac{M}{n} \right)$, it follows that:

$$\lim_{n\to+\infty} \mathbb{P}\left(E_n^{(M)} \right) = 0.$$

This implies $\mathbb{P}(F_M) = 0$ for any $M > 0$. As argued at the beginning of the proof, this is sufficient to conclude that $\mathbb{P}(L_{t_0} < +\infty) = 0$, which rigorously proves that $L_{t_0} = +\infty$ with probability one. □

A significant implication of Corollary 3.1 is the following:

COROLLARY 3.4.– *For every $H \in (0,1) \setminus \{1/2\}$, the ZgfBm is not a semimartingale*[1].

Proof. This result follows directly from Corollary 3.2 and Lemma 2.1 from Bojdecki et al. (2004b). Lemma 2.1 states that any real continuous centered Gaussian process $\zeta = (\zeta_t)_{t\in[0,1]}$ satisfying

$$\alpha|t-s|^h \le \mathbb{E}\left[(\zeta_t - \zeta_s)^2 \right] \le \beta|t-s|^h, \quad t,s \in [0,1],$$

for some $h \in (0,2)$, $h \neq 1$, and positive constants α and β, is not a semimartingale. □

3.2.1. *Covariance structure of ZgfBm increments*

A fundamental aspect of any stochastic process is the correlation between its increments. This property sheds light on its *memory* and how past changes influence future ones. For the generalized fractional Brownian motion, understanding the covariance of its increments is essential, particularly in discerning its departure from traditional processes, such as the Brownian motion.

Let $Z^H = (Z^H_t(a,b))_{t\ge0}$ be a ZgfBm. The covariance of its increments is given by the following proposition.

PROPOSITION 3.2.– *Let $0 \le s_1 < s_2 \le t_1 < t_2$ be four ordered time points. The covariance between the increments $Z^H_{t_2} - Z^H_{t_1}$ and $Z^H_{s_2} - Z^H_{s_1}$ is given by:*

$$\begin{aligned} &Cov(Z^H_{t_2} - Z^H_{t_1}, Z^H_{s_2} - Z^H_{s_1}) \\ &= -ab\left[(t_2+s_2)^{2H} - (t_2+s_1)^{2H} - (t_1+s_2)^{2H} + (t_1+s_1)^{2H} \right] \\ &\quad - \frac{a^2+b^2}{2}\left[|t_2-s_2|^{2H} - |t_2-s_1|^{2H} - |t_1-s_2|^{2H} + |t_1-s_1|^{2H} \right]. \end{aligned}$$

1 For the formal definition of a semimartingale and related concepts, please refer to Definition A.5.

Proof. Let $\Delta Z_t^H = Z_{t_2}^H - Z_{t_1}^H$ and $\Delta Z_s^H = Z_{s_2}^H - Z_{s_1}^H$. Using the bilinearity of the covariance operator, we expand the desired term:

$$\begin{aligned}\mathrm{Cov}(\Delta Z_t^H, \Delta Z_s^H) &= \mathrm{Cov}(Z^H(t_2), Z^H(s_2)) - \mathrm{Cov}(Z^H(t_2), Z^H(s_1)) \\ &\quad - \mathrm{Cov}(Z^H(t_1), Z^H(s_2)) + \mathrm{Cov}(Z^H(t_1), Z^H(s_1)).\end{aligned}$$

Next, we substitute the explicit covariance function for $Z^H(a,b)$ from Lemma 3.1, specifically item 2, into the expanded covariance expression. After algebraic manipulation and grouping terms by coefficients of a, b, it can be shown that the terms proportional to $(a+b)^2$ cancel out. This leads to:

$$\begin{aligned}&\mathrm{Cov}(\Delta Z_t^H, \Delta Z_s^H) \\ &= -ab\left[(t_2+s_2)^{2H} - (t_2+s_1)^{2H} - (t_1+s_2)^{2H} + (t_1+s_1)^{2H}\right] \\ &\quad - \frac{a^2+b^2}{2}\left[|t_2-s_2|^{2H} - |t_2-s_1|^{2H} - |t_1-s_2|^{2H} + |t_1-s_1|^{2H}\right].\end{aligned}$$

Given the ordering $0 \le s_1 < s_2 \le t_1 < t_2$, all differences within the absolute value signs are positive, allowing us to remove them. Combining these simplified terms, we arrive at the stated explicit expression for the covariance of increments. □

REMARK 3.1.– *The explicit covariance formula for the increments of $Z^H(a,b)$ immediately reveals their non-stationary nature, whenever both $a \neq 0$ and $b \neq 0$. This is because the expression depends not solely on the increment lengths but also on their absolute positions on the time axis, as evidenced by the term proportional to $-ab$ involving sums of time points like $(t_2+s_2)^{2H}$. When either $a = 0$ or $b = 0$, the process reduces to a scaled tsfBm, and its increments are stationary. The non-stationarity of increments is a unique feature of the ZgfBm that distinguishes it from fBm, which has stationary increments. This non-stationarity is further supported by Lemma 3.2, where the variance of increments is shown to depend on t and s.*

To quantify the deviation of the ZgfBm from a process with stationary increments, we will compare the covariance structure of its increments with that of the fBm. Recall that for fBm, which corresponds to the case $a = 1, b = 0$ in our framework, the covariance of its unit-length increments is given by: for any integer $p \ge 0$ and $n \ge 0$:

$$\begin{aligned}\mathbb{E}\left[(B_{p+1}^H - B_p^H)(B_{p+n+1}^H - B_{p+n}^H)\right] &= \mathbb{E}\left[(B_1^H - B_0^H)(B_{n+1}^H - B_n^H)\right] \\ &= \frac{1}{2}\left[(n+1)^{2H} - 2n^{2H} + (n-1)^{2H}\right] \\ &= R_B(0,n).\end{aligned}$$

This explicitly shows the stationary nature of fBm increments, as their covariance depends only on the lag n and not on the starting point p.

For the ZgfBm, we define the corresponding *autocovariance of unit-length increments, where* p *denotes the starting point of the first increment:*

$$R_Z(p, p+n) = \mathbb{E}\Big((Z^H_{p+1} - Z^H_p)(Z^H_{p+n+1} - Z^H_{p+n})\Big), \quad p \geq 1. \qquad [3.7]$$

In the following proposition, we characterize $R_Z(p, p+n)$ *and analyze its asymptotic behavior relative to* $R_B(0, n)$.

PROPOSITION 3.3.– *For every* $n \geq 1$, *as* $p \to \infty$, *it holds that:*

$$R_Z(p, p+n) = (a^2+b^2)R_B(0, n) - ab\Big(2^{2H-1}H(2H-1)\Big)p^{2(H-1)}\Big(1+o(1)\Big),$$

and consequently

$$\lim_{p\to\infty} R_Z(p, p+n) = (a^2 + b^2)R_B(0, n). \qquad [3.8]$$

Proof. A direct calculation yields:

$$R_Z(p, p+n) = \frac{a^2+b^2}{2}\left[(n+1)^{2H} - 2n^{2H} + (n-1)^{2H}\right]$$

$$-ab\left[(2p+n+2)^{2H} - 2(2p+n+1)^{2H} + (2p+n)^{2H}\right] \qquad [3.9]$$

$$= (a^2 + b^2)R_B(0, n) - abf_p(n),$$

for every $n \geq 1$, where

$$f_p(n) = (2p+n+2)^{2H} - 2(2p+n+1)^{2H} + (2p+n)^{2H}.$$

For $p \to \infty$, applying a Taylor expansion, we obtain:

$$f_p(n) = (2p)^{2H}\left[\Big(1+\frac{n+2}{2p}\Big)^{2H} - 2\Big(1+\frac{n+1}{2p}\Big)^{2H} + \Big(1+\frac{n}{2p}\Big)^{2H}\right]$$

$$= \Big(2^{2H-1}H(2H-1)\Big)p^{2(H-1)}\Big(1+o(1)\Big).$$

Since $H < 1$, the term $p^{2(H-1)}$ tends to 0 as $p \to \infty$. □

REMARK 3.2.– *If* $a \neq 0$ *and* $b \neq 0$, *the ZgfBm increments are not stationary. The proposition thus shows that, while the ZgfBm increments are not stationary, their covariance function asymptotically approaches a stationary one as* $p \to \infty$.

Now, we shift our focus to the behavior of the ZgfBm increments as the lag $n \to \infty$, specifically investigating the long-range dependence of the process Z^H.

DEFINITION 3.2.– *We say that the increments of a stochastic process X are long-range dependent (LRD) if for every integer $p \geq 1$, the sum of their autocovariances diverges:*

$$\sum_{n\geq 1} R_X(p, p+n) = \infty,$$

where $R_X(p, p+n) = \mathbb{E}\left[(X_{p+1} - X_p)(X_{p+n+1} - X_{p+n})\right]$. Conversely, we say the increments of X are short-range dependent (SRD) if the sum of their autocovariances converges:

$$\sum_{n\geq 1} R_X(p, p+n) < \infty.$$

We establish the following result concerning the long-range dependence of ZgfBm increments.

THEOREM 3.2.– *For every $(a, b) \in \mathbb{R}^2 \setminus \{(0, 0)\}$, the increments of $Z^H(a, b)$ are long-range dependent if and only if $H > \frac{1}{2}$ and $a \neq b$.*

Proof. For every integer $p \geq 1$, we use the expression for $R_Z(p, p+n)$ derived previously. Applying a Taylor expansion as $n \to \infty$, we have:

$$R_Z(p, p+n) = \frac{a^2+b^2}{2} n^{2H} \left[\left(1+\frac{1}{n}\right)^{2H} - 2 + \left(1-\frac{1}{n}\right)^{2H}\right]$$

$$-abn^{2H}\left[\left(1+\frac{2p+2}{n}\right)^{2H} - 2\left(1+\frac{2p+1}{n}\right)^{2H} + \left(1+\frac{2p}{n}\right)^{2H}\right]$$

$$= H(2H-1)(a-b)^2 n^{2H-2} - 4H(2H-1)(H-1)ab(2p+1)n^{2H-3}\Big(1+o(1)\Big).$$

Let us first analyze the specific case where $H = \frac{1}{2}$: in this case, we directly compute $R_Z(p, p+n)$. By formula [3.9], we get:

$$R_Z(0, n) = \frac{a^2+b^2}{2} n \left[\left(1+\frac{1}{n}\right) - 2 + \left(1-\frac{1}{n}\right)\right] = 0.$$

This implies that the increments are uncorrelated for non-overlapping intervals (similar to standard Brownian motion). Therefore, $\sum_{n\geq 1} R_Z(p, p+n) = 0$, which is a finite value. Hence, the increments are short-range dependent.

Now, we consider the case where $H \neq \frac{1}{2}$. In this scenario, $H(2H-1) \neq 0$. We analyze two distinct cases based on the values of a and b:

– *Case 1*: $a \neq b$. As $n \to \infty$, the asymptotic behavior of $R_Z(p, p+n)$ is dominated by the first term:

$$R_Z(p, p+n) \sim H(2H-1)(a-b)^2 n^{2H-2}.$$

- If $H > \frac{1}{2}$: the exponent $2H-2$ satisfies $2H-2 > -1$. Since $H(2H-1) > 0$ and $(a-b)^2 > 0$, the dominant term $R_Z(p, p+n)$ is positive for large n. Consequently, the sum $\sum_{n\geq 1} R_Z(p, p+n)$ diverges (as it behaves like $\sum n^{2H-2}$). Thus, the increments are long-range dependent.

- If $H < \frac{1}{2}$: the exponent $2H-2$ satisfies $2H-2 < -1$. Although $H(2H-1) < 0$, the sum of absolute values $\sum_{n\geq 1} |R_Z(p, p+n)|$ converges (as it behaves like $\sum n^{2H-2}$). This ensures that the sum $\sum_{n\geq 1} R_Z(p, p+n)$ itself converges. Therefore, the increments are short-range dependent.

– *Case 2*: $a = b$. In this case, the first term $H(2H-1)(a-b)^2 n^{2H-2}$ vanishes. The asymptotic behavior of $R_Z(p, p+n)$ as $n \to \infty$ is then dominated by the second term:

$$R_Z(p, p+n) \sim -4H(2H-1)(H-1)a^2(2p+1)n^{2H-3}.$$

For any $H \in (0, 1)$, the exponent $2H-3$ is always strictly less than -1.

Since $2H-3 < -1$ for all $H \in (0, 1)$, the series $\sum_{n\geq 1} n^{2H-3}$ converges. Consequently,

$$\sum_{n\geq 1} R_Z(p, p+n) < \infty.$$

Thus, if $a = b$, the increments are always short-range dependent.

Combining all cases, the increments of $Z^H(a, b)$ are long-range dependent if and only if $H > \frac{1}{2}$ and $a \neq b$. □

REMARK 3.3.– *It is well-known that the increments of the SfBm, which corresponds to* $Z^H\left(\frac{1}{\sqrt{2}}, \frac{1}{\sqrt{2}}\right)$ *in our notation (i.e.* $a = b = \frac{1}{\sqrt{2}}$*), are short-range dependent for every* $H \in (0, 1)$*. Our Theorem 3.2 provides an important motivation for investigating the general process* $Z^H(a, b)$ *with* a *and* b *not necessarily equal. This generalized framework allows us to model phenomena exhibiting not only short-range dependence but also long-range dependence in the increments, depending on the choice of parameters* a *and* b*.*

3.3. Integral representation of the generalized fractional Brownian motion

In this section, we establish a crucial integral representation for ZgfBm in terms of the two-sided Brownian motion, commonly known as a moving average expression. This representation is fundamental for understanding its underlying structure and is often used in stochastic analysis.

LEMMA 3.3.– *The ZgfBm $Z_t^H(a,b)$ for $H \in (1/2,1)$ has the following moving-average representation:*

$$Z_t^H(a,b) = \frac{H-1/2}{2C(H)} \int_{\mathbb{R}} \int_0^t \Big[(a+b)\left(((u-s)_+)^{H-3/2} - ((s-u)_+)^{H-3/2}\right)$$
$$+(a-b)\left(((u-s)_+)^{H-3/2} + ((s-u)_+)^{H-3/2}\right)\Big]\, du\, dB(s) \tag{3.10}$$

where $B = \{B(s), s \in \mathbb{R}\}$ is a two-sided Brownian motion (see Exercise 1.10), and $(x)_+ = \max(x,0)$. The constant $C(H)$ is defined by the following expressions:

$$C(H) = \sqrt{\int_0^\infty \left((1+x)^{H-1/2} - x^{H-1/2}\right)^2 dx + \frac{1}{2H}}$$
$$= \frac{\Gamma(H+1/2)}{\sqrt{2H\Gamma(2H)\sin \pi H}}.$$

Proof. Let us first recall that the tsfBm $\{B^H(t); t \in \mathbb{R}\}$ has the integral representation:

$$B^H(t) = \frac{1}{C(H)} \int_{\mathbb{R}} \left(((t-s)_+)^{H-1/2} - ((-s)_+)^{H-1/2}\right) dB(s), \quad t \in \mathbb{R}. \tag{3.11}$$

This representation is originally due to Mandelbrot and Van Ness (1968). For a detailed proof, readers may refer to Taqqu (1994, p. 321) or Mishura (2008, p. 9). The derivation of the second expression for the constant $C(H)$ is provided in Appendix A of Mishura (2008).

Applying equations [3.1] and [3.11], we obtain, for every $t \in [0, +\infty)$:

$$Z_t^H(a,b) =: \frac{1}{2C(H)}$$
$$\times \Big[\int_{\mathbb{R}} (a+b)\left(((t-s)_+)^{H-1/2} + ((t+s)_-)^{H-1/2} - 2((-s)_+)^{H-1/2}\right) \tag{3.12}$$
$$+(a-b)\left(((t-s)_+)^{H-1/2} - ((t+s)_-)^{H-1/2}\right) dB(s)\Big].$$

This, together with the identities:

$$((t-s)_+)^{H-1/2} + ((t+s)_-)^{H-1/2} - 2\left((-s)_+\right)^{H-1/2}$$

$$((u-s)_+)^{H-3/2} - ((u+s)_-)^{H-3/2}\, du$$

and

$$((t-s)_+)^{H-1/2} - ((t+s)_-)^{H-1/2} = \int_0^t ((u-s)_+)^{H-3/2} + ((u+s)_-)^{H-3/2}\, du,$$

for every $H \in (1/2, 1)$, complete the proof of Lemma 3.3. □

3.4. Alternative expression via odd and even fractional Brownian motions

The ZgfBm can also be expressed in an alternative form, linking it to the odd and even fractional Brownian motions. Recall from Chapter 2 that any tsfBm $B^H = (B_t^H)_{t\in\mathbb{R}}$ can be uniquely decomposed into an odd part $B^{H,o}$ and an even part $B^{H,e}$. Specifically, for $t \geq 0$, we have:

$$B_t^H = B_t^{H,o} + B_t^{H,e} \quad \text{and} \quad B_{-t}^H = -B_t^{H,o} + B_t^{H,e}.$$

Note that $B^{H,o}$ and $B^{H,e}$ are independent Gaussian processes.

PROPOSITION 3.4.– *Let $Z_t^H(a,b)$ be a ZgfBm. Then, for $t \geq 0$, $Z_t^H(a,b)$ can be expressed as a linear combination of the odd and even fractional Brownian motions:*

$$Z_t^H(a,b) = (a-b)B_t^{H,o} + (a+b)B_t^{H,e}. \qquad [3.13]$$

Proof. Substituting the decomposition of fBm into its odd and even parts, $B_t^H = B_t^{H,o} + B_t^{H,e}$ and $B_{-t}^H = -B_t^{H,o} + B_t^{H,e}$ (for $t \geq 0$), into the definition of Z_t^H (equation [3.1]):

$$\begin{aligned} Z_t^H &= aB_t^H + bB_{-t}^H \\ &= a(B_t^{H,o} + B_t^{H,e}) + b(-B_t^{H,o} + B_t^{H,e}) \\ &= (a-b)B_t^{H,o} + (a+b)B_t^{H,e}. \end{aligned}$$

This proves the proposition. The independence of $B^{H,o}$ and $B^{H,e}$ is a known property from their construction and covariance functions. □

REMARK 3.4.– *The decomposition of ZgfBm into its odd and even fBm components [3.13] offers a powerful alternative perspective for analyzing its properties. While the characteristics detailed in the previous sections are directly derived from the initial definition, this decomposition can simplify derivations or provide deeper insights into other aspects of ZgfBm, leveraging the well-understood properties and independence of odd and even fBms. Consequently, for the remainder of our investigation, we may utilize either the initial definition of ZgfBm or its representation in terms of* $B^{H,o}$ *and* $B^{H,e}$*, whichever is most suitable for the specific analysis.*

As a concrete example of the utility highlighted in the preceding remark, and building upon the ZgfBm decomposition presented in Proposition 3.4 and the series expansions for odd and even fBms established in the previous chapter, we now derive the corresponding series representation for ZgfBm. We maintain the notations introduced in section 2.6 of that chapter.

Let $X_1, X_2, \dots$ and $Y_1, Y_2, \dots$ be two independent sequences of independent, centered Gaussian random variables on a common probability space, with $\mathrm{Var}(X_n) = \sigma_n^2$ and $\mathrm{Var}(Y_n) = \tau_n^2$, respectively.

THEOREM 3.3.– *Let* $Z_t^H(a,b)$ *be a generalized fractional Brownian motion. Then, for* $t \in [0,1]$, $Z_t^H(a,b)$ *can be expressed as the following series:*

$$Z_t^H = (a-b)\sum_{n=1}^{\infty} \frac{\sin(x_n t)}{x_n} X_n + (a+b)\sum_{n=1}^{\infty} \frac{1-\cos(y_n t)}{y_n} Y_n.$$

Both series converge absolutely and uniformly with probability 1 in $t \in [0,1]$.

Proof. From Theorem 2.5, we have:

$$B_t^{H,o} = \sum_{n=1}^{\infty} \frac{\sin(x_n t)}{x_n} X_n \quad \text{and} \quad B_t^{H,e} = \sum_{n=1}^{\infty} \frac{1-\cos(y_n t)}{y_n} Y_n.$$

By substituting these series in the decomposition [3.13], we obtain:

$$\begin{aligned} Z_t^H &= (a-b)\left(\sum_{n=1}^{\infty} \frac{\sin(x_n t)}{x_n} X_n\right) + (a+b)\left(\sum_{n=1}^{\infty} \frac{1-\cos(y_n t)}{y_n} Y_n\right) \\ &= \sum_{n=1}^{\infty} (a-b)\frac{\sin(x_n t)}{x_n} X_n + \sum_{n=1}^{\infty} (a+b)\frac{1-\cos(y_n t)}{y_n} Y_n. \end{aligned}$$

The absolute and uniform convergence with probability 1 for both series, in $t \in [0,1]$, is directly inherited from the convergence properties established for the individual series expansions of $B_t^{H,o}$ and $B_t^{H,e}$ in Theorem 2.5. □

REMARK 3.5.– *The series expansion for ZgfBm presented in Theorem 3.3 offers significant advantages for both theoretical analysis and practical applications. From a theoretical standpoint, it provides a constructive definition of ZgfBm, explicitly revealing its structure as a sum of deterministic basis functions weighted by independent Gaussian random variables. This representation can simplify the derivation of certain properties or facilitate the study of ZgfBm in various functional spaces. Furthermore, this series expansion serves as the most practical and powerful basis for the numerical simulation of ZgfBm sample paths. Its key advantage is the ability to efficiently generate approximations of ZgfBm trajectories by truncating the infinite series to a finite number of terms, thereby circumventing the need for more computationally intensive methods like those that rely on large covariance matrices or circulant embedding. The established absolute and uniform convergence guarantees that increasing the number of terms leads to progressively more accurate representations of the true ZgfBm path.*

This was demonstrated in a study where the expansion was truncated at $n = 2,000$ (obtained in Theorem 3.3). Using Matlab, numerical values of the positive real zeros of the Bessel functions J_H and J_{1-H} were inserted for different Hurst parameters ($H = 0.25, 0.5,$ and 0.75). The resulting simulations generated $Z^H(100, 5)$ trajectories (Figure 3.1) and $sfBm\ \xi^H = Z^H\left(\frac{1}{\sqrt{2}}, \frac{1}{\sqrt{2}}\right)$ trajectories (Figure 3.2). These results illustrate the core property of ZgfBm: a larger H value corresponds to smoother sample paths, while smaller H values lead to more wildly fluctuating paths. Figure 3.3 further presented generated sample paths for the fBm $B^H = Z^H(1, 0)$, its even part $B^{H,e} = Z^H(0.5, 0.5)$ and its odd part $B^{H,o} = Z^H(0.5, -0.5)$, all for $H = 0.3$.

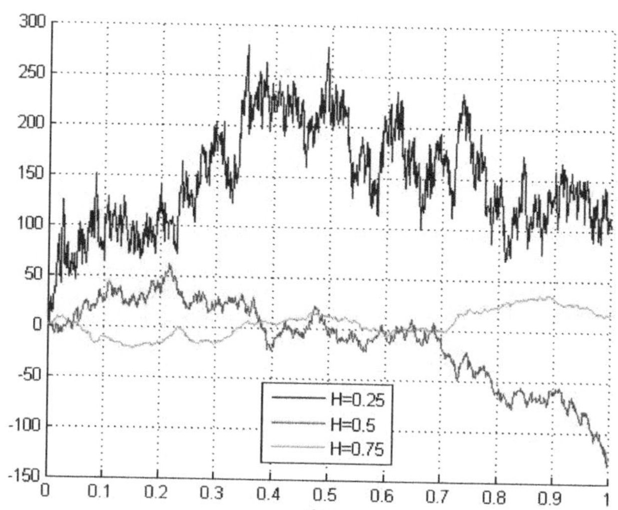

Figure 3.1. *Sample paths of Z^H (100,5) with different values of H. For a color version of this figure, see: www.iste.co.uk/zili/brownian.zip*

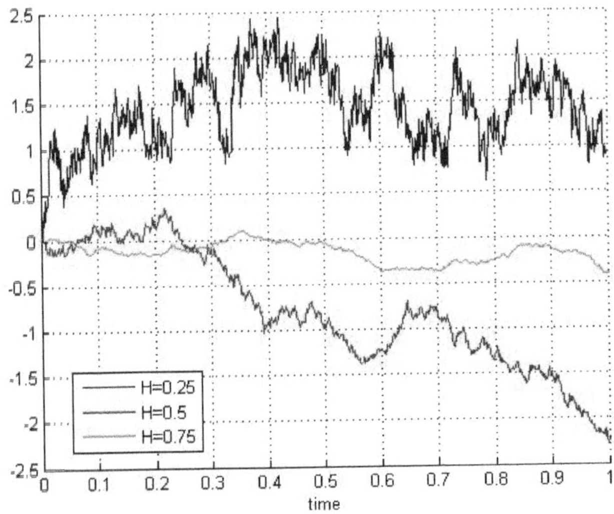

Figure 3.2. *Sample paths of sffBm ξ^H with different values of H. For a color version of this figure, see: www.iste.co.uk/zili/brownian.zip*

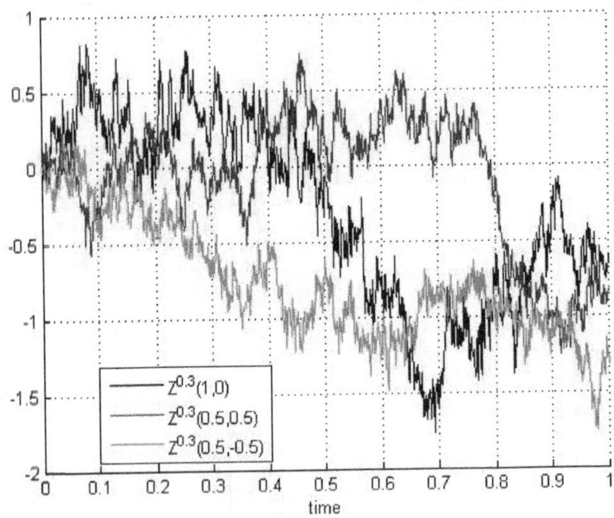

Figure 3.3. *Sample paths of $Z^{0.3}$(a, b) with different values of a and b. For a color version of this figure, see: www.iste.co.uk/zili/brownian.zip*

3.5. Exercises

EXERCISE 3.1.–

1) Using the definition $Z_t^H(a,b) = aB_t^H + bB_{-t}^H$, explicitly verify that if $a = 1$ and $b = 0$, then Z_t^H is indeed a standard fBm (B_t^H).

2) Use the ZgfBm covariance formula from Lemma 3.1 (item 2) to demonstrate that for $a = b = \frac{1}{\sqrt{2}}$, the ZgfBm covariance matches that of SfBm.

EXERCISE 3.2.–

1) Prove part 3 of Lemma 3.1:

$$\mathbb{E}[Z_t^H(a,b)^2] = (a^2 + b^2 - (2^{2H} - 2)ab)t^{2H}.$$

2) Justify that the condition $(a,b) \neq (0,0)$ and $H \in (0,1)$ guarantees that the variance $\mathbb{E}[Z_t^H(a,b)^2]$ is strictly positive for $t > 0$.

3) Corollary 3.2 asserts that ZgfBm paths are almost surely Hölder continuous of any order strictly less than H. State the precise conditions of Kolmogorov's continuity theorem. Then, explain how Corollary 3.1 (which provides moment bounds for ZgfBm increments) enables the application of this theorem to prove the Hölder continuity of ZgfBm paths. You do not need to re-derive Corollary 3.1, but specify the form of the inequality it must provide to satisfy Kolmogorov's condition.

EXERCISE 3.3.– *Theorem 3.2 states that ZgfBm increments are LRD if and only if $H > 1/2$ and $a \neq b$.*

1) Why does the case $H = 1/2$ always lead to short-range dependence for ZgfBm, regardless of a and b? Refer to the direct calculation in the proof.

2) If $a = b$, the increments are also always short-range dependent. Explain why the asymptotic behavior of $R_Z(p, p+n)$ in this case (n^{2H-3}) ensures short-range dependence for all $H \in (0,1)$.

EXERCISE 3.4.– *The covariance of the ZgfBm was derived in Lemma 3.1. Using the decomposition $Z_t^H = (a-b)B_t^{H,o} + (a+b)B_t^{H,e}$ and leveraging the known properties of independent odd and even fBms ($B^{H,o}$ and $B^{H,e}$), re-derive the expression for Cov($Z_t^H(a,b), Z_s^H(a,b)$). Verify that your result matches the formula provided in Lemma 3.1.*

EXERCISE 3.5.– *The variance of ZgfBm was established in Lemma 3.1.*

1) Using the decomposition $Z_t^H = (a-b)B_t^{H,o} + (a+b)B_t^{H,e}$, calculate $\mathbb{E}[Z_t^H(a,b)^2]$.

2) Consider the case of sfBm, which corresponds to $a = b = \frac{1}{\sqrt{2}}$ in the ZgfBm framework. How does the decomposition of Z_t^H simplify in this specific case?

Compute its variance using this simplified form and discuss the implications for the contributions of the odd and even components.

EXERCISE 3.6.– *The self-similarity of ZgfBm with index H was demonstrated in Proposition 3.1. Provide an alternative proof of this property by utilizing the decomposition of Z_t^H into its odd and even fBm components.*

4

Parameter Estimation for the Generalized Fractional Brownian Motion

This chapter presents a comprehensive methodology for the estimation of all key parameters of the ZgfBm: the Hurst parameter H and the scaling coefficients a and b. Accurately estimating these parameters is crucial for characterizing the long/short-range dependence of process and its asymmetric properties, which are essential for modeling various phenomena in fields like finance and hydrology.

4.1. Estimation of the Hurst parameter H when a and b are known

We begin this chapter by addressing the problem of estimating the Hurst parameter H for the ZgfBm under the foundational assumption that the deterministic coefficients a and b are known constants. While in many practical scenarios these coefficients are unknown, this case provides a crucial starting point for understanding the statistical properties of the ZgfBm. The approach presented here is directly applicable to estimating H for various important special cases of ZgfBm, such as fractional Brownian motion, odd and even fractional Brownian motion and standard Brownian motion which correspond to specific, known values of a and b. We will present a method based on increment-based statistics and establish the consistency of the resulting estimators, laying the groundwork for more complex estimation problems to follow.

Consider the ZgfBm $Z^H = \{Z_t^H, t \in [0,T]\}$ as defined in Definition 3.1. To facilitate parameter estimation from discrete observations, we analyze the properties of its increments. We discretize the interval $[0,T]$ by considering time instants $t_i = i/N$, and we denote the increments as $\Delta Z_i^H = Z_{t_i}^H - Z_{t_{i-1}}^H$ for $i = 1, \ldots, N$.

The following proposition presents a foundational result for our parameter estimation methodology: the explicit formula for the expected value of the squared discrete increments of the ZgfBm.

PROPOSITION 4.1.– *For discrete observations of the ZgfBm $Z^H(a,b)$ at time instants $t_i = i/N$, we have:*

$$\mathbb{E}\left[\left(Z^H_{\frac{i}{N}}(a,b) - Z^H_{\frac{i-1}{N}}(a,b)\right)^2\right]$$

$$= (a^2+b^2)N^{-2H} - ab\left(\frac{2}{N}\right)^{2H}\left[i^{2H} - 2\left(i-\frac{1}{2}\right)^{2H} + (i-1)^{2H}\right].$$

Proof. Substituting $t = i/N$ and $s = (i-1)/N$ into formula [3.2]:

$$\mathbb{E}\left[\left(Z^H_{\frac{i}{N}}(a,b) - Z^H_{\frac{i-1}{N}}(a,b)\right)^2\right]$$

$$= (a^2+b^2)\left(\frac{1}{N}\right)^{2H} - 2ab\left(2^{2H-1}\left(\left(\frac{i}{N}\right)^{2H} + \left(\frac{i-1}{N}\right)^{2H}\right)\right.$$

$$\left. - \left(\frac{2i-1}{N}\right)^{2H}\right)$$

$$= (a^2+b^2)N^{-2H} - abN^{-2H}\left[2^{2H}\left(i^{2H} + (i-1)^{2H}\right) - 2(2i-1)^{2H}\right].$$

Now, we rearrange the terms in the bracket:

$$2^{2H}\left(i^{2H} + (i-1)^{2H}\right) - 2(2i-1)^{2H} = 2^{2H}\left[i^{2H} - 2\left(i-\frac{1}{2}\right)^{2H} + (i-1)^{2H}\right].$$

Substituting this back into the expression for the expected squared increment, and factoring out $2^{2H}N^{-2H} = (2/N)^{2H}$, we obtain the stated result. This completes the proof. □

The expected value of the squared increment derived in Proposition 4.1 provides the theoretical foundation for constructing a practical estimator for H.

We can build an estimator by considering the empirical average of these squared increments, which converges to the expected value as the number of observations grows.

4.1.1. *Parameter estimation via squared increments: an asymptotic study*

We define a statistic based on the squared increments of the ZgfBm. For observations taken at $t_i = i/N$ for $i = 0, \ldots, N$:

$$S_N(Z^H(a,b)) = \frac{1}{N}\sum_{i=1}^{N}\left(Z^H_{\frac{i}{N}}(a,b) - Z^H_{\frac{i-1}{N}}(a,b)\right)^2. \qquad [4.1]$$

This statistic is a natural choice for estimating the Hurst parameter, as its asymptotic behavior is directly related to H.

Our next step is to analyze its asymptotic behavior as the number of observations N becomes large. The following lemma provides the asymptotic mean of this statistic, which is a crucial first step toward establishing its consistency as an estimator for H.

LEMMA 4.1 (asymptotic mean of $S_N(Z^H(a,b))$).– *As $N \to \infty$, the expected value of the statistic $S_N(Z^H(a,b))$ satisfies:*

$$\mathbb{E}[S_N(Z^H(a,b))] = (a^2+b^2)N^{-2H}\left(1 + O\left(\frac{1}{N}\right)\right).$$

This implies that $\mathbb{E}[S_N(Z^H(a,b))] \sim (a^2+b^2)N^{-2H}$ as $N \to \infty$.[1]

Proof. From the definition of $S_N(Z^H(a,b))$ in [4.1] and linearity of expectation, we have:

$$\mathbb{E}[S_N(Z^H(a,b))] = \frac{1}{N}\sum_{i=1}^{N}\mathbb{E}\left[\left(Z^H_{\frac{i}{N}}(a,b) - Z^H_{\frac{i-1}{N}}(a,b)\right)^2\right].$$

Using Proposition 4.1:

$$\mathbb{E}[S_N(Z^H(a,b))]$$
$$= \frac{1}{N}\sum_{i=1}^{N}\left((a^2+b^2)N^{-2H} - ab\left(\frac{2}{N}\right)^{2H}\left[i^{2H} - 2\left(i-\frac{1}{2}\right)^{2H} + (i-1)^{2H}\right]\right)$$

---

1 For two deterministic sequences of real numbers $(x_n)_{n\geq 1}$ and $(y_n)_{n\geq 1}$, such that for all $n \geq n_0$, $y_n \neq 0$ for some integer n_0, their asymptotic equivalence as $n \to \infty$ is denoted by $x_n \sim y_n$, and is defined by the limit:

$$\lim_{n\to\infty}\frac{xn}{yn} = 1.$$

$$= (a^2 + b^2)N^{-2H} - \frac{ab2^{2H}}{N} \sum_{i=1}^{N} \left[\left(\frac{i}{N}\right)^{2H} - 2\left(\frac{i-1/2}{N}\right)^{2H} + \left(\frac{i-1}{N}\right)^{2H} \right]$$

$$= (a^2 + b^2)N^{-2H} - \frac{ab2^{2H}}{N^{1+2H}} \sum_{i=1}^{N} \left[i^{2H} - 2\left(i - \frac{1}{2}\right)^{2H} + (i-1)^{2H} \right]. \tag{4.2}$$

Let us use the Euler–Maclaurin formula (see Lemma A.2). For a sufficiently smooth function $f(x)$ on $[0, 1]$, the formula for a sum starting at $i = 1$ is:

$$\frac{1}{N} \sum_{i=1}^{N} f\left(\frac{i}{N}\right) = \int_0^1 f(x)dx + \frac{f(1) - f(0)}{2N} + O\left(\frac{1}{N^2}\right).$$

Applying this formula to the terms in the summations with $f(x) = x^{2H}$, and using the fact that $\int_0^1 x^{2H} dx = \frac{1}{2H+1}$, we get:

$$\frac{1}{N} \sum_{i=1}^{N} \left(\frac{i}{N}\right)^{2H} = \frac{1}{2H+1} + \frac{1}{2N} + O\left(\frac{1}{N^2}\right).$$

$$\begin{aligned}
\frac{1}{N} \sum_{i=1}^{N} \left(\frac{i-1}{N}\right)^{2H} &= \frac{1}{N} \sum_{j=0}^{N-1} f\left(\frac{j}{N}\right) \\
&= \frac{1}{N} \sum_{i=1}^{N} \left(\frac{i}{N}\right)^{2H} + \frac{f(0) - f(1)}{N} \\
&= \int_0^1 f(x)dx + \frac{f(0) - f(1)}{2N} + O\left(\frac{1}{N^2}\right) \\
&= \frac{1}{2H+1} - \frac{1}{2N} + O\left(\frac{1}{N^2}\right).
\end{aligned}$$

Now, for $\frac{1}{N} \sum_{i=1}^{N} \left(\frac{2i-1}{N}\right)^{2H}$, this sum can be expressed as $\frac{1}{N} \sum_{i=1}^{N} f\left(\frac{i-1/2}{N}\right)$, where $f(x) = (2x)^{2H}$. Applying Euler–Maclaurin for midpoint sum formula (see Lemma A.3), and using the fact that

$$\int_0^1 (2x)^{2H} dx = \frac{2^{2H}}{2H+1},$$

we get

$$\frac{1}{N} \sum_{i=1}^{N} \left(\frac{2i-1}{N}\right)^{2H} = \frac{2^{2H}}{2H+1} + O\left(\frac{1}{N^2}\right).$$

Substituting these into the expression [4.2]:

$$\begin{aligned}
&\mathbb{E}[S_N(Z^H(a,b))]\\
&= (a^2+b^2)N^{-2H} - 2^{2H}abN^{-2H}\left(\frac{1}{N}\sum_{i=1}^{N}\left(\frac{i}{N}\right)^{2H} + \frac{1}{N}\sum_{i=1}^{N}\left(\frac{i-1}{N}\right)^{2H}\right)\\
&\quad + 2abN^{-2H}\left(\frac{1}{N}\sum_{i=1}^{N}\left(\frac{2i-1}{N}\right)^{2H}\right)\\
&= (a^2+b^2)N^{-2H}\\
&\quad - 2^{2H}abN^{-2H}\left(\frac{1}{2H+1}+O\left(\frac{1}{N}\right)+\frac{1}{2H+1}+O\left(\frac{1}{N}\right)\right)\\
&\quad + 2abN^{-2H}\left(\frac{2^{2H}}{2H+1}+O\left(\frac{1}{N}\right)\right)\\
&= (a^2+b^2)N^{-2H} - \frac{2^{2H+1}ab}{2H+1}N^{-2H} + O\left(N^{-2H-1}\right)\\
&\quad + \frac{2^{2H+1}ab}{2H+1}N^{-2H} + O\left(N^{-2H-1}\right)\\
&= (a^2+b^2)N^{-2H} + O\left(N^{-2H-1}\right)\\
&= (a^2+b^2)N^{-2H}\left(1+O\left(\frac{1}{N}\right)\right).
\end{aligned}$$

This completes the proof of Lemma 4.1. □

To facilitate the estimation of the Hurst parameter H and enable a rigorous analysis of our estimator's properties, we first establish the explicit covariance structure of the discrete increments of the ZgfBm in the following lemma.

LEMMA 4.2.– *We have:*

$$\begin{aligned}
&Cov(\Delta Z_i^H, \Delta Z_j^H) = -ab\left(\frac{1}{N}\right)^{2H}\left[(i+j)^{2H} - 2(i+j-1)^{2H}\right.\\
&\left.+(i+j-2)^{2H}\right]\\
&-\frac{(a^2+b^2)}{2}\left(\frac{1}{N}\right)^{2H}\left[2|i-j|^{2H} - |i-j+1|^{2H} - |i-j-1|^{2H}\right]
\end{aligned}$$

for every $i,j \in \{1,\ldots,N\}$.

Proof. Recall the covariance function of the ZgfBm:

$$\mathrm{Cov}(Z_u^H, Z_v^H) = \frac{(a+b)^2}{2}(|u|^{2H} + |v|^{2H}) - \frac{a^2+b^2}{2}|u-v|^{2H} - ab|u+v|^{2H}.$$

The covariance of the increments ΔZ_i^H and ΔZ_j^H is given by the following expansion:

$$\mathrm{Cov}(\Delta Z_i^H, \Delta Z_j^H) = \mathrm{Cov}(Z_{t_i}^H - Z_{t_{i-1}}^H, Z_{t_j}^H - Z_{t_{j-1}}^H)$$
$$= \mathrm{Cov}(Z_{t_i}^H, Z_{t_j}^H) - \mathrm{Cov}(Z_{t_i}^H, Z_{t_{j-1}}^H) - \mathrm{Cov}(Z_{t_{i-1}}^H, Z_{t_j}^H) + \mathrm{Cov}(Z_{t_{i-1}}^H, Z_{t_{j-1}}^H).$$

We now substitute the explicit covariance function into each term and group the terms by their coefficients $\frac{(a+b)^2}{2}$, $-ab$ and $-\frac{a^2+b^2}{2}$.

$$= \frac{(a+b)^2}{2}\left[(t_i^{2H} + t_j^{2H}) - (t_i^{2H} + t_{j-1}^{2H}) - (t_{i-1}^{2H} + t_j^{2H}) + (t_{i-1}^{2H} + t_{j-1}^{2H})\right]$$
$$- ab\left[|t_i + t_j|^{2H} - |t_i + t_{j-1}|^{2H} - |t_{i-1} + t_j|^{2H} + |t_{i-1} + t_{j-1}|^{2H}\right]$$
$$- \frac{a^2+b^2}{2}\left[|t_i - t_j|^{2H} - |t_i - t_{j-1}|^{2H} - |t_{i-1} - t_j|^{2H} + |t_{i-1} - t_{j-1}|^{2H}\right]$$

Let us evaluate each bracketed term separately.

First bracket (coefficient $\frac{(a+b)^2}{2}$): the terms in the first bracket cancel out:

$$(t_i^{2H} + t_j^{2H}) - (t_i^{2H} + t_{j-1}^{2H}) - (t_{i-1}^{2H} + t_j^{2H}) + (t_{i-1}^{2H} + t_{j-1}^{2H}) = 0$$

Second bracket (coefficient $-ab$): substituting $t_k = k/N$, we simplify this bracket as follows:

$$|t_i + t_j|^{2H} - |t_i + t_{j-1}|^{2H} - |t_{i-1} + t_j|^{2H} + |t_{i-1} + t_{j-1}|^{2H}$$
$$= \left|\frac{i+j}{N}\right|^{2H} - \left|\frac{i+(j-1)}{N}\right|^{2H} - \left|\frac{(i-1)+j}{N}\right|^{2H} + \left|\frac{(i-1)+(j-1)}{N}\right|^{2H}$$
$$= \left(\frac{1}{N}\right)^{2H}\left[(i+j)^{2H} - (i+j-1)^{2H} - (i+j-1)^{2H} + (i+j-2)^{2H}\right]$$
$$= \left(\frac{1}{N}\right)^{2H}\left[(i+j)^{2H} - 2(i+j-1)^{2H} + (i+j-2)^{2H}\right]$$

Third bracket (coefficient $-\frac{a^2+b^2}{2}$): substituting $t_k = k/N$, we simplify the third bracket:

$$|t_i - t_j|^{2H} - |t_i - t_{j-1}|^{2H} - |t_{i-1} - t_j|^{2H} + |t_{i-1} - t_{j-1}|^{2H}$$
$$= \left|\frac{i-j}{N}\right|^{2H} - \left|\frac{i-(j-1)}{N}\right|^{2H} - \left|\frac{(i-1)-j}{N}\right|^{2H} + \left|\frac{(i-1)-(j-1)}{N}\right|^{2H}$$
$$= \left(\frac{1}{N}\right)^{2H} \left[|i-j|^{2H} - |i-j+1|^{2H} - |i-j-1|^{2H} + |i-j|^{2H}\right]$$
$$= \left(\frac{1}{N}\right)^{2H} \left[2|i-j|^{2H} - |i-j+1|^{2H} - |i-j-1|^{2H}\right]$$

Finally, we substitute the simplified bracketed terms back into the grouped sum, and we get:

$$\mathrm{Cov}(\Delta Z_i^H, \Delta Z_j^H)$$
$$= -ab\left(\frac{1}{N}\right)^{2H} \left[(i+j)^{2H} - 2(i+j-1)^{2H} + (i+j-2)^{2H}\right]$$
$$- \frac{a^2+b^2}{2}\left(\frac{1}{N}\right)^{2H} \left[2|i-j|^{2H} - |i-j+1|^{2H} - |i-j-1|^{2H}\right]$$

This result matches the stated lemma, thus completing the proof. □

The following interesting theorem is crucial for constructing a consistent estimator for H.

THEOREM 4.1.– *Let $S_N(Z^H(a,b))$ be the statistic defined by equation [4.1]. As $N \to +\infty$, the variance of $S_N(Z^H(a,b))$, denoted Var(S_N), exhibits the following convergence properties:*

1) Asymptotic behavior of variance:

- *If $H \in (0, 3/4) \setminus \{1/2\}$, then Var$(S_N) = O(N^{-1-4H})$.*
- *If $H = 3/4$, then Var$(S_N) = O(N^{-4}\ln N)$.*
- *If $H \in (3/4, 1)$, then Var$(S_N) = O(N^{-4})$.*

2) Convergence to zero:

$$\lim_{N\to\infty} \mathit{Var}(S_N(a,b)) = 0.$$

Proof. For brevity, we denote the statistic as $S_N = S_N(Z^H(a,b))$. By definition, the variance of the statistic S_N is:

$$\begin{aligned}\text{Var}(S_N) &= \text{Var}\left(\frac{1}{N}\sum_{i=1}^{N}(\Delta Z_i^H)^2\right)\\ &= \frac{1}{N^2}\text{Var}\left(\sum_{i=1}^{N}(\Delta Z_i^H)^2\right)\\ &= \frac{1}{N^2}\left[\sum_{i=1}^{N}\text{Var}((\Delta Z_i^H)^2) + \sum_{i\neq j}\text{Cov}((\Delta Z_i^H)^2,(\Delta Z_j^H)^2)\right].\end{aligned} \quad [4.3]$$

Since the increments ΔZ_i^H are zero-mean Gaussian random variables, we can use the following properties for zero-mean Gaussian variables X and Y:

$$\text{Var}(X^2) = 2(\mathbb{E}[X^2])^2 \text{ and } \text{Cov}(X^2,Y^2) = 2(\text{Cov}(X,Y))^2.$$

First, we analyze the sum of the variances. Let $\sigma_i^2 = \mathbb{E}[(\Delta Z_i^H)^2]$. From Proposition 4.1, we know that:

$$\sigma_i^2 = (a^2+b^2)N^{-2H} - ab\left(\frac{2}{N}\right)^{2H}\left[i^{2H} - 2\left(i-\frac{1}{2}\right)^{2H} + (i-1)^{2H}\right].$$

As discussed in the proof of Lemma 4.1, the second term is of a smaller order. Specifically, the bracketed term is a discrete second-order central difference for the function $f(x) = x^{2H}$. Using Taylor's theorem, for large i, it behaves as:

$$i^{2H} - 2\left(i-\frac{1}{2}\right)^{2H} + (i-1)^{2H} \sim \frac{1}{4}\cdot f''(i) = \frac{1}{4}\cdot 2H(2H-1)i^{2H-2} = O(i^{2H-2}).$$

This implies that

$$\sigma_i^2 = (a^2+b^2)N^{-2H} + O(N^{-2H}i^{2H-2}).$$

Thus, the leading order of σ_i^2 is N^{-2H} for all $i \in \{1,\dots,N\}$, so $\sigma_i^2 = O(N^{-2H})$.

The variance of the squared increment is:

$$\text{Var}((\Delta Z_i^H)^2) = 2(\sigma_i^2)^2 = O(N^{-4H}).$$

The sum of these variances is:

$$\sum_{i=1}^{N}\text{Var}((\Delta Z_i^H)^2) = \sum_{i=1}^{N} O(N^{-4H}) = N\cdot O(N^{-4H}) = O(N^{1-4H}).$$

Next, we analyze the sum of the off-diagonal covariances:

$$\sum_{i\neq j} \text{Cov}((\Delta Z_i^H)^2, (\Delta Z_j^H)^2) = 2\sum_{i\neq j} (\text{Cov}(\Delta Z_i^H, \Delta Z_j^H))^2.$$

From Lemma 4.2, the covariance of the increments is:

$$\begin{aligned}&\text{Cov}(\Delta Z_i^H, \Delta Z_j^H)\\&= \frac{1}{2}\left(\frac{1}{N}\right)^{2H} \left[-2ab\left((i+j)^{2H} - 2(i+j-1)^{2H} + (i+j-2)^{2H}\right)\right.\\&\quad\left. -(a^2+b^2)\left(2|i-j|^{2H} - |i-j+1|^{2H} - |i-j-1|^{2H}\right)\right].\end{aligned}$$

The asymptotic behavior of the double sum of these squared covariances, as $N \to \infty$, is a known result. Let us denote this sum as:

$$\Sigma_N^{cov} = 2\sum_{i\neq j} (\text{Cov}(\Delta Z_i^H, \Delta Z_j^H))^2.$$

If $H = 1/2$, the increments are independent and uncorrelated. Thus,

$$\text{Cov}(\Delta Z_i^H, \Delta Z_j^H) = 0 \text{ for } i \neq j, \text{ and } \Sigma_N^{cov} = 0.$$

If $H = 3/4$, the sum of the squared covariances leads to a logarithmic term, with:

$$\Sigma_N^{cov} = O(N^{-2} \ln N).$$

If $H \in (0,1) \setminus \{1/2, 3/4\}$, the sum of the squared covariances is dominated by a term of order $O(N^{-2})$.

Now, we combine the asymptotic orders of the diagonal and off-diagonal terms to find the order of Var $\left(\sum_{i=1}^N (\Delta Z_i^H)^2\right)$. We compare the orders of the two summations: $O(N^{1-4H})$ and the respective orders of Σ_N^{cov}.

– *Case 1*: $H \in (0, 3/4) \setminus \{1/2\}$: In this case, $1 - 4H > 1 - 4(3/4) = -2$. The term $O(N^{1-4H})$ dominates the term $O(N^{-2})$. Therefore:

$$\text{Var}\left(\sum_{i=1}^N (\Delta Z_i^H)^2\right) = O(N^{1-4H}).$$

– *Case 2*: $H = 3/4$: In this case, $1 - 4H = 1 - 3 = -2$. The two terms have orders $O(N^{-2})$ and $O(N^{-2} \ln N)$. The logarithmic term dominates, so:

$$\text{Var}\left(\sum_{i=1}^N (\Delta Z_i^H)^2\right) = O(N^{-2} \ln N).$$

– *Case 3*: $H \in (3/4, 1)$: In this case, $1 - 4H < 1 - 4(3/4) = -2$. The term $O(N^{-2})$ dominates the term $O(N^{1-4H})$. Therefore:

$$\text{Var}\left(\sum_{i=1}^{N}(\Delta Z_i^H)^2\right) = O(N^{-2}).$$

Finally, we obtain the asymptotic order of $\text{Var}(S_N)$ stated in part 1 of the theorem by dividing the order of the total variance by N^2.

For the second part, since all of the exponents of N in the asymptotic orders are negative, we have:

$$\lim_{N\to\infty} \text{Var}(S_N) = 0.$$

By Chebyshev's inequality, this implies that the statistic S_N converges in probability to its expected value, that is:

$$S_N - \mathbb{E}[S_N] \xrightarrow{P} 0.$$

This completes the proof. □

COROLLARY 4.1.– *The statistic $S_N(Z^H(a,b))$ has the following convergence property as $N \to \infty$:*

$$\frac{S_N(Z^H(a,b))}{\mathbb{E}[S_N(Z^H(a,b))]} \xrightarrow{P} 1.$$

Proof. Let us denote $S_N = S_N(Z^H(a,b))$ and $\mu_N = \mathbb{E}[S_N(Z^H(a,b))]$. We aim to show that the ratio $\frac{S_N}{\mu_N}$ converges to 1 in probability. We can express the ratio as:

$$\frac{S_N}{\mu_N} = 1 + \frac{S_N - \mu_N}{\mu_N}.$$

By Slutsky's theorem (Theorem A.1), if we can show that the second term on the right-hand side, $\frac{S_N - \mu_N}{\mu_N}$, converges to 0 in probability, the proof will be complete.

A sufficient condition for convergence in probability to a constant (in this case, 0) is that the variance of the random variable converges to zero (by Chebyshev's inequality). We consider the variance of the term:

$$\text{Var}\left(\frac{S_N - \mu_N}{\mu_N}\right) = \frac{1}{\mu_N^2}\text{Var}(S_N - \mu_N) = \frac{1}{\mu_N^2}\text{Var}(S_N),$$

since μ_N is a deterministic sequence.

From Lemma 4.1, we know that the asymptotic behavior of the mean is:

$$\mu_N = \mathbb{E}[S_N(Z^H(a,b))] \sim (a^2+b^2)N^{-2H}.$$

Since $(a,b) \neq (0,0)$, it follows that $a^2+b^2>0$. Therefore, the square of the mean behaves as:

$$\mu_N^2 = O(N^{-4H}).$$

Next, we combine this result with the asymptotic orders of Var(S_N) established in Theorem 4.1 and analyze the convergence of the variance of the ratio on a case-by-case basis.

– *Case 1*: $H \in (0,3/4)\setminus\{1/2\}$: from Theorem 4.1, we have Var$(S_N) = O(N^{-1-4H})$. The variance of the ratio is:

$$\text{Var}\left(\frac{S_N-\mu_N}{\mu_N}\right) = \frac{O(N^{-1-4H})}{O(N^{-4H})} = O(N^{-1}).$$

As $N\to\infty$, this term converges to 0.

– *Case 2*: $H = 3/4$: for this case, $4H=3$, so $\mu_N^2 = O(N^{-4H}) = O(N^{-3})$. From Theorem 4.1, Var$(S_N) = O(N^{-4}\ln N)$. The variance of the ratio is:

$$\text{Var}\left(\frac{S_N-\mu_N}{\mu_N}\right) = \frac{O(N^{-4}\ln N)}{O(N^{-3})} = O(N^{-1}\ln N).$$

As $N\to\infty$, this term also converges to 0.

– *Case 3*: $H\in(3/4,1)$: in this final case, from Theorem 4.1, we have Var$(S_N) = O(N^{-4})$. The variance of the ratio is:

$$\text{Var}\left(\frac{S_N-\mu_N}{\mu_N}\right) = \frac{O(N^{-4})}{O(N^{-4H})} = O(N^{-4+4H}).$$

Since $H\in(3/4,1)$, we have $4H-4<0$. Therefore, the exponent is negative, and the term N^{-4+4H} converges to 0 as $N\to\infty$.

In all sub-cases, the variance of the term $\dfrac{S_N-\mu_N}{\mu_N}$ converges to zero. By Chebyshev's inequality, this implies convergence in probability:

$$\frac{S_N-\mu_N}{\mu_N} \xrightarrow{P} 0.$$

Substituting this back into our expression for the ratio, we conclude that:

$$\frac{S_N}{\mu_N} = 1 + \frac{S_N-\mu_N}{\mu_N} \xrightarrow{P} 1. \qquad \square$$

Building upon the variance convergence established in Theorem 4.1, the following corollary details the essential asymptotic behavior of $S_N(Z^H(a,b))$ itself.

COROLLARY 4.2.– *As $N \to \infty$, the statistic $S_N(Z^H(a,b))$ satisfies the following asymptotic approximations in probability:*

1) $S_N(Z^H(a,b)) \approx (a^2+b^2)N^{-2H}$.[2]

2) $\log\left(S_N(Z^H(a,b))\right) \approx \log\left((a^2+b^2)N^{-2H}\right)$.

Proof. Let us first show that $\dfrac{S_N(Z^H(a,b))}{(a^2+b^2)N^{-2H}} \xrightarrow{P} 1$. We can express the ratio as:

$$\frac{S_N(Z^H(a,b))}{(a^2+b^2)N^{-2H}} = \frac{S_N(Z^H(a,b))}{\mathbb{E}[S_N(Z^H(a,b))]} \cdot \frac{\mathbb{E}[S_N(Z^H(a,b))]}{(a^2+b^2)N^{-2H}}.$$

From the preceding Corollary 4.1, we know that $\dfrac{S_N(Z^H(a,b))}{\mathbb{E}[S_N(Z^H(a,b))]} \xrightarrow{P} 1$ as $N \to \infty$. From Lemma 4.1, we have the asymptotic behavior of the expected value:

$$\mathbb{E}[S_N(Z^H(a,b))] = (a^2+b^2)N^{-2H}\left(1+O\left(\frac{1}{N}\right)\right).$$

This implies that:

$$\lim_{N\to\infty} \frac{\mathbb{E}[S_N(Z^H(a,b))]}{(a^2+b^2)N^{-2H}} = \lim_{N\to\infty}\left(1+O\left(\frac{1}{N}\right)\right) = 1.$$

By Slutsky's theorem (Theorem A.1), if a sequence of random variables converges in probability to a constant, and a deterministic sequence converges to a constant, their product converges in probability to the product of the constants. Thus,

$$\frac{S_N(Z^H(a,b))}{(a^2+b^2)N^{-2H}} \xrightarrow{P} 1,$$

which proves the first part of the corollary.

Concerning the second part, it suffices to show that $\dfrac{\log\left(S_N(Z^H(a,b))\right)}{\log\left((a^2+b^2)N^{-2H}\right)} \xrightarrow{P} 1$. Let $Y_N = (a^2+b^2)N^{-2H}$. Consider the following ratio.

---

2 For two sequences of random variables $(X_n)_{n\geq 1}$ and $(Y_n)_{n\geq 1}$, we denote their asymptotic equivalence in probability as $X_n \approx Y_n$, if $\mathbb{P}(Y_n \neq 0) \to 1$ as $n \to \infty$, and for every $\delta > 0$:

$$\lim_{n\to\infty} \mathbb{P}\left(\left|\frac{X_n}{Y_n} - 1\right| > \delta\right) = 0.$$

This is equivalent to stating that $\dfrac{X_n}{Y_n} \xrightarrow{P} 1$ as $n \to \infty$.

$$
\begin{aligned}
\frac{\log(S_N(Z^H(a,b)))}{\log(Y_N)} &= \frac{\log\left(\frac{S_N(Z^H(a,b))}{Y_N} \cdot Y_N\right)}{\log(Y_N)} \\
&= \frac{\log\left(\frac{S_N(Z^H(a,b))}{Y_N}\right) + \log(Y_N)}{\log(Y_N)} \\
&= 1 + \frac{\log\left(\frac{S_N(Z^H(a,b))}{Y_N}\right)}{\log(Y_N)}.
\end{aligned}
$$

Let $R_N = \dfrac{S_N(Z^H(a,b))}{Y_N}$. From Part 1, we know that $R_N \xrightarrow{P} 1$ as $N \to \infty$. By the continuous mapping theorem, since $\log(x)$ is continuous for $x > 0$, it follows that:

$$\log(R_N) \xrightarrow{P} \log(1) = 0.$$

Now consider the denominator $\log(Y_N) = \log(a^2 + b^2) - 2H \log(N)$.

As $N \to \infty$, since $H \in (0,1)$, we have $2H \log(N) \to \infty$, and thus

$$|\log(Y_N)| \to \infty.$$

Since $\log(R_N) \xrightarrow{P} 0$ and $|\log(Y_N)| \to \infty$ deterministically, it implies that:

$$\frac{\log(R_N)}{\log(Y_N)} \xrightarrow{P} 0.$$

Thus,

$$\frac{\log(S_N(Z^H(a,b)))}{\log(Y_N)} = 1 + \frac{\log(R_N)}{\log(Y_N)} \xrightarrow{P} 1.$$

This completes the proof. □

4.1.2. *Estimators for the Hurst parameter* H

Building upon the consistency results for the statistic $S_N(Z^H(a,b))$, particularly the asymptotic equivalence established in Corollary 4.2, we can now derive explicit estimators for the Hurst parameter H.

From Corollary 4.2, we have the following asymptotic relationship in probability for the logarithm of the statistic:

$$\log(S_N(Z^H(a,b))) \approx \log(a^2 + b^2) - 2H \log(N).$$

This relationship forms the basis of our estimation strategy. By rearranging this expression to isolate H, we arrive at a natural estimator that utilizes the known values of the deterministic coefficients a and b:

$$\widehat{H}_N^{(1)}(a,b) = \frac{\log(a^2+b^2) - \log(S_N(Z^H(a,b)))}{2\log(N)}. \qquad [4.4]$$

Furthermore, as $N \to \infty$, the term $\log(a^2+b^2)/(2\log N)$ converges to zero. This crucial observation leads to a simplified estimator that is asymptotically equivalent to the first, thereby maintaining its consistency properties. This simplified estimator is given by:

$$\widehat{H}_N^{(2)}(a,b) = -\frac{\log S_N(Z^H(a,b))}{2\log N}. \qquad [4.5]$$

Although $\widehat{H}_N^{(2)}(a,b)$ does not explicitly include a^2+b^2 in its numerator, its consistency, as we will demonstrate, inherently relies on the fact that (a^2+b^2) is a nonzero constant, ensuring the validity of the logarithmic approximations.

Now, we formally establish the consistency of these estimators for H.

THEOREM 4.2.– *The statistics* $\widehat{H}_N^{(1)}(a,b)$ *and* $\widehat{H}_N^{(2)}(a,b)$ *are consistent estimators of the Hurst parameter* H.

Proof. Let us prove the consistency of both estimators by showing their convergence in probability to H. For simplicity, let $S_N = S_N(Z^H(a,b))$.

The first estimator is defined as:

$$\widehat{H}_N^{(1)}(a,b) = \frac{\log(a^2+b^2) - \log(S_N)}{2\log(N)}.$$

Our goal is to show $\widehat{H}_N^{(1)}(a,b) \xrightarrow{P} H$ as $N \to \infty$.

From part 2 of Corollary 4.2, we have established the following asymptotic equivalence in probability:

$$\log(S_N) \approx \log\left((a^2+b^2)N^{-2H}\right).$$

This is equivalent to the ratio of the two quantities converging to 1 in probability. Let

$$L_N = \log(S_N) \text{ and } M_N = \log((a^2+b^2)N^{-2H}).$$

We have:

$$\frac{L_N}{M_N} \xrightarrow{P} 1 \; as \; N \to \infty.$$

We can express our estimator in terms of L_N and M_N. By definition,

$$M_N = \log(a^2 + b^2) - 2H\log(N).$$

We add and subtract the term M_N in the numerator of the estimator's expression:

$$\begin{aligned}\widehat{H}_N^{(1)}(a,b) &= \frac{\log(a^2+b^2) - L_N}{2\log(N)} \\ &= \frac{(\log(a^2+b^2) - 2H\log(N)) + 2H\log(N) - L_N}{2\log(N)} \\ &= \frac{M_N + 2H\log(N) - L_N}{2\log(N)} \\ &= H + \frac{M_N - L_N}{2\log(N)}.\end{aligned}$$

To prove consistency, we must show that the remainder term $\dfrac{M_N - L_N}{2\log(N)}$ converges to 0 in probability.

We can analyze the remainder term by factoring the numerator:

$$\frac{M_N - L_N}{2\log(N)} = \frac{M_N}{2\log(N)}\left(1 - \frac{L_N}{M_N}\right).$$

Let us analyze the two factors: The first factor is a deterministic sequence:

$$\lim_{N\to\infty} \frac{M_N}{2\log(N)} = \lim_{N\to\infty} \frac{\log(a^2+b^2) - 2H\log(N)}{2\log(N)}$$

$$= \lim_{N\to\infty}\left(\frac{\log(a^2+b^2)}{2\log(N)} - H\right) = -H.$$

The second factor is a sequence of random variables. Since $\dfrac{L_N}{M_N} \xrightarrow{P} 1$ and the function $f(x) = 1 - x$ is continuous, we can apply the continuous mapping theorem to get:

$$1 - \frac{L_N}{M_N} \xrightarrow{P} 1 - 1 = 0 \quad \text{as } N \to \infty.$$

By Slutsky's theorem, the product of a sequence converging deterministically to a constant $(-H)$ and a sequence converging in probability to 0 converges in probability to 0. Thus,

$$\frac{M_N - L_N}{2\log(N)} \xrightarrow{P} (-H)\cdot 0 = 0 \quad \text{as } N \to \infty.$$

Substituting this back into our expression for $\widehat{H}_N^{(1)}(a,b)$, we have:

$$\widehat{H}_N^{(1)}(a,b) = H + \frac{M_N - L_N}{2\log(N)} \xrightarrow{P} H + 0 = H.$$

This demonstrates that $\widehat{H}_N^{(1)}(a,b)$ is a consistent estimator for H.

Now, we can establish the consistency of the second estimator by relating it to the first. From their definitions, we have:

$$\begin{aligned}\widehat{H}_N^{(2)}(a,b) &= -\frac{\log S_N}{2\log N} \\ &= \frac{\log(a^2+b^2) - \log S_N}{2\log N} - \frac{\log(a^2+b^2)}{2\log N} \\ &= \widehat{H}_N^{(1)}(a,b) - \frac{\log(a^2+b^2)}{2\log(N)}.\end{aligned}$$

We have already shown that $\widehat{H}_N^{(1)}(a,b) \xrightarrow{P} H$ as $N \to \infty$. The second term, $\frac{\log(a^2+b^2)}{2\log(N)}$, is a deterministic sequence that converges to 0. By Slutsky's theorem, the difference of a sequence converging in probability to H and a deterministic sequence converging to 0 converges in probability to $H - 0$. Therefore:

$$\widehat{H}_N^{(2)}(a,b) \xrightarrow{P} H - 0 = H \quad \text{as } N \to \infty.$$

This completes the proof that both estimators are consistent. □

4.2. Maximum likelihood estimation of a and b, when H is known

Having established a methodology for estimating the Hurst parameter H under the assumption that the scaling coefficients a and b are known, we now turn to the inverse problem. In this section, we develop an approach to estimate the deterministic coefficients a and b from discrete observations of the ZgfBm process, assuming that H is a known quantity. This scenario is particularly relevant in applications where the underlying long-range dependence structure is well established, and the goal is to quantify the specific asymmetric properties of the process. We will derive the maximum likelihood estimation for a and b by leveraging the Gaussian nature of the process and introducing a suitable reparameterization.

The observation vector for the process remains $\mathbf{Z} = (Z_{t_1}, Z_{t_2}, \ldots, Z_{t_n})'$, where $t_k = kh$ for a fixed step size $h > 0$. The process can be decomposed as:

$$Z_t^H = (a-b)B_t^{H,o} + (a+b)B_t^{H,e},$$

where $B^{H,o}$ and $B^{H,e}$ are the odd and even fractional Brownian motions, respectively.

4.2.1. *Estimating* a *and* b *when* $a \neq b$ *and* $a \neq -b$

To simplify the estimation process, we reparameterize the problem by defining α and ϵ as:

$$\alpha = a - b \quad \text{and} \quad \epsilon = \frac{a+b}{a-b}. \tag{4.6}$$

This reparameterization is valid for $a \neq b$, ensuring $\alpha \neq 0$. Conversely, the original parameters a and b can be expressed in terms of α and ϵ as:

$$a = \frac{\alpha(1+\epsilon)}{2} \quad \text{and} \quad b = \frac{\alpha(\epsilon - 1)}{2}. \tag{4.7}$$

The estimation plan is to first estimate the unknown parameters α and ϵ from observations $\mathbf{Z}$, and then deduce the estimated values of a and b using the inverse relationships in equation [4.7].

In vector form, the process can be written as:

$$\mathbf{Z} = \alpha\Big(\mathbf{B^o} + \epsilon\mathbf{B^e}\Big),$$

where

$$\mathbf{B^o} = \Big(B_h^{H,o}, B_{2h}^{H,o}, \dots, B_{nh}^{H,o}\Big)' \quad \text{and} \quad \mathbf{B^e} = \Big(B_h^{H,e}, B_{2h}^{H,e}, \dots, B_{nh}^{H,e}\Big)'. \tag{4.8}$$

Since $\mathbf{Z}$ is a Gaussian process, the joint probability density function of the observation vector is:

$$f(\mathbf{Z}) = (2\pi\alpha^2)^{-n/2} \mid \Gamma \mid^{-1/2} \exp\Big(-\frac{1}{2\alpha^2}\mathbf{Z}'\Gamma^{-1}\mathbf{Z}\Big),$$

where $\mathbf{\Gamma}$ is the covariance matrix of the rescaled process $\mathbf{B^o} + \epsilon\mathbf{B^e}$:

$$\begin{aligned}
\Gamma &= \mathrm{Cov}(\mathbf{B^o} + \epsilon\mathbf{B^e}) = \mathrm{Cov}(\mathbf{B^o}) + \epsilon^2\mathrm{Cov}(\mathbf{B^e}) \\
&= \mathbf{\Sigma_O} + \epsilon^2\mathbf{\Sigma_E} \\
&= \Bigg(\frac{h^{2H}}{4}\Big((i+j)^{2H} - \mid i-j \mid^{2H}\Big) \\
&\quad + \epsilon^2\frac{h^{2H}}{2}\Big(i^{2H} + j^{2H} - \frac{1}{2}\Big((i+j)^{2H} + \mid i-j \mid^{2H}\Big)\Big)\Bigg)_{i,j=1,\dots,n}.
\end{aligned} \tag{4.9}$$

LEMMA 4.3.– *Let* $\mathbf{\Gamma} = \mathbf{\Sigma_O} + \epsilon^2\mathbf{\Sigma_E}$. *Then* $\mathbf{\Gamma}$ *is a symmetric positive definite matrix, and thus invertible, for any real value of* ϵ.

Proof. First, let us establish the symmetry of $\mathbf{\Gamma}$. Since $\mathbf{\Sigma_O}$ and $\mathbf{\Sigma_E}$ are covariance matrices, they are by definition symmetric. The sum of symmetric matrices is also symmetric, and multiplying a symmetric matrix by a scalar ϵ^2 preserves symmetry. Therefore, $\mathbf{\Gamma}$ is a symmetric matrix.

Next, we prove that $\mathbf{\Gamma}$ is positive definite. A symmetric matrix $\mathbf{M}$ is positive definite if for any nonzero real vector $\mathbf{x} \in \mathbb{R}^n$, the quadratic form $\mathbf{x}'\mathbf{M}\mathbf{x} > 0$.

Consider the quadratic form for Γ:

$$\mathbf{x}'\mathbf{\Gamma}\mathbf{x} = \mathbf{x}'(\mathbf{\Sigma_O} + \epsilon^2\mathbf{\Sigma_E})\mathbf{x} = \mathbf{x}'\mathbf{\Sigma_O}\mathbf{x} + \epsilon^2\mathbf{x}'\mathbf{\Sigma_E}\mathbf{x}.$$

From Corollaries 2.1 and 2.2, we know that $\mathbf{\Sigma_O}$ and $\mathbf{\Sigma_E}$ are symmetric positive definite matrices. By definition of positive definiteness, this means that for all nonzero vectors $\mathbf{x} \in \mathbb{R}^n$, $\mathbf{x}'\mathbf{\Sigma_O}\mathbf{x} > 0$ and $\mathbf{x}'\mathbf{\Sigma_E}\mathbf{x} > 0$.

Furthermore, ϵ^2 is always nonnegative. Thus, for any nonzero vector $\mathbf{x}$:

$$\mathbf{x}'\mathbf{\Gamma}\mathbf{x} = (\text{a positive value}) + (\text{a nonnegative value}).$$

Since the first term $(\mathbf{x}'\mathbf{\Sigma_O}\mathbf{x})$ is strictly positive, and the second term $(\epsilon^2\mathbf{x}'\mathbf{\Sigma_E}\mathbf{x})$ is nonnegative, their sum must always be strictly positive.

In all cases where $\mathbf{x} \neq \mathbf{0}$, we have $\mathbf{x}'\Gamma\mathbf{x} > 0$. Therefore, Γ is a positive definite matrix. □

A fundamental property of positive definite matrices is that they are always invertible. Thus, $\mathbf{\Gamma}^{-1}$ exists, which is essential for the formulation of the Gaussian probability density function and the subsequent maximum likelihood estimation procedure. This also ensures that the determinant $\mid \mathbf{\Gamma} \mid$ is strictly positive. So, the log-likelihood function is:

$$\ln f(\mathbf{Z}; \alpha^2, \epsilon^2) = -\frac{n}{2}\ln(2\pi) - \frac{n}{2}\ln \alpha^2 - \frac{1}{2}\ln \mid \mathbf{\Gamma} \mid - \frac{1}{2\alpha^2}\mathbf{Z}'\mathbf{\Gamma}^{-1}\mathbf{Z} \quad [4.10]$$

We find maximum likelihood estimators by maximizing the log-likelihood function [6.9] over its parameter space:

$$(\hat{\alpha}^2, \hat{\epsilon}^2) = \arg\max_{\alpha^2, \epsilon^2} \ln f(\mathbf{Z}; \alpha^2, \epsilon^2). \quad [4.11]$$

The first-order conditions for maximum likelihood estimation require the gradient of the log-likelihood to be equal to zero. In other words, the maximum likelihood estimators of α^2 and ϵ^2 are obtained by maximizing the log-likelihood function $\ln f(\mathbf{Z}; \alpha^2, \epsilon^2)$ with respect to α^2 and ϵ^2, respectively. By the relationships, we have, for every matrix A depending on a parameter x:

$$\frac{\partial}{\partial x}A^{-1} = -A^{-1}\Big(\frac{\partial}{\partial x}A\Big)A^{-1}, \quad [4.12]$$

$$\frac{\partial}{\partial x} \ln | A |= \mathrm{Tr}\Big(A^{-1}\frac{\partial}{\partial x}A\Big), \qquad [4.13]$$

and consequently,

$$\begin{cases} -\dfrac{n}{2\alpha^2} + \dfrac{1}{2\alpha^4}\mathbf{Z}'\Gamma^{-1}\mathbf{Z} = 0, \\ -\dfrac{1}{2}\mathrm{Tr}(\Gamma^{-1}\Sigma_E) + \dfrac{1}{2\alpha^2}\mathbf{Z}'\Gamma^{-1}\Sigma_E\Gamma^{-1}\mathbf{Z} = 0. \end{cases} \qquad [4.14]$$

where

$$\begin{aligned} \Sigma_E &= \Big(Cov\Big(B^{H,e}_{ih}, B^{H,e}_{jh}\Big)\Big)_{i,j=1,\dots,n} \\ &= \left(\frac{h^{2H}}{2}\left(i^{2H} + j^{2H} - \frac{1}{2}\Big((i+j)^{2H} + | i-j |^{2H}\Big)\right)\right)_{i,j=1,\dots,n}. \end{aligned} \qquad [4.15]$$

The solution to this nonlinear system of equations may be found by a variety of numerical methods. However, we will present a hybrid approach that simplifies the numerical search by exploiting some convenient analytical manipulations of the first-order conditions.

First, we observe that the first line of equation [6.1] can be rewritten as:

$$\hat{\alpha}^2 = \frac{1}{n}\mathbf{Z}'\Gamma^{-1}\mathbf{Z}. \qquad [4.16]$$

Substituting equation [6.3] into equation [6.9], we get the profile log-likelihood function:

$$L_n(\mathbf{Z}, \epsilon^2) = -\frac{n}{2}\ln(2\pi) + \frac{n}{2}\ln n - \frac{n}{2}\ln(\mathbf{Z}'\Gamma^{-1}\mathbf{Z}) - \frac{1}{2}\ln | \Gamma | -\frac{n}{2}. \qquad [4.17]$$

The maximum likelihood estimator of ϵ^2 is obtained by maximizing $L_n(\mathbf{Z}, \epsilon^2)$ with respect to ϵ^2, or, equivalently, by minimizing the function:

$$M_n(\mathbf{Z}, \epsilon^2) = \frac{1}{2}\ln | \Gamma | + \frac{n}{2}\ln(\mathbf{Z}'\Gamma^{-1}\mathbf{Z}). \qquad [4.18]$$

This minimizing problem can be reformulated in terms of the first derivatives. Hence, the maximum likelihood estimator of ϵ^2 is the solution of:

$$(\mathbf{Z}'\Gamma^{-1}\mathbf{Z})\mathrm{Tr}(\Gamma^{-1}\Sigma_E) - n(\mathbf{Z}'\Gamma^{-1}\Sigma_E\Gamma^{-1}\mathbf{Z}) = 0. \qquad [4.19]$$

Indeed, the solution for this equation is algebraically complex and would not lead to a closed form. Hence, we turn to solve the optimization problem [6.5]. For solving this optimization problem, we also have to rely on the numerical solution. Now, the hybrid algorithm can be summarized as follows:

– *Step 1*: obtain the optimal value of the estimator $\hat{\epsilon}^2$ by numerically minimizing the function $M_n(\mathbf{Z}, \epsilon^2)$ (from equation [6.5]). This can be done using optimization routines such as MATLAB's `fminsearch`, which implements a quasi-Newton method.

– *Step 2*: calculate the valuation of $\hat{\alpha}^2$ by substituting the valuation of the estimator $\hat{\epsilon}^2$ into equation [6.3].

– *Step 3*: obtain the valuation of the estimators $\hat{a}$ and $\hat{b}$ based on the inverse relationships given in equation [4.7]:

$$\hat{a} = \frac{\hat{\alpha}(1+\hat{\epsilon})}{2} \quad \text{and} \quad \hat{b} = \frac{\hat{\alpha}(\hat{\epsilon}-1)}{2}.$$

Note that the signs of $\hat{\alpha}$ and $\hat{\epsilon}$ may need careful consideration if taking square roots. Typically, we estimate $\hat{\alpha}^2$ and $\hat{\epsilon}^2$ and then infer $\hat{\alpha}$ and $\hat{\epsilon}$ based on assumptions (e.g. $a, b > 0$). If a and b can be negative, the sign of α and ϵ would need to be determined by further conditions or by exploring both positive and negative roots.

The advantage of this approach is that the problem in [6.5] requires a one-dimensional search, which is computationally more efficient than the two-dimensional search required for simultaneously solving the first-order conditions for both α^2 and ϵ^2.

The practical implementation of this hybrid maximum likelihood estimation is detailed in Exercise 4.5, which provides explicit and comprehensive guidance for the numerical simulation and estimation steps, ensuring clarity in applying the one-dimensional search procedure.

With an explicit expression for the estimator $\hat{\alpha}^2$, we can now rigorously discuss its statistical properties. We begin by formally establishing its consistency.

THEOREM 4.3.– *The maximum likelihood estimator $\hat{\alpha}^2$ (defined by equation [6.3]) is unbiased and converges in mean square to α^2 as $n \to \infty$.*

Proof. Note that $\alpha^{-1}\mathbf{Z}$ is a zero-mean Gaussian vector, so $\alpha^{-1}\mathbf{Z} \sim \mathcal{N}_n(0, \Gamma)$ where the covariance matrix Γ is defined by equation [4.9]. As established in Lemma 4.3, Γ is an $n \times n$ positive-definite symmetric matrix.

Applying the results of Lemmas 1.9 and 1.10 for the expectation and variance of a quadratic form of a zero-mean Gaussian vector, we have:

$$\mathbb{E}\left[\mathbf{Z}'\Gamma^{-1}\mathbf{Z}\right] = \alpha^2 \text{Tr}(\Gamma^{-1}\Gamma) = \alpha^2 n \qquad [4.20]$$

$$\mathbb{E}\left[(\mathbf{Z}'\Gamma^{-1}\mathbf{Z})^2\right] = \alpha^4\Big(\text{Tr}(\Gamma^{-1}\Gamma)\Big)^2 + 2\alpha^4\text{Tr}(\Gamma^{-1}\Gamma\Gamma^{-1}\Gamma) = \alpha^4(n^2+2n). \qquad [4.21]$$

Using the definition of the estimator $\hat{\alpha}^2$ from equation [6.3], we can now compute its expectation and variance.

For the expectation, we get:

$$\mathbb{E}\left[\hat{\alpha}^2\right] = \mathbb{E}\left[\frac{1}{n}\mathbf{Z}'\Gamma^{-1}\mathbf{Z}\right] = \frac{1}{n}\mathbb{E}\left[\mathbf{Z}'\Gamma^{-1}\mathbf{Z}\right] = \frac{1}{n}(\alpha^2 n) = \alpha^2. \qquad [4.22]$$

This shows that $\hat{\alpha}^2$ is an unbiased estimator of α^2.

For the variance, we use the formula $\mathrm{Var}(X) = \mathbb{E}\left[X^2\right] - (\mathbb{E}[X])^2$:

$$\mathrm{Var}(\hat{\alpha}^2) = \mathrm{Var}\left(\frac{1}{n}\mathbf{Z}'\Gamma^{-1}\mathbf{Z}\right) = \frac{1}{n^2}\mathrm{Var}(\mathbf{Z}'\Gamma^{-1}\mathbf{Z}) \qquad [4.23]$$

$$= \frac{1}{n^2}\left[\mathbb{E}\left[(\mathbf{Z}'\Gamma^{-1}\mathbf{Z})^2\right] - \left(\mathbb{E}\left[\mathbf{Z}'\Gamma^{-1}\mathbf{Z}\right]\right)^2\right] \qquad [4.24]$$

$$= \frac{1}{n^2}\left[\alpha^4(n^2 + 2n) - (\alpha^2 n)^2\right] \qquad [4.25]$$

$$= \frac{1}{n^2}\left[\alpha^4 n^2 + 2\alpha^4 n - \alpha^4 n^2\right] = \frac{2\alpha^4}{n}. \qquad [4.26]$$

Since $\mathrm{Var}(\hat{\alpha}^2) \to 0$ as $n \to \infty$, the estimator converges in mean square to α^2. As mean square convergence implies convergence in probability, this also establishes the consistency of $\hat{\alpha}^2$. □

4.2.2. *Special cases: when* $a = b$ *or* $a = -b$

When $a = b$ or $a = -b$, the ZgfBm simplifies to a form that is a scaled version of either an even or odd fractional Brownian motion, warranting a dedicated and direct analysis for parameter estimation.

Case 1: $a = b$

When $a = b$, the reparameterization $\epsilon = \dfrac{a+b}{a-b}$ is undefined. In this specific case, the decomposition from Proposition 3.4 simplifies significantly:

$$Z_t^H = (a-b)B_t^{H,o} + (a+b)B_t^{H,e} = 2aB_t^{H,e}.$$

Thus, Z_t^H is solely a constant multiple of an even fBm. The observed data vector becomes $\mathbf{Z} = 2a\mathbf{B}^{\mathbf{e}}$, where $\mathbf{B}^{\mathbf{e}} = (B_{t_1}^{H,e}, \ldots, B_{t_n}^{H,e})'$.

Since $\mathbf{B}^{\mathbf{e}}$ is a zero-mean Gaussian vector with a known covariance matrix $\mathbf{\Sigma_E}$ (where $(\mathbf{\Sigma_E})_{ij} = \mathrm{Cov}(B_{t_i}^{H,e}, B_{t_j}^{H,e})$), the joint probability density function of $\mathbf{Z}$ simplifies to:

$$f(\mathbf{Z}; a) = (2\pi(2a)^2)^{-n/2} \mid \mathbf{\Sigma_E} \mid^{-1/2} \exp\left(- \frac{1}{2(2a)^2}\mathbf{Z}'\mathbf{\Sigma_E}^{-1}\mathbf{Z}\right).$$

The log-likelihood function is then expressed as:

$$\ln f(\mathbf{Z}; a) = -\frac{n}{2}\ln(2\pi) - \frac{n}{2}\ln(4a^2) - \frac{1}{2}\ln|\,\Sigma_{\mathbf{E}}\,| - \frac{1}{8a^2}\mathbf{Z}'\Sigma_{\mathbf{E}}{}^{-1}\mathbf{Z}.$$

Maximizing this log-likelihood with respect to a^2 directly yields the maximum likelihood estimator for a^2:

$$\hat{a}^2 = \frac{1}{4n}\mathbf{Z}'\Sigma_{\mathbf{E}}{}^{-1}\mathbf{Z}.$$

This estimator for a^2 can be shown to possess properties of unbiasedness and L^2-consistency. Its derivation is analogous to the $\hat{\alpha}^2$ case in the general maximum likelihood estimation problem, simply by replacing α with $2a$ and the general covariance matrix Γ with the specific matrix $\Sigma_{\mathbf{E}}$.

Case 2: $a = -b$

When $a = -b$, it follows that $a + b = 0$. In the context of the reparameterization, this implies $\epsilon = \dfrac{a+b}{a-b} = 0$. The decomposition from Proposition 3.4 then simplifies to:

$$Z_t^H = (a-(-a))B_t^{H,o} + (a+(-a))B_t^{H,e} = 2aB_t^{H,o}.$$

Consequently, Z_t^H is simply a constant multiple of an odd fBm. The observation vector is $\mathbf{Z} = 2a\mathbf{B}^{\mathbf{o}}$, where $\mathbf{B}^{\mathbf{o}} = (B_{t_1}^{H,o}, \ldots, B_{t_n}^{H,o})'$.

Given that $\mathbf{B}^{\mathbf{o}}$ is a zero-mean Gaussian vector with a known covariance matrix $\Sigma_{\mathbf{O}}$ (where $(\Sigma_{\mathbf{O}})_{ij} = \mathrm{Cov}(B_{t_i}^{H,o}, B_{t_j}^{H,o})$), the joint probability density function of $\mathbf{Z}$ simplifies to:

$$f(\mathbf{Z}; a) = (2\pi(2a)^2)^{-n/2}\,|\,\Sigma_{\mathbf{O}}\,|^{-1/2}\exp\Big(-\frac{1}{2(2a)^2}\mathbf{Z}'\Sigma_{\mathbf{O}}{}^{-1}\mathbf{Z}\Big).$$

The corresponding log-likelihood function is:

$$\ln f(\mathbf{Z}; a) = -\frac{n}{2}\ln(2\pi) - \frac{n}{2}\ln(4a^2) - \frac{1}{2}\ln|\,\Sigma_{\mathbf{O}}\,| - \frac{1}{8a^2}\mathbf{Z}'\Sigma_{\mathbf{O}}{}^{-1}\mathbf{Z}.$$

Maximizing this expression with respect to a^2 directly yields the MLE for a^2:

$$\hat{a}^2 = \frac{1}{4n}\mathbf{Z}'\Sigma_{\mathbf{O}}{}^{-1}\mathbf{Z}.$$

Similar to the previous case, this estimator for a^2 is also unbiased and consistent in L^2, owing to its analogous structure to the general MLE derived for $\hat{\alpha}^2$.

For the remainder of this section, we will focus on the general case where $a \neq b$ and $a \neq -b$.

4.2.3. *Asymptotic normality of* $\hat{\alpha}^2$

To establish the asymptotic normality of our estimator, we will show that a properly normalized version of $\hat{\alpha}^2$ converges in distribution to a standard normal random variable. This property is crucial for constructing confidence intervals and performing hypothesis tests.

THEOREM 4.4.– *As $n \to \infty$, the normalized estimator $\hat{\alpha}^2$ converges in distribution to a standard normal random variable:*

$$\frac{\hat{\alpha}^2 - \alpha^2}{\alpha^2\sqrt{2/n}} \xrightarrow{D} \mathcal{N}(0,1).$$

Proof. The observation vector $\mathbf{Z}$ follows a multivariate Gaussian distribution with mean vector $\mathbf{0}$ and covariance matrix $\alpha^2\mathbf{\Gamma}$. Thus, we have $\mathbf{Z} \sim \mathcal{N}_n(0, \alpha^2\mathbf{\Gamma})$.

To simplify the quadratic form, let us define a new random vector $\mathbf{Y}$:

$$\mathbf{Y} = \alpha^{-1}\mathbf{Z}.$$

The vector $\mathbf{Y}$ is also Gaussian, with mean $\mathbf{0}$ and covariance matrix:

$$\mathbb{E}[\mathbf{Y}\mathbf{Y}'] = \mathbb{E}[(\alpha^{-1}\mathbf{Z})(\alpha^{-1}\mathbf{Z})'] = \alpha^{-2}\mathbb{E}[\mathbf{Z}\mathbf{Z}'] = \alpha^{-2}(\alpha^2\mathbf{\Gamma}) = \mathbf{\Gamma}.$$

Thus, $\mathbf{Y} \sim \mathcal{N}_n(0, \Gamma)$.

Now, we can express the estimator $\hat{\alpha}^2$ in terms of $\mathbf{Y}$:

$$\hat{\alpha}^2 = \frac{1}{n}(\alpha\mathbf{Y})'\Gamma^{-1}(\alpha\mathbf{Y}) = \frac{\alpha^2}{n}\mathbf{Y}'\Gamma^{-1}\mathbf{Y}.$$

Let $\mathbf{Q}_n = \mathbf{Y}'\Gamma^{-1}\mathbf{Y}$. Our goal is to establish the asymptotic normality of $\hat{\alpha}^2$, which is equivalent to establishing the asymptotic normality of $\mathbf{Q}_n$ since $\frac{\alpha^2}{n}$ is a deterministic scaling factor.

From Lemma 4.3, we know that $\mathbf{\Gamma}$ is a symmetric positive definite matrix. Therefore, $\mathbf{\Gamma}$ can be spectrally decomposed as:

$$\mathbf{\Gamma} = U\Lambda U',$$

where U is an $n \times n$ orthogonal matrix ($UU' = I_n$) and Λ is an $n \times n$ diagonal matrix containing the positive eigenvalues of Γ. From this decomposition, we can write:

$$\mathbf{\Gamma}^{-1} = U\Lambda^{-1}U'.$$

Now, let us introduce another transformation by defining $\mathbf{V}$:

$$\mathbf{V} = \Lambda^{-1/2}U'\mathbf{Y}.$$

We check the mean and covariance of $\mathbf{V}$:

$$\mathbb{E}[\mathbf{V}] = \Lambda^{-1/2}U'\mathbb{E}[\mathbf{Y}] = \Lambda^{-1/2}U'\mathbf{0} = \mathbf{0}.$$

$$\begin{aligned}\mathrm{Cov}(\mathbf{V}) &= \mathbb{E}[\mathbf{V}\mathbf{V}'] = \mathbb{E}[(\Lambda^{-1/2}U'\mathbf{Y})(\Lambda^{-1/2}U'\mathbf{Y})'] \\ &= \Lambda^{-1/2}U'\mathbb{E}[\mathbf{Y}\mathbf{Y}']U(\Lambda^{-1/2})' = \Lambda^{-1/2}U'\mathbf{\Gamma}U\Lambda^{-1/2} \\ &= \Lambda^{-1/2}U'(U\Lambda U')U\Lambda^{-1/2} = \Lambda^{-1/2}\Lambda\Lambda^{-1/2} = I_n.\end{aligned}$$

Since $\mathbf{Y}$ is a Gaussian vector, $\mathbf{V}$ is also a Gaussian vector. Therefore, $\mathbf{V} \sim \mathcal{N}_n(0, I_n)$, which means that the components $V_1, V_2, \ldots, V_n$ of $\mathbf{V}$ are independent and identically distributed standard normal random variables.

Next, we express the quadratic form $\mathbf{Q}_n$ in terms of $\mathbf{V}$. From the transformation above, we have $\mathbf{Y} = U\Lambda^{1/2}\mathbf{V}$. Substituting this, we get:

$$\begin{aligned}\mathbf{Q}_n &= \mathbf{Y}'\mathbf{\Gamma}^{-1}\mathbf{Y} = (U\Lambda^{1/2}\mathbf{V})'(U\Lambda^{-1}U')(U\Lambda^{1/2}\mathbf{V}) \\ &= \mathbf{V}'\Lambda^{1/2}U'(U\Lambda^{-1}U')(U\Lambda^{1/2}\mathbf{V}).\end{aligned}$$

Using the property $U'U = I_n$:

$$\mathbf{Q}_n = \mathbf{V}'\Lambda^{1/2}\Lambda^{-1}\Lambda^{1/2}\mathbf{V} = \mathbf{V}'I_n\mathbf{V} = \sum_{i=1}^{n} V_i^2.$$

This is the sum of n independent squared standard normal random variables. By definition, this sum follows a chi-squared distribution with n degrees of freedom (see Definition 1.2). Therefore, $\mathbf{Q}_n \sim \chi^2(n)$.

For a chi-squared random variable X with n degrees of freedom, its mean is $\mathbb{E}[X] = n$ and its variance is $\mathrm{Var}(X) = 2n$. A fundamental result in probability theory states that as the degrees of freedom $n \to \infty$, a chi-squared distributed random variable, when properly standardized, converges in distribution to a standard normal random variable. Specifically:

$$\frac{\mathbf{Q}_n - \mathbb{E}(\mathbf{Q}_n)}{\sqrt{\mathrm{Var}(\mathbf{Q}_n)}} = \frac{\mathbf{Q}_n - n}{\sqrt{2n}} \xrightarrow{D} \mathcal{N}(0, 1).$$

Finally, we substitute this back into the expression for $\hat{\alpha}^2$. We want to show that:

$$\frac{\hat{\alpha}^2 - \alpha^2}{\alpha^2\sqrt{2/n}} \xrightarrow{D} \mathcal{N}(0, 1).$$

Let us manipulate this expression using the relationship $\hat{\alpha}^2 = \dfrac{\alpha^2}{n}\mathbf{Q}_n$:

$$\frac{\hat{\alpha}^2 - \alpha^2}{\alpha^2\sqrt{2/n}} = \frac{\frac{\alpha^2}{n}\mathbf{Q}_n - \alpha^2}{\alpha^2\sqrt{2/n}} = \frac{\alpha^2\left(\frac{\mathbf{Q}_n}{n} - 1\right)}{\alpha^2\sqrt{2/n}} = \frac{\frac{\mathbf{Q}_n - n}{n}}{\sqrt{2/n}} = \frac{\mathbf{Q}_n - n}{\sqrt{2n}}.$$

Since we have already shown that $\frac{\mathbf{Q}_n - n}{\sqrt{2n}} \xrightarrow{D} \mathcal{N}(0,1)$, the conclusion follows directly. This completes the proof. □

4.3. Estimation of all three parameters (a, b, H)

In the preceding sections, we developed methodologies for estimating either the Hurst parameter H (assuming a and b were known) or the scaling coefficients a and b (assuming H was known). We now address the most general and challenging case: estimating all three parameters (a, b, H) simultaneously from a single realization of the ZgfBm process.

When all parameters are unknown, an explicit, closed-form solution for the MLEs is no longer available. The problem transforms from a direct calculation to a high-dimensional nonlinear optimization task. The core challenge lies in the complex dependence of the covariance matrix of process on the Hurst parameter H and the coefficients a and b.

The general approach is to find the values $(\hat{a}, \hat{b}, \hat{H})$ that maximize the joint log-likelihood function. We will focus on the case where $a \neq b$, which allows us to use a reparameterized form of the problem. This lets us estimate (α, ϵ, H) where the parameter space is more tractable for numerical optimization.

The observation vector $\mathbf{Z} = (Z_{t_1}, \ldots, Z_{t_n})'$ is a zero-mean Gaussian vector with a covariance matrix $\mathbf{\Sigma}$ that depends on all three unknown parameters. From the previous sections, we know that $\mathbf{\Sigma} = \alpha^2\mathbf{\Gamma}$, where $\alpha = a - b$ and the matrix $\mathbf{\Gamma}$ itself depends on the reparameterized coefficient $\epsilon = (a+b)/(a-b)$ and the Hurst parameter H.

The joint log-likelihood function for the observed data is given by:

$$L(\mathbf{Z}; \alpha, \epsilon, H) = -\frac{n}{2}\ln(2\pi) - \frac{n}{2}\ln(\det(\alpha^2\mathbf{\Gamma}(H,\epsilon))) - \frac{1}{2}\mathbf{Z}'(\alpha^2\mathbf{\Gamma}(H,\epsilon))^{-1}\mathbf{Z}.$$

Simplifying this expression, we obtain:

$$L(\mathbf{Z}; \alpha, \epsilon, H) = -\frac{n}{2}\ln(2\pi) - \frac{n}{2}\ln(\alpha^2) - \frac{1}{2}\ln(\det(\mathbf{\Gamma}(H,\epsilon))) - \frac{1}{2\alpha^2}\mathbf{Z}'\mathbf{\Gamma}(H,\epsilon)^{-1}\mathbf{Z}. \quad [4.27]$$

Since a direct analytical solution is not feasible, we will employ a numerical strategy. A particularly effective method for this type of problem is the *profile likelihood* approach. This method reduces the dimensionality of the optimization problem, transforming a three-dimensional search into a more manageable two-dimensional search.

The core idea of the profile likelihood method is to analytically maximize the log-likelihood with respect to a subset of parameters while holding the others fixed. This allows us to define a new, lower dimensional function, the profile log-likelihood, which can then be numerically maximized with respect to the remaining parameters.

Let's apply this method to our problem. We begin by maximizing the log-likelihood with respect to the scaling parameter α^2, holding ϵ and H constant. The log-likelihood function can be written as:

$$L(\mathbf{Z}; \alpha^2, \epsilon, H) = \text{const} - \frac{n}{2}\ln(\alpha^2) - \frac{1}{2\alpha^2}\mathbf{Z}'\mathbf{\Gamma}(H, \epsilon)^{-1}\mathbf{Z}. \quad [4.28]$$

To find the value of α^2 that maximizes this expression, we take the partial derivative with respect to α^2 and set it to zero:

$$\frac{\partial L}{\partial(\alpha^2)} = -\frac{n}{2\alpha^2} + \frac{1}{2(\alpha^2)^2}\mathbf{Z}'\mathbf{\Gamma}(H, \epsilon)^{-1}\mathbf{Z} = 0. \quad [4.29]$$

Solving for α^2, we obtain the conditional MLE for α^2 given ϵ and H:

$$\hat{\alpha}^2(\epsilon, H) = \frac{1}{n}\mathbf{Z}'\mathbf{\Gamma}(H, \epsilon)^{-1}\mathbf{Z}. \quad [4.30]$$

This expression is analogous to the one we found in the previous section for estimating a and b when H was known.

Now, we define the *profile log-likelihood function* for the remaining parameters (ϵ, H) by substituting this conditional estimator $\hat{\alpha}^2(\epsilon, H)$ back into the full log-likelihood function:

$$\begin{aligned}
&L_P(\mathbf{Z}; \epsilon, H) = L(\mathbf{Z}; \hat{\alpha}^2(\epsilon, H), \epsilon, H) \\
&= -\frac{n}{2}\ln(2\pi) - \frac{n}{2}\ln(\hat{\alpha}^2(\epsilon, H)) - \frac{1}{2}\ln(\det(\mathbf{\Gamma}(H, \epsilon))) - \frac{1}{2\hat{\alpha}^2(\epsilon, H)}\mathbf{Z}'\mathbf{\Gamma}(H, \epsilon)^{-1}\mathbf{Z} \\
&= -\frac{n}{2}\ln(2\pi) - \frac{n}{2}\ln\left(\frac{1}{n}\mathbf{Z}'\mathbf{\Gamma}^{-1}\mathbf{Z}\right) - \frac{1}{2}\ln(\det(\mathbf{\Gamma})) - \frac{1}{2\left(\frac{1}{n}\mathbf{Z}'\mathbf{\Gamma}^{-1}\mathbf{Z}\right)}\mathbf{Z}'\mathbf{\Gamma}^{-1}\mathbf{Z} \\
&= -\frac{n}{2}\ln(2\pi) - \frac{n}{2}\ln\left(\frac{1}{n}\mathbf{Z}'\mathbf{\Gamma}^{-1}\mathbf{Z}\right) - \frac{1}{2}\ln(\det(\mathbf{\Gamma})) - \frac{n}{2}.
\end{aligned}$$

Ignoring the constant terms for the purpose of optimization, the function to be maximized is:

$$L_P(\mathbf{Z}; \epsilon, H) = -\frac{n}{2}\ln\left(\frac{1}{n}\mathbf{Z}'\mathbf{\Gamma}(H, \epsilon)^{-1}\mathbf{Z}\right) - \frac{1}{2}\ln(\det(\mathbf{\Gamma}(H, \epsilon))). \quad [4.31]$$

The problem is now reduced to a two-dimensional maximization:

$$(\hat{\epsilon}, \hat{H}) = \arg\max_{\epsilon, H} L_P(\mathbf{Z}; \epsilon, H). \quad [4.32]$$

This is a significant simplification, as it replaces a three-parameter search with a two-parameter search. This two-dimensional optimization can still be computationally intensive but is far more tractable than the original problem.

The full procedure is as follows:

1) *Step 1 (numerical search)*: numerically maximize the profile log-likelihood function $L_P(\mathbf{Z}; \epsilon, H)$ with respect to ϵ and H. This can be done using standard optimization algorithms like grid search, Nelder-Mead, or other quasi-Newton methods. The result is the estimated pair $(\hat{\epsilon}, \hat{H})$.

2) *Step 2 (analytical calculation)*: substitute the estimated values $(\hat{\epsilon}, \hat{H})$ into the conditional MLE for α^2 from equation [4.30] to obtain the estimate for $\hat{\alpha}^2$:

$$\hat{\alpha}^2 = \frac{1}{n}\mathbf{Z}'\mathbf{\Gamma}(\hat{H}, \hat{\epsilon})^{-1}\mathbf{Z}.$$

3) *Step 3 (back-transformation)*: use the inverse relationships from the reparameterization to find the final estimates for a and b:

$$\hat{a} = \frac{\hat{\alpha}(1+\hat{\epsilon})}{2} \quad \text{and} \quad \hat{b} = \frac{\hat{\alpha}(\hat{\epsilon}-1)}{2}.$$

This profile likelihood approach provides a robust and practical method for estimating all three parameters simultaneously. It combines the analytical elegance of MLE with the necessity of numerical optimization, making it well-suited for this complex problem.

4.4. Exercises

EXERCISE 4.1.–

1) Let $f(x)$ be a function that is twice continuously differentiable on an open interval containing $[x-1, x]$. For $x \geq 1$, prove that the second-order finite difference expression $f(x) - 2f(x-1/2) + f(x-1)$ can be exactly related to the second derivative of $f(x)$ as:

$$f(x) - 2f(x-1/2) + f(x-1) = \frac{1}{4}f''(c)$$

for some $c \in (x-1, x)$.

Hint: consider applying Taylor's theorem with Lagrange remainder for $f(x)$ expanded around the central point $x - 1/2$. Expand $f(x)$ and $f(x-1)$ around $x-1/2$ with a step size of $1/2$.

2) Now, consider the specific function $f(x) = x^{2H}$, where $H \in (0, 1)$. Using the result from the previous part, demonstrate that for $i \geq 1$, the term $K_{i,H} = i^{2H} - 2(i-1/2)^{2H} + (i-1)^{2H}$ is bounded by a constant independent of i and N.

EXERCISE 4.2.–

1) Let $f(x)$ be a function that is at least four times continuously differentiable. For sufficiently large x, show that the second-order central finite difference $f(x+1) - 2f(x) + f(x-1)$ can be expressed as $f''(x)+O(f^{(4)}(x))$. Subsequently, for the specific function $f(x) = x^{2H}$ (where $H \in (0,1)$), demonstrate that $f(x+1) - 2f(x) + f(x-1) = O(x^{2H-2})$.

Hint: use Taylor's theorem to expand $f(x+1)$ and $f(x-1)$ around x up to the fourth-order term with remainder.

2) Using the result from part 1, demonstrate that for large $i, j, |i-j|, i+j$ (where applicable):

a) $(i+j)^{2H} - 2(i+j-1)^{2H} + (i+j-2)^{2H} = O((i+j)^{2H-2})$.

b) $2|i-j|^{2H} - |i-j+1|^{2H} - |i-j-1|^{2H} = O(|i-j|^{2H-2})$.

EXERCISE 4.3 (asymptotic variance of the increment statistic for odd fractional Brownian motion (OFBM)).– *Consider the odd fractional Brownian motion (OFBM), denoted $B^{H,o} = \{B_t^{H,o}, t \in [0,T]\}$. Let $S_N(B^{H,o})$ be the statistic defined as:*

$$S_N(B^{H,o}) = \frac{1}{N}\sum_{i=1}^{N}\left(B^{H,o}_{\frac{i}{N}} - B^{H,o}_{\frac{i-1}{N}}\right)^2.$$

Derive the asymptotic behavior of Var($S_N(B^{H,o})$) for OFBM directly from its definition.

1) Begin by establishing the covariance of the increments for OFBM. That is, for $\Delta B_i^{H,o} = B_{i/N}^{H,o} - B_{(i-1)/N}^{H,o}$ and $\Delta B_j^{H,o} = B_{j/N}^{H,o} - B_{(j-1)/N}^{H,o}$, show that:

$$Cov(\Delta B_i^{H,o}, \Delta B_j^{H,o}) = \frac{1}{2N^{2H}}\left[|i-j+1|^{2H} - 2|i-j|^{2H} + |i-j-1|^{2H}\right].$$

2) Determine the expected value of the squared increment $\mathbb{E}[(\Delta B_i^{H,o})^2]$.

3) Using these results, derive the asymptotic behavior of Var($S_N(B^{H,o})$) as $N \to \infty$.

EXERCISE 4.4 (consistency of the Hurst parameter estimator for sub-fractional Brownian motion (SFBM)).– *Consider the sub-fractional Brownian motion (SFBM) $Z_t^H = \sqrt{2}B_t^{H,e}$.*

1) Calculate $\mathbb{E}[(\Delta Z_i^H)^2] = \mathbb{E}[(Z_{i/N}^H - Z_{(i-1)/N}^H)^2]$.

2) Derive the asymptotic expected value of the statistic $S_N(Z^H) = \frac{1}{N}\sum_{i=1}^{N}(\Delta Z_i^H)^2$.

3) Propose a consistent estimator for H, say $\widehat{H}_{SFBM,N}$, based on this asymptotic expectation.

4) Rigorously prove the consistency of $\widehat{H}_{SFBM,N}$ for H.

EXERCISE 4.5 (numerical implementation and simulation).– *This exercise requires implementing the hybrid maximum likelihood estimation algorithm for the scaling coefficients a and b using simulated data.*

Parameters: sample size: $n = 50$; Hurst parameter (known): $H = 0.7$; step size: $h = 1$; true coefficients (for simulation): $a = 2$, $b = 0.5$; minimization function: $M_n(\mathbf{Z}, \epsilon^2) = \frac{1}{2}\ln|\,\mathbf{\Gamma}\,| + \frac{n}{2}\ln(\mathbf{Z}'\mathbf{\Gamma}^{-1}\mathbf{Z})$.

1) Simulate the ZgfBm observation vector $\mathbf{Z}$ of size $n = 50$ using the true parameters.

- Hint: first, calculate the true $\alpha = a - b$ and $\epsilon = (a+b)/(a-b)$. Then, generate the $n \times n$ true covariance matrix of the data, $\mathbf{C} = \alpha^2\mathbf{\Gamma}_{true}$. Use a Cholesky decomposition ($\mathbf{C} = \mathbf{LL}'$) on $\mathbf{C}$ to generate the Gaussian vector $\mathbf{Z} = \mathbf{L}\cdot\mathbf{W}$, where $\mathbf{W} \sim \mathcal{N}_n(0, \mathbf{I}_n)$.

2) Write a function that accepts ϵ^2 as an input and returns the value of $M_n(\mathbf{Z}, \epsilon^2)$. The simulated data vector $\mathbf{Z}$ (from question 1) should be held constant (fixed in the function's scope).

- Hint: fix H and recalculate $\mathbf{\Gamma}$: the base covariance matrices $\mathbf{\Sigma_O}$ and $\mathbf{\Sigma_E}$ must be calculated using the known $H = 0.7$. Crucially, the full covariance matrix $\mathbf{\Gamma} = \mathbf{\Sigma_O} + \epsilon^2\mathbf{\Sigma_E}$ must be recomputed and inverted for every ϵ^2 value in the optimization loop. Use numerically stable routines for matrix determinant and inversion.

3) Use an optimization routine (e.g. `fminsearch`) to minimize the function $M_n(\mathbf{Z}, \epsilon^2)$ and find the maximum likelihood estimate $\hat{\epsilon}^2$.

- Hint: a good starting value for the search could be $\epsilon^2 = 1$. Ensure the search domain is constrained to $\epsilon^2 \in [0, \infty)$.

4) Using the resulting $\hat{\epsilon}^2$ from question 3 complete the estimation for a and b.

- Hint: calculate $\hat{\alpha}^2$ using $\hat{\alpha}^2 = \frac{1}{n}\mathbf{Z}'\mathbf{\Gamma}(\hat{\epsilon}^2)^{-1}\mathbf{Z}$. Find $\hat{\alpha}$ and $\hat{\epsilon}$ by taking the positive square roots (assuming $a, b > 0$), and then apply the inverse transformation to get $\hat{a}$ and $\hat{b}$. Compare your final estimates $(\hat{a}, \hat{b})$ against the true values $(2, 0.5)$.

EXERCISE 4.6 (profile likelihood estimation for ZgfBm parameters $(\hat{a}, \hat{b}, \hat{H})$).–

– Part 1: data simulation and setup. Objective: generate the synthetic data set $\mathbf{Z}$ based on known parameters; set the following true values for the simulation ($t_k = k$); Hurst parameter: $\mathbf{H}_{true} = \mathbf{0.75}$; scaling coefficients: $\mathbf{a}_{true} = \mathbf{1.5}$, $\mathbf{b}_{true} = \mathbf{0.5}$; sample size: $\mathbf{n} = \mathbf{1{,}000}$.

1) Record the corresponding true reparameterized values α_{true} and ϵ_{true}.

2) Implement the full ZgfBm covariance function to build the $\mathbf{n} \times \mathbf{n}$ *matrix* Σ_{true} *using the true parameters.*

3) Generate a single realization $\mathbf{Z} \sim \mathcal{N}_n(\mathbf{0}, \Sigma_{true})$. *This* $\mathbf{Z}$ *is the observed data for estimation.*

– Part 2: numerical optimization $(\hat{\epsilon}, \hat{H})$ *objective: implement the profile log-likelihood and perform the two-dimensional numerical maximization (step 1).*

1) Write a function that calculates and returns the $n \times n$ *matrix* $\mathbf{\Gamma}(H, \epsilon)$ *for any input* H *and* ϵ.

2) Write a function that calculates the profile log-likelihood value. This function must handle the following computational tasks robustly. Computation of $\mathbf{\Gamma}(H, \epsilon)$; *computation of* $\mathbf{\Gamma}(H, \epsilon)^{-1}$ *and computation of* $\det(\mathbf{\Gamma}(H, \epsilon))$.

3) Perform two-dimensional maximization (step 1): use a numerical optimization routine to find the maximum of L_P *over the search space (*$H \in [0.05, 0.95]$ *and* $\epsilon \in [1.01, 5.0]$*):*

$$(\hat{\epsilon}, \hat{H}) = \arg\max_{\epsilon, H} L_P(\mathbf{Z}; \epsilon, H)$$

4) Record results: record the resulting estimated pair $\hat{\epsilon}$ *and* $\hat{H}$.

– Part 3: analytical calculation and final estimates. Objective: complete the estimation by calculating $\hat{\alpha}^2$ *and transforming back to* $\hat{a}$ *and* $\hat{b}$ *(steps 2 and 3).*

1) Calculate $\hat{\alpha}^2$ *(step 2): using the numerically optimized values* $(\hat{\epsilon}, \hat{H})$ *from part 2, compute the conditional maximum likelihood estimate for* $\hat{\alpha}^2$:

$$\hat{\alpha}^2 = \frac{1}{n}\mathbf{Z}'\mathbf{\Gamma}(\hat{H}, \hat{\epsilon})^{-1}\mathbf{Z}.$$

Calculate $\hat{\alpha} = \sqrt{\hat{\alpha}^2}$.

2) Back-transform to $\hat{a}$ *and* $\hat{b}$ *(step 3): use the inverse relationships to find the final estimates* $\hat{a}$ *and* $\hat{b}$:

$$\hat{a} = \frac{\hat{\alpha}(1 + \hat{\epsilon})}{2} \quad \text{and} \quad \hat{b} = \frac{\hat{\alpha}(\hat{\epsilon} - 1)}{2}.$$

3) Evaluation and reporting: summarize all estimated parameters $(\hat{H}, \hat{a}, \hat{b})$ *and compare them to the true values. Present the results in a table, including the relative error for each parameter. Discuss the computational stability and convergence properties observed during the optimization phase.*

5

Stochastic Calculus with Generalized Fractional Brownian Motion

5.1. Introduction

Building on the introduction of Zili generalized fractional Brownian motion (ZgfBm) in previous chapters, this chapter develops the stochastic calculus for this more general class of Gaussian processes. For a Hurst index $H \in (1/2, 1)$, we begin by examining the unique characteristics of its covariance function and proving key probabilistic properties, such as exponential integrability.

The core of this chapter is dedicated to deriving generalized Itô formulas for ZgfBm, utilizing both Skorokhod and symmetric integrals. These results are not only fundamental for the ZgfBm itself, but also provide a powerful resource for those working with more specific cases, such as standard fBm and sub-fractional Brownian motion (sfBm). Since ZgfBm is a generalization of these processes, the formulas and techniques derived here are directly applicable to them. This systematic exposition aims to provide a robust mathematical framework for theoretical and applied contexts, exemplified by the derivation of a generalized fractional Black–Scholes equation.

5.2. Useful ZgfBm's characteristics

We will first establish key properties of the covariance function $R^{H,a,b}(s,t)$ of ZgfBm, which are crucial for the subsequent development in this book. Specifically, we require its second mixed partial derivative. Note that for $H \in (1/2, 1)$, the covariance function $R^{H,a,b}(s,t)$ is C^1 (continuously differentiable) on the entire domain. However, due to the exponent $2H - 2 \in (-1, 0)$ in the second derivative of the term $|t - s|^{2H}$, the function lacks a classical second-order partial derivative on the diagonal $s = t$. Consequently, the mixed partial derivative $\dfrac{\partial^2 R^{H,a,b}(s,t)}{\partial s \partial t}$ must be defined in the rigorous sense of distributions (or generalized functions) across the entire domain. As is standard, for $s \neq t$, the distributional derivative coincides with the classical pointwise derivative.

LEMMA 5.1.– *For every $H \in (1/2, 1)$, the second mixed partial derivative of the covariance function $R^{H,a,b}(s,t)$, in the sense of distributions, is given by:*

$$\frac{\partial^2 R^{H,a,b}(s,t)}{\partial s \partial t} = H(2H-1)\left[(a^2+b^2)|t-s|^{2H-2} - 2ab(t+s)^{2H-2}\right]. \qquad [5.1]$$

Proof. Let $R^{H,a,b}(s,t)$ be the covariance function, defined for $(s,t) \in [0,T]^2$:

$$R^{H,a,b}(t,s) = \frac{1}{2}(a+b)^2\left(s^{2H}+t^{2H}\right) - ab(t+s)^{2H} - \frac{a^2+b^2}{2}|t-s|^{2H}.$$

A distribution $T \in \mathcal{D}'(\mathbb{R}^2)$ acts on a test function $\phi \in C_c^\infty(\mathbb{R}^2)$ (smooth functions with compact support). The partial derivative $\frac{\partial T}{\partial s}$ is defined by:

$$\left\langle \frac{\partial T}{\partial s}, \phi \right\rangle = -\left\langle T, \frac{\partial \phi}{\partial s} \right\rangle.$$

Let us examine each term of $R^{H,a,b}(s,t)$:

– Term 1: $T_1(s,t) = \frac{1}{2}(a+b)^2(s^{2H}+t^{2H})$. For $s,t \in [0,T]$, s^{2H} and t^{2H} are C^∞ functions since $2H > 1$. Thus, their classical derivatives exist and are smooth. The mixed partial derivative is given as:

$$\frac{\partial^2 T_1}{\partial s \partial t} = 0.$$

This holds in the classical sense and, consequently, in the sense of distributions.

– Term 2: $T_2(s,t) = -ab(t+s)^{2H}$. For $s,t \in [0,T]$, $(t+s)^{2H}$ is a C^∞ function since $t+s \geq 0$ and $2H > 1$. Its classical derivatives exist and are smooth. Calculating the classical mixed partial derivative, we get:

$$\frac{\partial T_2}{\partial s} = -2abH(t+s)^{2H-1}.$$

$$\frac{\partial^2 T_2}{\partial s \partial t} = -2abH(2H-1)(t+s)^{2H-2}.$$

This equality also holds in the sense of distributions.

– Term 3: $T_3(s,t) = -\frac{a^2+b^2}{2}|t-s|^{2H}$.

This is the term that introduces a singularity on the diagonal $s = t$. We consider the function of one variable $f(x) = |x|^{2H}$. It is a known result in distribution theory that its second distributional derivative is given by $f''(x) = 2H(2H-1)|x|^{2H-2}$ for $x \neq 0$ and the singularity is absorbed by the distributional definition.

Let us apply the chain rule for distributional derivatives. Let the distribution be $U(s,t) = f(t-s) = |t-s|^{2H}$. The mixed partial derivative can be computed as:

$$\frac{\partial^2 U}{\partial s \partial t} = \frac{\partial}{\partial s}\left(\frac{\partial U}{\partial t}\right).$$

By the chain rule for distributions applied to $t \mapsto t-s$, we have:

$$\frac{\partial U}{\partial t} = \frac{\partial}{\partial t} f(t-s) = f'(t-s) \cdot \frac{\partial(t-s)}{\partial t} = f'(t-s).$$

Similarly, for the derivative with respect to s:

$$\frac{\partial}{\partial s} f(t-s) = f'(t-s) \cdot \frac{\partial(t-s)}{\partial s} = -f'(t-s).$$

Therefore:

$$\frac{\partial^2 U}{\partial s \partial t} = \frac{\partial}{\partial s}\left(f'(t-s)\right) = f''(t-s) \cdot \frac{\partial(t-s)}{\partial s} = f''(t-s) \cdot (-1).$$

Thus, we have:

$$\frac{\partial^2 |t-s|^{2H}}{\partial s \partial t} = -\frac{d^2}{dx^2}|x|^{2H}\bigg|_{x=t-s} = -\left(2H(2H-1)|t-s|^{2H-2}\right).$$

Substituting this result back into the expression for T_3:

$$\frac{\partial^2 T_3}{\partial s \partial t} = -\frac{a^2+b^2}{2}\left(-2H(2H-1)|t-s|^{2H-2}\right) = (a^2+b^2)H(2H-1)|t-s|^{2H-2}.$$

Combining the derivatives of all three terms, we obtain the total mixed partial derivative of the covariance function $R^{H,a,b}(s,t)$:

$$\frac{\partial^2 R^{H,a,b}(s,t)}{\partial s \partial t} = 0 - 2abH(2H-1)(t+s)^{2H-2} + (a^2+b^2)H(2H-1)|t-s|^{2H-2}.$$

Factoring out the common term $H(2H-1)$, we arrive at the desired result:

$$\frac{\partial^2 R^{H,a,b}(s,t)}{\partial s \partial t} = H(2H-1)\left[(a^2+b^2)|t-s|^{2H-2} - 2ab(t+s)^{2H-2}\right].$$

This completes the proof. □

COROLLARY 5.1.– *Let* $\frac{1}{2} < H < 1$, *and denote* $\alpha_H = H(2H-1)$. *For the pointwise values of the second mixed partial derivative of* $R^{H,a,b}(s,t)$ *for* $s \neq t$, *we have the following bounds:*

$$C_1|t-s|^{2H-2} \leq \frac{\partial^2 R^{H,a,b}(s,t)}{\partial s \partial t} \leq C_2|t-s|^{2H-2}, \qquad [5.2]$$

for every $s, t \in [0,T]$; $s \neq t$, *with:*

$$C_1 = \alpha_H \min((a-b)^2, a^2+b^2) \quad and \quad C_2 = \alpha_H(|a|+|b|)^2.$$

Proof. From Lemma 5.1, we know that for $s \neq t$, the pointwise value of the second mixed partial derivative is given by:

$$\frac{\partial^2 R^{H,a,b}(s,t)}{\partial s \partial t} = H(2H-1)\left[(a^2+b^2)|t-s|^{2H-2} - 2ab(t+s)^{2H-2}\right].$$

Let $\alpha_H = H(2H-1)$. Since $H \in (1/2, 1)$, $\alpha_H > 0$. Also, $2H-2 \in (-1,0)$, which implies that the function $f(x) = x^{2H-2}$ is strictly decreasing for $x > 0$.

We can rewrite the expression as:

$$\frac{\partial^2 R^{H,a,b}(s,t)}{\partial s \partial t} = \alpha_H|t-s|^{2H-2}\left[(a^2+b^2) - 2ab\left(\frac{t+s}{|t-s|}\right)^{2H-2}\right].$$

Let

$$M(s,t) = (a^2+b^2) - 2ab\left(\frac{t+s}{|t-s|}\right)^{2H-2}.$$

For $s, t \geq 0, s \neq t$, we have $\frac{t+s}{|t-s|} \geq 1$. Since $x \mapsto x^{2H-2}$ is a decreasing function, it follows that:

$$0 < \left(\frac{t+s}{|t-s|}\right)^{2H-2} \leq 1.$$

For the upper bound C_2: we seek the maximum value of $M(s,t)$.

$$M(s,t) = (a^2+b^2) - 2ab\left(\frac{t+s}{|t-s|}\right)^{2H-2}.$$

We know that $-2ab \leq 2|a||b|$. Since $\left(\frac{t+s}{|t-s|}\right)^{2H-2} \leq 1$ and $2|a||b| \geq 0$, we have:

$$\begin{aligned} M(s,t) &\leq (a^2+b^2) + 2|a||b|\left(\frac{t+s}{|t-s|}\right)^{2H-2} \\ &\leq (a^2+b^2) + 2|a||b|(1) \\ &= (|a|+|b|)^2. \end{aligned}$$

This inequality holds for all $a, b \in \mathbb{R}$. Thus,

$$\frac{\partial^2 R^{H,a,b}(s,t)}{\partial s \partial t} \leq \alpha_H(|a|+|b|)^2|t-s|^{2H-2},$$

so we set $C_2 = \alpha_H(|a|+|b|)^2$.

For the lower bound C_1: we seek the minimum value of $M(s,t)$.

1) If $ab \geq 0$: then $-2ab \leq 0$. The term $-2ab\left(\frac{t+s}{|t-s|}\right)^{2H-2}$ is nonpositive. Its minimum (most negative) value occurs when $\left(\frac{t+s}{|t-s|}\right)^{2H-2}$ is maximized, which is 1.

This is achieved, for example, when $s = 0$ (and $t > 0$), where:

$$\frac{t+s}{|t-s|} = \frac{t}{t} = 1.$$

In this case, we have:

$$M(s,t) \geq (a^2+b^2) - 2ab(1) = (a-b)^2.$$

2) If $ab < 0$: then $-2ab > 0$. The term $-2ab\left(\frac{t+s}{|t-s|}\right)^{2H-2}$ is positive.

Its minimum value (closest to 0) occurs when $\left(\frac{t+s}{|t-s|}\right)^{2H-2}$ is minimized, which approaches 0.

This minimum is approached as $\frac{t+s}{|t-s|} \to \infty$ (i.e. as $s \to t$). In this case, $M(s,t)$ approaches a^2+b^2. More precisely:

$$M(s,t) = (a^2+b^2) - 2ab\left(\frac{t+s}{|t-s|}\right)^{2H-2} \geq (a^2+b^2) + 0 = a^2+b^2$$

(since $-2ab > 0$ and the term in parenthesis is nonnegative).

Therefore, the minimum value of $M(s,t)$ for $s \neq t$ is $\min((a-b)^2, a^2+b^2)$. Thus,

$$\frac{\partial^2 R^{H,a,b}(s,t)}{\partial s \partial t} \geq \alpha_H \min((a-b)^2, a^2+b^2)|t-s|^{2H-2},$$

so we set $C_1 = \alpha_H \min((a-b)^2, a^2+b^2)$. Both constants C_1 and C_2 are positive because $\alpha_H > 0$ (since $H \in (1/2, 1)$) and $(a,b) \neq (0,0)$. □

The following characteristic of the second partial derivative of $R^{H,a,b}$ will also be useful for the remainder of this book.

LEMMA 5.2.– *When* $\frac{1}{2} < H < 1$, *there exists a positive constant* C_3, *such that:*

$$\frac{1}{2\epsilon}\int_{-\epsilon}^{\epsilon}\left|\frac{\partial^2 R^{H,a,b}(s+\sigma,t)}{\partial s\partial t}\right| d\sigma \leq C_3|t-s|^{2H-2}, \qquad [5.3]$$

for every $\epsilon > 0$ *and* $s, t \in [0,T]; s \neq t$.

Proof. From Lemma 5.1, for $s' \neq t$, we have the pointwise bound:

$$\left|\frac{\partial^2 R^{H,a,b}(s',t)}{\partial s'\partial t}\right| \leq C_2|t-s'|^{2H-2},$$

where $C_2 = \alpha_H(|a|+|b|)^2$ and $\alpha_H = H(2H-1)$. Thus,

$$\frac{1}{2\epsilon}\int_{-\epsilon}^{\epsilon}\left|\frac{\partial^2 R^{H,a,b}(s+\sigma,t)}{\partial s\partial t}\right| d\sigma \leq \frac{1}{2\epsilon}\int_{-\epsilon}^{\epsilon} C_2|t-(s+\sigma)|^{2H-2}d\sigma.$$

Let $x = t-s$. We are evaluating the integral $\frac{C_2}{2\epsilon}\int_{-\epsilon}^{\epsilon}|x-\sigma|^{2H-2}d\sigma$. To simplify, let $u = x-\sigma$, so $du = -d\sigma$.

The integration limits change from $\sigma = -\epsilon$ to $u = x+\epsilon$, and from $\sigma = \epsilon$ to $u = x-\epsilon$.

The expression becomes $\frac{C_2}{2\epsilon}\int_{x-\epsilon}^{x+\epsilon}|u|^{2H-2}du$. To evaluate this integral and establish a uniform bound, we distinguish two cases based on the relationship between $|x|$ and ϵ.

Case 1: if $|x| \geq 2\epsilon$

In this case, the interval $[x-\epsilon, x+\epsilon]$ does not contain 0. Without loss of generality, assume $x > 0$, so $x \geq 2\epsilon$.

Since $u > 0$ for all $u \in [x-\epsilon, x+\epsilon]$, we have $J(\epsilon) = \int_{x-\epsilon}^{x+\epsilon} u^{2H-2}du$.

Since $2H-2 \in (-1,0)$, the function $f(u) = u^{2H-1}$ is differentiable, and its derivative is $(2H-1)u^{2H-2}$. By the mean value theorem, there exists some $\xi \in (x-\epsilon, x+\epsilon)$ such that:

$$J(\epsilon) = \frac{1}{2H-1}\left((x+\epsilon)^{2H-1} - (x-\epsilon)^{2H-1}\right) = 2\epsilon\xi^{2H-2}.$$

Now we need to find an upper bound for ξ^{2H-2}. Since $\xi \in (x-\epsilon, x+\epsilon)$ and $x-\epsilon \geq 2\epsilon-\epsilon = \epsilon$, we know that $\xi \geq x-\epsilon \geq \epsilon$.

Furthermore, since $x \geq 2\epsilon$, we have $x - \epsilon \geq x/2$. So, $\xi \geq x - \epsilon \geq x/2$. Since $2H - 2 < 0$, the function $y \mapsto y^{2H-2}$ is decreasing for $y > 0$. Thus, applying the decreasing property to $\xi \geq x/2$:

$$\xi^{2H-2} \leq (x/2)^{2H-2} = 2^{2-2H} x^{2H-2} = 2^{2-2H} |x|^{2H-2}.$$

Substituting this into the expression for $J(\epsilon)$:

$$J(\epsilon) \leq 2\epsilon \cdot 2^{2-2H} |x|^{2H-2}.$$

Now, consider the term we want to bound:

$$\frac{C_2}{2\epsilon} J(\epsilon) \leq \frac{C_2}{2\epsilon} \left(2\epsilon \cdot 2^{2-2H} |x|^{2H-2}\right) = C_2 2^{2-2H} |x|^{2H-2}.$$

This provides the upper bound for this case.

Case 2: if $|x| < 2\epsilon$

In this case, the interval $[x-\epsilon, x+\epsilon]$ is either very close to 0 or contains 0. Without loss of generality, assume $x \geq 0$. The integral is $J(\epsilon) = \displaystyle\int_{x-\epsilon}^{x+\epsilon} |u|^{2H-2} du$.

Since $|x| < 2\epsilon$, the interval $[x - \epsilon, x + \epsilon]$ is entirely contained within $[-3\epsilon, 3\epsilon]$. Therefore:

$$J(\epsilon) \leq \int_{-3\epsilon}^{3\epsilon} |u|^{2H-2} du.$$

Since $|u|^{2H-2}$ is an even function, we have:

$$J(\epsilon) \leq 2 \int_{0}^{3\epsilon} u^{2H-2} du.$$

The antiderivative of u^{2H-2} is $\frac{u^{2H-1}}{2H-1}$. Since $2H - 1 > 0$:

$$J(\epsilon) \leq 2 \left[\frac{u^{2H-1}}{2H-1}\right]_0^{3\epsilon} = 2\frac{(3\epsilon)^{2H-1}}{2H-1} = \frac{2 \cdot 3^{2H-1} \epsilon^{2H-1}}{2H-1}.$$

Now, consider the term we want to bound:

$$\frac{C_2}{2\epsilon} J(\epsilon) \leq \frac{C_2}{2\epsilon} \left(\frac{2 \cdot 3^{2H-1} \epsilon^{2H-1}}{2H-1}\right) = \frac{C_2 3^{2H-1}}{2H-1} \epsilon^{2H-2}.$$

To express this in terms of $|x|^{2H-2}$, we use the condition $|x| < 2\epsilon$, which implies $\epsilon > \frac{|x|}{2}$. Since $2H - 2 < 0$, the function $y \mapsto y^{2H-2}$ is decreasing. Thus,

$$\epsilon^{2H-2} < \left(\frac{|x|}{2}\right)^{2H-2} = 2^{2-2H} |x|^{2H-2}.$$

So, for this case:

$$\frac{C_2}{2\epsilon}J(\epsilon) < \frac{C_2 3^{2H-1}}{2H-1}2^{2-2H}|x|^{2H-2}.$$

The universal constant C_3 will be the maximum of the coefficients from both cases.

$$C_3 = \max\left(C_2 2^{2-2H}\ ,\ C_2\frac{3^{2H-1}}{2H-1}2^{2-2H}\right).$$

Since $H \in (1/2, 1)$, we have $2H-1 \in (0,1)$. This implies $\frac{1}{2H-1} > 1$. Also, $3^{2H-1} > 1$. Therefore, the term $\dfrac{3^{2H-1}}{2H-1}$ is strictly greater than 1. This means $\dfrac{3^{2H-1}}{2H-1}2^{2-2H}$ is greater than 2^{2-2H}. Thus, the second term dominates:

$$C_3 = C_2\frac{3^{2H-1}}{2H-1}2^{2-2H}.$$

This provides the desired constant C_3 such that $\dfrac{C_2}{2\epsilon}J(\epsilon) \leq C_3|x|^{2H-2}$ for all $x \neq 0$. □

We will complete this section by the following useful lemma.

LEMMA 5.3.– *For every integer* $p > 0$ *and real* $\lambda > 0$ *satisfying* $p\lambda < \dfrac{1}{2C_H(a,b)T^{2H}}$, *we have:*

$$\mathbb{E}\left[\exp\left(p\lambda \sup_{0\leq t\leq T}|Z_t^H(a,b)|^2\right)\right] < \infty.$$

Proof. Let $Z_t \equiv Z_t^H(a,b)$ for brevity. Since Z_t is a centered Gaussian process, we can use known concentration inequalities for Gaussian processes. According to Theorem 4.2 in Nourdin (2012), for a centered Gaussian process X_t with finite variance $\sup\limits_{u\in[0,1]}\mathrm{Var}(X_u)$, if $m := \mathbb{E}\left[\sup\limits_{u\in[0,1]}X_u\right]$ is finite, then for all $x > m$:

$$\mathbb{P}\left(\sup_{u\in[0,1]}X_u \geq x\right) \leq \exp\left(-\frac{(x-m)^2}{2\sup\limits_{u\in[0,1]}\mathrm{Var}(X_u)}\right). \qquad [5.4]$$

For our process $Z_u^H(a,b)$, Lemma 3.1 provides the explicit variance:

$$E[Z_u^2] = [(a+b)^2 - ab2^{2H}]u^{2H}.$$

Since $u \in [0,1]$:

$$\sup_{u\in[0,1]} \mathrm{Var}(Z_u^H(a,b)) = [(a+b)^2 - ab2^{2H}] = C_H(a,b). \tag{5.5}$$

For $Z_u^H(a,b)$, $m := \mathbb{E}\left[\sup_{u\in[0,1]} Z_u^H(a,b)\right]$ is finite because its sample paths are almost surely continuous on the compact set $[0,1]$, which implies its supremum is finite.

Using the fact that $Z^H(a,b)$ and $-Z^H(a,b)$ have the same law (due to symmetry of Gaussian distributions around zero), we have:

$$\begin{aligned}&\mathbb{P}\left(\sup_{u\in[0,1]} |Z_u^H(a,b)| \geq y\right)\\ &= \mathbb{P}\left(\sup_{u\in[0,1]} Z_u^H(a,b) \geq y\right) + \mathbb{P}\left(\sup_{u\in[0,1]} (-Z_u^H(a,b)) \geq y\right)\\ &= 2\mathbb{P}\left(\sup_{u\in[0,1]} Z_u^H(a,b) \geq y\right).\end{aligned}$$

Let $y = \sqrt{x}$. For $x > m^2$, this implies $\sqrt{x} > m$.

Applying equation [5.4] with $X_u = Z_u^H(a,b)$ and $y = \sqrt{x}$:

$$\begin{aligned}\mathbb{P}\left(\sup_{u\in[0,1]} \left|Z_u^H(a,b)\right|^2 \geq x\right) &= \mathbb{P}\left(\sup_{u\in[0,1]} |Z_u^H(a,b)| \geq \sqrt{x}\right)\\ &\leq 2\mathbb{P}\left(\sup_{u\in[0,1]} Z_u^H(a,b) \geq \sqrt{x}\right)\\ &\leq 2\exp\left(-\frac{(\sqrt{x}-m)^2}{2C_H(a,b)}\right).\end{aligned} \tag{5.6}$$

This inequality holds for $x > m^2$. For $0 \leq x \leq m^2$, the probability on the left side is less than or equal to 1.

Now, by the self-similarity property (cf. Assertion 1 in Lemma 3.1), for $t \in [0,T]$:

$$Z_t^H(a,b) \stackrel{\mathcal{D}}{=} T^H Z_{t/T}^H(a,b).$$

Therefore:

$$\sup_{0\leq t\leq T} |Z_t^H(a,b)|^2 \stackrel{\mathcal{D}}{=} \sup_{0\leq t\leq T} |T^H Z_{t/T}^H(a,b)|^2 = T^{2H} \sup_{0\leq u\leq 1} |Z_u^H(a,b)|^2.$$

We want to evaluate the expectation $\mathbb{E}\left[e^{p\lambda \sup_{0\le t\le T}|Z_t^H(a,b)|^2}\right]$. For a nonnegative random variable Y, the expectation of e^{kY} $(k > 0)$ can be computed using the following formula:

$$\mathbb{E}[e^{kY}] = \int_0^{\infty} k e^{kx}\mathbb{P}(Y \ge x)dx.$$

Let $Y = \sup_{0\le u\le 1} |Z_u^H(a,b)|^2$.

$$\mathbb{E}\left[e^{p\lambda \sup_{0\le t\le T}|Z_t^H(a,b)|^2}\right] = \mathbb{E}\left[e^{p\lambda T^{2H}Y}\right]$$

$$= \int_0^{+\infty} p\lambda T^{2H} e^{p\lambda T^{2H}x}\mathbb{P}\Big(\sup_{u\in[0,1]}\left|Z_u^H(a,b)\right|^2 \ge x\Big)dx$$

$$\le 2p\lambda T^{2H}\int_{m^2}^{+\infty} e^{p\lambda T^{2H}x}\exp\left(-\frac{(\sqrt{x}-m)^2}{2C_H(a,b)}\right)dx$$

$$+p\lambda T^{2H}\int_0^{m^2} e^{p\lambda T^{2H}x}dx.$$

The second integral is clearly finite:

$$p\lambda T^{2H}\int_0^{m^2} e^{p\lambda T^{2H}x}dx = \left[e^{p\lambda T^{2H}x}\right]_0^{m^2} = e^{p\lambda T^{2H}m^2} - 1 < \infty.$$

Now consider the first integral. Let $\gamma = p\lambda T^{2H}$ and $C = 2C_H(a,b)$. We need to evaluate $\int_{m^2}^{+\infty} e^{\gamma x}\exp\left(-\frac{(\sqrt{x}-m)^2}{C}\right)dx$. The exponent is:

$$\gamma x - \frac{(\sqrt{x}-m)^2}{C} = \gamma x - \frac{x - 2m\sqrt{x} + m^2}{C} = \left(\gamma - \frac{1}{C}\right)x + \frac{2m}{C}\sqrt{x} - \frac{m^2}{C}.$$

Let $\gamma_H = \gamma - \frac{1}{C} = p\lambda T^{2H} - \frac{1}{2C_H(a,b)}$.

The condition $p\lambda < \frac{1}{2C_H(a,b)T^{2H}}$ implies $p\lambda T^{2H} < \frac{1}{2C_H(a,b)}$, so $\gamma_H < 0$.

Let $\delta = -\gamma_H > 0$. The exponent becomes $-\delta x + \frac{2m}{C}\sqrt{x} - \frac{m^2}{C}$. To show the integral converges, we observe the asymptotic behavior of the exponent as $x \to \infty$.

The term $-\delta x$ dominates $\frac{2m}{C}\sqrt{x}$. Specifically, for sufficiently large x, say $x > x_0$, we have $\frac{2m}{C}\sqrt{x} < \frac{\delta}{2}x$. This is true because $\frac{\sqrt{x}}{x} = \frac{1}{\sqrt{x}} \to 0$ as $x \to \infty$.

Thus, for $x > x_0$, $\quad -\delta x + \frac{2m}{C}\sqrt{x} - \frac{m^2}{C} < -\delta x + \frac{\delta}{2}x - \frac{m^2}{C} = -\frac{\delta}{2}x - \frac{m^2}{C}.$

So the integrand is bounded by $K\exp\left(-\frac{\delta}{2}x\right)$ for some constant $K = \exp\left(-\frac{m^2}{C}\right)$.

The integral $\int_{x_0}^{\infty} Ke^{-(\delta/2)x}dx$ converges because $\delta/2 > 0$. Therefore, the first integral is finite, and the sum of two finite integrals is finite. This concludes the proof. □

5.3. Stochastic integration with respect to ZgfBm when $\frac{1}{2} < H < 1$

We begin this section by recalling fundamental concepts from Malliavin calculus and adapting them to our particular Gaussian process $Z^H(a,b)$.

Let $(\Omega, \mathcal{F}, \mathbb{P})$ be a complete probability space where the σ-algebra $\mathcal{F}$ is generated by the two-sided fractional Brownian motion $(B_t^H)_{t\in\mathbb{R}}$ with parameter H. For any fixed $(a,b) \in \mathbb{R}^2 \setminus \{(0,0)\}$, we denote by $\mathcal{H}_{a,b}$ the *canonical Hilbert space* associated with the Gaussian process $Z^H(a,b)$. This Hilbert space is defined as the closure of the linear span generated by the indicator functions $\mathbf{1}_{[0,t]}$, $t > 0$, with respect to the inner product:

$$< \mathbf{1}_{[0,t]}, \mathbf{1}_{[0,s]} >_{\mathcal{H}_{a,b}} = \mathcal{R}^{H,a,b}(t,s) = \mathrm{Cov}(Z_t^H(a,b), Z_s^H(a,b)).$$

The isometry between $\mathcal{H}_{a,b}$ and the Gaussian space associated with $Z^H(a,b)$ is denoted by $\varphi \longmapsto Z^H(a,b)(\varphi)$. This isometry is extended from the mapping:

$$\mathbf{1}_{[0,t]} \longmapsto Z_t^H(a,b).$$

For any $\varphi, \psi \in \mathcal{H}_{a,b}$, the inner product can be expressed as:

$$< \varphi, \psi >_{\mathcal{H}_{a,b}} = \int_0^T \int_0^T \varphi(u)\psi(v)\frac{\partial^2 \mathcal{R}^{H,a,b}}{\partial u \partial v}(u,v)dudv.$$

Consider the subspace $|\mathcal{H}_{a,b}|$ of $\mathcal{H}_{a,b}$, which is defined as the set of measurable functions f on $[0,T]$ such that:

$$\|f\|^2_{|\mathcal{H}_{a,b}|} = \int_0^T \int_0^T |f(u)|\,|f(v)| \left|\frac{\partial^2 R^{H,a,b}(u,v)}{\partial u \partial v}\right| dudv < +\infty.$$

Denoting $\mathcal{H}_{a,b}^{\otimes 2}$ as the *second tensor product* of $\mathcal{H}_{a,b}$, we also define $|\mathcal{H}_{a,b}|^{\otimes 2} \subset \mathcal{H}_{a,b}^{\otimes 2}$ as the linear space of measurable functions f on $[0,T]^2$ such that:

$$\|f\|^2_{|\mathcal{H}_{a,b}|^{\otimes 2}} = \int_{[0,T]^4} |f(r,s)f(t,u)| \\ \times \left| \frac{\partial^2 R^{H,a,b}(r,t)}{\partial r \partial t} \frac{\partial^2 R^{H,a,b}(s,u)}{\partial s \partial u} \right| drdsdtdu < +\infty.$$

NOTE.– Here, $f(r,s)$ refers to an element of the space $|\mathcal{H}_{a,b}|^{\otimes 2}$, which is a function of two variables $(r,s) \in [0,T]^2$.

We denote by $C_b^\infty(\mathbb{R}^n)$ the set of all infinitely continuously differentiable functions $f : \mathbb{R}^n \longrightarrow \mathbb{R}$ such that f and all its derivatives are bounded.

Let $\mathcal{S}_{a,b}$ be the set of *smooth random variables* $F_{a,b}$ of the form:

$$F_{a,b} = f(Z^H(a,b)(h_1), \ldots, Z^H(a,b)(h_n)), \quad [5.7]$$

where $f \in C_b^\infty(\mathbb{R}^n)$, $h_1, \ldots, h_n \in \mathcal{H}_{a,b}$, and $n \geq 1$.

The *Malliavin derivative* of a random variable $F_{a,b}$ of the form [5.7] is defined by:

$$D^{a,b}F_{a,b} = \sum_{i=1}^{n} \frac{\partial f}{\partial x_i}(Z^H(a,b)(h_1), \ldots, Z^H(a,b)(h_n))h_i.$$

To make the notation less cluttered, $D^{a,b}$ will be simply denoted as D.

EXAMPLE 5.1.– *Consider a simple case where $F_{a,b}$ is directly one of the Gaussian random variables, say $Z^H(a,b)(\varphi)$ for some $\varphi \in \mathcal{H}_{a,b}$. This fits the form of equation [5.7] with $n = 1$, $f(x) = x$ and $h_1 = \varphi$.*

Applying the definition, the Malliavin derivative of $Z^H(a,b)(\varphi)$ with respect to time r is simply given as:

$$D_r(Z^H(a,b)(\varphi)) = \frac{\partial x}{\partial x}(Z^H(a,b)(\varphi)) \cdot \varphi(r) = 1 \cdot \varphi(r) = \varphi(r).$$

This tells us that the Malliavin derivative of a fundamental Gaussian random variable $Z^H(a,b)(\varphi)$ is precisely the function $\varphi(r)$ that defines it within the Hilbert space.

Now, consider our specific process $Z^H_{t_{j-1}}(a,b)$. As established by the isometry, $Z^H_{t_{j-1}}(a,b)$ corresponds to $Z^H(a,b)(\mathbf{1}_{[0,t_{j-1}]})$. Therefore, applying the result above, its Malliavin derivative would be:

$$D_r Z^H_{t_{j-1}}(a,b) = D_r(Z^H(a,b)(\mathbf{1}_{[0,t_{j-1}]})) = \mathbf{1}_{[0,t_{j-1}]}(r). \quad [5.8]$$

This means the derivative at time r is 1 if $r \in [0,t_{j-1}]$ and 0 otherwise, indicating how much $Z^H_{t_{j-1}}(a,b)$ is influenced by the noise up to its evaluation time t_{j-1}.

5.3.1. *Chain rule formula*

A crucial property of the Malliavin derivative, analogous to the classical chain rule in ordinary calculus, allows us to differentiate composite functions of random variables. For a continuously differentiable function $\phi : \mathbb{R} \to \mathbb{R}$ and a random variable $F \in D^{1,2}(\mathcal{H}_{a,b})$ (a Malliavin-differentiable random variable), the Malliavin derivative of $\phi(F)$ is given by:

$$D_t(\phi(F)) = \phi'(F) D_t F, \quad \text{for almost all } t \in [0,T].$$

This rule extends naturally to functions of multiple variables or, as in our context, functions of time and a stochastic process. For a sufficiently smooth function $f(s,x)$ and a stochastic process $X = \{X_s\}_{s\in[0,T]}$, where each X_s is Malliavin-differentiable, an example of this chain rule application is given by:

$$\begin{aligned} D_r\left(f(t_{j-1}, Z^H_{t_{j-1}}(a,b))\right) &= \frac{\partial f}{\partial x}(t_{j-1}, Z^H_{t_{j-1}}(a,b)) D_r Z^H_{t_{j-1}}(a,b) \\ &= \frac{\partial f}{\partial x}(t_{j-1}, Z^H_{t_{j-1}}(a,b)) \mathbf{1}_{[0,t_{j-1}]}(r). \end{aligned} \qquad [5.9]$$

5.3.2. *The divergence operator*

The derivative operator D is a closable unbounded operator from $L^p(\Omega)$ into $L^p(\Omega; \mathcal{H}_{a,b})$, for any $p \geq 1$.

We define the iteration of the operator D such that the iterated derivative operator D^k is a random variable with values in $\mathcal{H}_{a,b}^{\otimes k}$. The operator D^k maps $L^p(\Omega)$ into $L^p(\Omega; \mathcal{H}_{a,b}^{\otimes k})$.

For any positive integer k and any real $p \geq 1$, we denote by $D^{k,p}(\mathcal{H}_{a,b})$ the *domain of* D^k *in* $L^p(\Omega)$, which is the closure of $\mathcal{S}_{a,b}$ with respect to the norm defined by:

$$\|F\|_{D^{k,p}(\mathcal{H}_{a,b})} = \left[\mathbb{E}(|F|^p) + \sum_{j=1}^{k} \mathbb{E}\left(\|D^j F\|^p_{\mathcal{H}_{a,b}^{\otimes j}}\right)\right]^{1/p}.$$

In a similar way, given a Hilbert space V, we denote by $D^{k,p}(V)$ the corresponding *Sobolev space of* V*-valued random variables*.

We denote $D^{1,2}(|\mathcal{H}_{a,b}|)$ as the subspace of $D^{1,2}(\mathcal{H}_{a,b})$ consisting of elements u such that $u \in |\mathcal{H}_{a,b}|$ a.s., $Du \in |\mathcal{H}_{a,b}|^{\otimes 2}$ a.s., and:

$$\mathbb{E}\left[\|u\|^2_{|\mathcal{H}_{a,b}|} + \|Du\|^2_{|\mathcal{H}_{a,b}|^{\otimes 2}}\right] < \infty.$$

We denote by δ the *adjoint of the derivative operator* D. That is, δ is an unbounded operator on $L^2(\Omega; \mathcal{H}_{a,b})$ with values in $L^2(\Omega)$ such that:

1) The *domain of* δ, denoted by $Dom\delta$, is the set of $\mathcal{H}_{a,b}$-valued square integrable random variables $u \in L^2(\Omega; \mathcal{H}_{a,b})$ such that

$$|\mathbb{E}[< DF, u >_{\mathcal{H}_{a,b}}]| \leq c\|F\|_{L^2(\Omega)},$$

for all $F \in D^{1,2}(\mathcal{H}_{a,b})$, where c is some constant depending on u.

2) If $u \in Dom\delta$, then $\delta(u)$ is the unique element of $L^2(\Omega)$ characterized by the duality relationship:

$$\mathbb{E}[F\delta(u)] = \mathbb{E}[< DF, u >_{\mathcal{H}_{a,b}}]$$

for any $F \in D^{1,2}(\mathcal{H}_{a,b})$.

Note that $D^{1,2}(|\mathcal{H}_{a,b}|) \subset D^{1,2}(\mathcal{H}_{a,b}) \subset Dom\delta$. The operator δ is commonly called the *divergence operator*.

REMARK 5.1.– *Let* $|\mathcal{H}| = |\mathcal{H}_{1,0}|$ *be the set of measurable functions* f *on* $[0,T]$ *such that:*

$$\|f\|^2_{|\mathcal{H}|} = \int_0^T \int_0^T |f(u)|\,|f(v)|\,|u-v|^{2H-2}dudv < +\infty.$$

From Corollary 5.1, the following properties hold:

1) For all $(a,b) \in \mathbb{R}^2 \setminus \{(0,0)\}$, *we have the norm inequality:*

$$\|\cdot\|_{|\mathcal{H}_{a,b}|} \leq C_2\|\cdot\|_{|\mathcal{H}|},$$

where $C_2 = \alpha_H C_{2,H}$. *This implies the continuous embedding* $|\mathcal{H}| \subset |\mathcal{H}_{a,b}|$. *Consequently, for the corresponding Malliavin Sobolev spaces, we have:*

$$D^{1,2}(|\mathcal{H}|) \subset D^{1,2}(\mathcal{H}_{a,b}) \subset Dom\delta.$$

2) For all $(a,b) \in \mathbb{R}^2 \setminus \{(0,0)\}$ *with* $a \neq b$, *we have the equivalent norm inequalities:*

$$C_1\|\cdot\|_{|\mathcal{H}|} \leq \|\cdot\|_{|\mathcal{H}_{a,b}|} \leq C_2\|\cdot\|_{|\mathcal{H}|},$$

where

$$C_1 = \alpha_H C^{\min}_{1,H} \text{ and } C_2 = \alpha_H C_{2,H}.$$

These inequalities imply that $|\mathcal{H}| = |\mathcal{H}_{a,b}|$ *as sets, meaning they consist of the same functions, and their norms are equivalent.*

Consequently, their associated Sobolev spaces are also equivalent:

$$D^{1,2}(|\mathcal{H}|) = D^{1,2}(\mathcal{H}_{a,b}) \subset Dom\delta.$$

Moreover, according to Aloès and Nualart (2003), $|\mathcal{H}|$ *is a Banach space with respect to both norms* $\|\cdot\|_{|\mathcal{H}|}$ *and* $\|\cdot\|_{|\mathcal{H}_{a,b}|}$. *Furthermore, the set* $\mathcal{E}$ *of step functions on* $[0,T]$ *is dense in* $|\mathcal{H}|$.

The following proposition, which provides the *integration by parts formula for the Skorokhod integral,* will be useful. Its proof can be found, for example, in Nualart (2006).

PROPOSITION 5.1.– *For every* $F \in D^{1,2}(\mathcal{H}_{a,b})$ *and* $u \in \operatorname{Dom}\delta$ *such that* $Fu \in L^2(\Omega;\mathcal{H}_{a,b})$, *we have that:*

1) $Fu \in \operatorname{Dom}\delta$;

2) $F(\delta u) = \delta(Fu) + \langle DF, u\rangle_{\mathcal{H}_{a,b}}$.

5.3.3. Divergence and symmetric integrals for ZgfBm

Now we are able to define a first type of integral with respect to the ZgfBm.

DEFINITION 5.1.– *Let* u *be an* $\mathcal{H}_{a,b}$*-valued random variable. The divergence integral (also known as the Skorokhod integral) of a process* $u \in D^{1,2}(|\mathcal{H}_{a,b}|)$ *with respect to* $Z^H(a,b)$ *is defined as* $\delta(u)$, *where* δ *is the divergence operator adjoint to the Malliavin derivative* D. *It is denoted by:*

$$\int_0^T u_s\, \delta Z_s^H(a,b) = \delta(u).$$

REMARK 5.2.– *Several key properties of the divergence integral* $\delta(u)$ *are crucial for its application:*

1) The Skorokhod integral operator δ *is a continuous operator from its domain* $Dom\delta \subset L^2(\Omega;\mathcal{H}_{a,b})$ *to* $L^2(\Omega)$. *More specifically, if a sequence* $u_n \in Dom\delta$ *converges to* $u \in Dom\delta$ *in* $L^2(\Omega;\mathcal{H}_{a,b})$, *then* $\delta(u_n)$ *converges to* $\delta(u)$ *in* $L^2(\Omega)$.

2) The notation $\int_0^T u_s\, \delta Z_s^H(a,b)$ *emphasizes that this integral is a formal adjoint and does not, in general, coincide with a pathwise Riemann–Stieltjes or Ito-type integral. Its definition is rooted in the duality relation with the Malliavin derivative.*

We can ask in which sense the divergence operator with respect to $Z^H(a,b)$ can be interpreted as a more classical stochastic integral. In order to address this question, we will provide a second type of integration with respect to $Z^H(a,b)$, which under certain conditions, can coincide with the divergence integral. This is the *symmetric integral* introduced by Russo and Vallois (1993). By convention, we will assume that all processes and functions vanish outside the interval $[0,T]$.

DEFINITION 5.2.– *Let* $u = \{u_t, t \in [0,T]\}$ *be a stochastic process with almost surely integrable trajectories. The symmetric integral of* u *with respect to the ZgfBm* $Z^H(a,b)$ *is defined as the limit in probability, as* ϵ *tends to zero, of:*

$$\int_0^T u_s \frac{Z^H_{s+\epsilon}(a,b) - Z^H_{s-\epsilon}(a,b)}{2\epsilon} ds \qquad [5.10]$$

provided this limit exists. It is denoted by $\int_0^T u_s dZ^H_s(a,b)$.

The following theorem establishes a sufficient condition for the existence of the symmetric integral and provides the crucial relationship between the divergence and symmetric integrals. This relationship, in turn, offers a representation of the divergence operator as a more classical stochastic integral.

THEOREM 5.1.– *Let* $u = \{u_t; t \in [0,T]\}$ *be a stochastic process in the space* $D^{1,2}(|\mathcal{H}|)$ *such that:*

$$\int_0^T \int_0^T |D_s u_t||t-s|^{2H-2} dsdt < \infty \ a.s. \qquad [5.11]$$

Then, for every $(a,b) \in \mathbb{R}^2 \setminus \{(0,0)\}$, *the symmetric integral exists, and we have the following relation:*

$$\begin{aligned}\int_0^T u_t dZ^H_t(a,b) &= \int_0^T u_t \delta Z^H_t(a,b) \\ &+\alpha_H \int_0^T \int_0^T D_s u_t \left[(a^2+b^2)|t-s|^{2H-2} - 2ab(t+s)^{2H-2}\right] dsdt.\end{aligned} \qquad [5.12]$$

The proof of Theorem 5.1 largely follows the steps presented in Proposition 3 of Aloès and Nualart (2003), and it relies on the following preliminary lemmas:

LEMMA 5.4.– *Let* $u = \{u_t; t \in [0,T]\}$ *be a stochastic process in the space* $D^{1,2}(|\mathcal{H}_{a,b}|)$ *and define the approximating process* u^ϵ *by:*

$$u^\epsilon_t := \frac{1}{2\epsilon} \int_{t-\epsilon}^{t+\epsilon} u_s ds, \qquad [5.13]$$

where we use the convention $u_s = 0$ *for* $s \notin [0,T]$. *Then, for every* $(a,b) \in \mathbb{R}^2$:

1) When $a \neq b$, *there exists a positive constant* $e_{H,a,b}$ *such that:*

$$\|u^\epsilon\|^2_{D^{1,2}(|\mathcal{H}_{a,b}|)} \leq e_{H,a,b} \|u\|^2_{D^{1,2}(|\mathcal{H}_{a,b}|)}. \qquad [5.14]$$

2) When $a = b$ *and* $(a,b) \neq (0,0)$, *there exists a positive constant* $e_{H,a}$ *such that:*

$$\|u^\epsilon\|^2_{D^{1,2}(|\mathcal{H}_{a,b}|)} \leq e_{H,a} \|u\|^2_{D^{1,2}(|\mathcal{H}|)}. \qquad [5.15]$$

Proof. The proof distinguishes between two cases based on the parameters (a,b).

Case 1: $a \neq b$

In this case, from Remark 5.1 (specifically item 2), the norms $\|\cdot\|_{|\mathcal{H}|}$ and $\|\cdot\|_{|\mathcal{H}_{a,b}|}$ are equivalent. This means there exist positive constants C_1 and C_2 such that for any measurable function f:

$$C_1\|f\|_{|\mathcal{H}|} \leq \|f\|_{|\mathcal{H}_{a,b}|} \leq C_2\|f\|_{|\mathcal{H}|}.$$

The Sobolev norm for $F \in D^{1,2}(|\mathcal{H}_{a,b}|)$ is defined as:

$$\|F\|^2_{D^{1,2}(|\mathcal{H}_{a,b}|)} = \mathbb{E}|[F|^2] + \mathbb{E}[\|DF\|^2_{|\mathcal{H}_{a,b}|}].$$

Similarly, for $F \in D^{1,2}(|\mathcal{H}|)$:

$$\|F\|^2_{D^{1,2}(|\mathcal{H}|)} = \mathbb{E}[|F|^2] + \mathbb{E}[\|DF\|^2_{|\mathcal{H}|}].$$

Since $\|g\|^2_{|\mathcal{H}_{a,b}|} \leq C_2^2\|g\|^2_{|\mathcal{H}|}$ for any function g, we can apply this to u^ϵ and Du^ϵ:

$$\|u^\epsilon\|^2_{D^{1,2}(|\mathcal{H}_{a,b}|)} = \mathbb{E}[|u^\epsilon|^2] + \mathbb{E}[\|Du^\epsilon\|^2_{|\mathcal{H}_{a,b}|}] \leq \mathbb{E}[|u^\epsilon|^2] + C_2^2\mathbb{E}[\|Du^\epsilon\|^2_{|\mathcal{H}|}].$$

From Aloès and Nualart (2003) (Proposition 3 for fractional Brownian motion (fBm), which corresponds to $Z^H(1,0)$ and thus the $|\mathcal{H}|$ space), there exists a positive constant e_H such that for a process $u \in D^{1,2}(|\mathcal{H}|)$:

$$\|u^\epsilon\|^2_{D^{1,2}(|\mathcal{H}|)} \leq e_H\|u\|^2_{D^{1,2}(|\mathcal{H}|)}. \quad [5.16]$$

Now, let $K = \max(1, C_2^2)$. Then, using the inequality derived from norm equivalence:

$$\begin{aligned}\|u^\epsilon\|^2_{D^{1,2}(|\mathcal{H}_{a,b}|)} &\leq \mathbb{E}[|u^\epsilon|^2] + C_2^2\mathbb{E}[\|Du^\epsilon\|^2_{|\mathcal{H}|}]\\ &\leq K\left(\mathbb{E}[|u^\epsilon|^2] + \mathbb{E}[\|Du^\epsilon\|^2_{|\mathcal{H}|}]\right)\\ &= K\|u^\epsilon\|^2_{D^{1,2}(|\mathcal{H}|)}.\end{aligned}$$

Combining this with equation [5.16]:

$$\|u^\epsilon\|^2_{D^{1,2}(|\mathcal{H}_{a,b}|)} \leq Ke_H\|u\|^2_{D^{1,2}(|\mathcal{H}|)}.$$

Since $a \neq b$, we also have the reverse inequality from Remark 5.1:

$$\|f\|_{|\mathcal{H}|} \leq \frac{1}{C_1}\|f\|_{|\mathcal{H}_{a,b}|},$$

which implies:

$$\|F\|^2_{D^{1,2}(|\mathcal{H}|)} \leq \max(1, 1/C_1^2)\|F\|^2_{D^{1,2}(|\mathcal{H}_{a,b}|)}.$$

Let $K' = \max(1, 1/C_1^2)$. Thus,

$$\|u\|^2_{D^{1,2}(|\mathcal{H}|)} \leq K'\|u\|^2_{D^{1,2}(|\mathcal{H}_{a,b}|)}.$$

Substituting this back:

$$\|u^\epsilon\|^2_{D^{1,2}(|\mathcal{H}_{a,b}|)} \leq Ke_H K' \|u\|^2_{D^{1,2}(|\mathcal{H}_{a,b}|)}.$$

So, for $a \neq b$, $e_{H,a,b} = Ke_H K' = \max(1, C_2^2) e_H \max(1, 1/C_1^2)$. This establishes inequality [5.14].

Case 2: $a = b$ and $(a,b) \neq (0,0)$

When $a = b$, the covariance structure of $Z^H(a,a)$ is given as:

$$R^{H,a,a}(t,s) = a^2(t^{2H} + s^{2H} - |t-s|^{2H}).$$

The second mixed partial derivative is:

$$\left| \frac{\partial^2 R^{H,a,a}(u,v)}{\partial u \partial v} \right| = \alpha_H (2a^2)|u-v|^{2H-2}.$$

This means the norm in $|\mathcal{H}_{a,a}|$ is related to the norm in $|\mathcal{H}|$ by:

$$\|f\|^2_{|\mathcal{H}_{a,a}|} = 2a^2 \|f\|^2_{|\mathcal{H}|}.$$

Consequently, for the Sobolev norms:

$$\|F\|^2_{D^{1,2}(|\mathcal{H}_{a,a}|)} = \mathbb{E}[|F|^2] + \mathbb{E}[(\|DF\|^2_{|\mathcal{H}_{a,a}|}] = \mathbb{E}[|F|^2] + 2a^2\mathbb{E}[\|DF\|^2_{|\mathcal{H}|}].$$

We know from equation [5.16] that smoothing works in $|\mathcal{H}|$:

$$\|u^\epsilon\|^2_{D^{1,2}(|\mathcal{H}|)} \leq e_H \|u\|^2_{D^{1,2}(|\mathcal{H}|)}.$$

We want to find a constant $e_{H,a}$ such that $\|u^\epsilon\|^2_{D^{1,2}(|\mathcal{H}_{a,a}|)} \leq e_{H,a}\|u\|^2_{D^{1,2}(|\mathcal{H}|)}$.

Let us analyze the components of the Sobolev norm for u^ϵ in $|\mathcal{H}_{a,a}|$:

$$\|u^\epsilon\|^2_{D^{1,2}(|\mathcal{H}_{a,a}|)} = \mathbb{E}[|u^\epsilon|^2] + 2a^2\mathbb{E}[\|Du^\epsilon\|^2_{|\mathcal{H}|}].$$

The result in equation [5.16] implies individual bounds (assuming such constants exist and are part of e_H): let $e_H^{(0)}$ be a constant for the L^2 part and $e_H^{(D)}$ for the derivative part, such that $e_H = \max(e_H^{(0)}, e_H^{(D)})$. Then:

$$\mathbb{E}[|u^\epsilon|^2] \leq e_H^{(0)} \mathbb{E}[|u|^2]$$

$$\mathbb{E}[\|Du^\epsilon\|^2_{|\mathcal{H}|}] \leq e_H^{(D)} \mathbb{E}[\|Du\|^2_{|\mathcal{H}|}]$$

Substituting these into the expression for $\|u^\epsilon\|^2_{D^{1,2}(|\mathcal{H}_{a,a}|)}$:

$$\|u^\epsilon\|^2_{D^{1,2}(|\mathcal{H}_{a,a}|)} \leq e_H^{(0)}\mathbb{E}[|u|^2] + 2a^2 e_H^{(D)}\mathbb{E}[\|Du\|^2_{|\mathcal{H}|}].$$

To obtain the desired form $e_{H,a}\|u\|^2_{D^{1,2}(|\mathcal{H}|)} = e_{H,a}\left(\mathbb{E}[|u|^2] + \mathbb{E}[\|Du\|^2_{|\mathcal{H}|}]\right)$, we need $e_{H,a} \geq e_H^{(0)}$ and $e_{H,a} \geq 2a^2 e_H^{(D)}$.

Therefore, we can set $e_{H,a} = \max(e_H^{(0)}, 2a^2 e_H^{(D)})$. This constant can be bounded by $e_H \max(1, 2a^2)$, since $e_H^{(0)} \leq e_H$ and $e_H^{(D)} \leq e_H$. This establishes inequality [5.15]. □

We now present a crucial lemma that relates the symmetric approximation of an integral to the divergence operator and establishes the convergence of the smoothed process.

LEMMA 5.5.– *Let $u = \{u_t; t \in [0,T]\}$ be a stochastic process in the space $D^{1,2}(|\mathcal{H}|)$. Denoting*

$$A^\epsilon u = \frac{1}{2\epsilon}\int_0^T \langle Du_s, \mathbf{1}_{[s-\epsilon,s+\epsilon]}\rangle_{|\mathcal{H}_{a,b}|} ds,$$

for every $\epsilon > 0$, we have:

$$\frac{1}{2\epsilon}\int_0^T u_s(Z^H_{s+\epsilon}(a,b) - Z^H_{s-\epsilon}(a,b))ds = \delta(u^\epsilon) + A^\epsilon u, \qquad [5.17]$$

and $\delta(u^\epsilon)$ converges in $L^2(\Omega)$ to $\delta(u)$ as ϵ tends to 0.

Proof. Let S_T be the set of smooth step processes of the form:

$$u = \sum_{j=0}^{m-1} F_j \mathbf{1}_{[t_j,t_{j+1}]},$$

where $F_j \in \mathcal{S}_{a,b}$, and $0 = t_0 < t_1 < \cdots < t_m = T$. We recall that by Remark 5.1, when $a \neq b$, $D^{1,2}(|\mathcal{H}_{a,b}|) = D^{1,2}(|\mathcal{H}|)$.

Since S_T is dense in $D^{1,2}(|\mathcal{H}|)$, there exists a sequence $\{u^n\}_{n\geq 1} \subset S_T$ such that $u^n \to u$ in the norm of the space $D^{1,2}(|\mathcal{H}|)$ as $n \to \infty$.

For $u^n \in S_T$, applying the product rule for the divergence operator (Proposition 5.1, assertion 2) and Fubini's theorem (as shown in similar proofs, e.g. in Nualart (2006) for divergence integrals), we get:

$$\frac{1}{2\epsilon}\int_0^T u_s^n(Z_{s+\epsilon}^H(a,b) - Z_{s-\epsilon}^H(a,b))ds$$
$$= \frac{1}{2\epsilon}\int_0^T \delta(u_s^n \mathbf{1}_{[s-\epsilon,s+\epsilon]}(.))ds + A^\epsilon u^n$$
$$= \int_0^T \frac{1}{2\epsilon}\left(\int_{r-\epsilon}^{r+\epsilon} u_s^n ds\right)\delta Z_r^H(a,b) + A^\epsilon u^n$$
$$= \int_0^T u_r^{\epsilon,n}\delta Z_r^H(a,b) + A^\epsilon u^n$$
$$= \delta(u^{\epsilon,n}) + A^\epsilon u^n.$$

Here, $u_r^{\epsilon,n} = \frac{1}{2\epsilon}\int_{r-\epsilon}^{r+\epsilon} u_s^n ds$ is the smoothed process.

Now we need to show that this relation extends to $u \in D^{1,2}(|\mathcal{H}|)$. This involves showing that the left-hand side converges to its desired limit as $n \to \infty$ for fixed ϵ, and that $A^\epsilon u^n$ converges to $A^\epsilon u$ as $n \to \infty$.

Convergence of the left-hand side: let

$$LHS(u^\epsilon) = \frac{1}{2\epsilon}\int_0^T u_s(Z_{s+\epsilon}^H(a,b) - Z_{s-\epsilon}^H(a,b))ds.$$

We want to show that $\mathbb{E}[|LHS(u^{n,\epsilon}) - LHS(u^\epsilon)|^2] \to 0$ as $n \to \infty$.

$$LHS(u^{n,\epsilon}) - LHS(u^\epsilon) = \frac{1}{2\epsilon}\int_0^T (u_s^n - u_s)(Z_{s+\epsilon}^H(a,b) - Z_{s-\epsilon}^H(a,b))ds.$$

Applying the Cauchy–Schwarz inequality in $L^2([0,T])$:

$$\mathbb{E}\left[\left|\frac{1}{2\epsilon}\int_0^T (u_s^n - u_s)(Z_{s+\epsilon}^H(a,b) - Z_{s-\epsilon}^H(a,b))ds\right|^2\right]$$
$$\leq \frac{1}{4\epsilon^2}\mathbb{E}\left[\left(\int_0^T |u_s^n - u_s|^2 ds\right)\left(\int_0^T |Z_{s+\epsilon}^H(a,b) - Z_{s-\epsilon}^H(a,b)|^2 ds\right)\right].$$

By the Cauchy–Schwarz inequality for expectations:

$$\leq \frac{1}{4\epsilon^2}\left(\mathbb{E}\left[\left(\int_0^T |u_s^n - u_s|^2 ds\right)^2\right]\right)^{1/2}$$
$$\times\left(\mathbb{E}\left[\left(\int_0^T |Z_{s+\epsilon}^H(a,b) - Z_{s-\epsilon}^H(a,b)|^2 ds\right)^2\right]\right)^{1/2}.$$

Since $u^n \to u$ in $D^{1,2}(|\mathcal{H}|)$, it is a standard result that this implies $u^n \to u$ in $L^2(\Omega; L^2([0,T]))$.

Therefore, the first term $\mathbb{E}\left[\left(\int_0^T |u_s^n - u_s|^2 ds\right)^2\right]$ converges to 0 as $n \to \infty$.

For fixed $\epsilon > 0$, $(Z_{s+\epsilon}^H(a,b) - Z_{s-\epsilon}^H(a,b))$ is a Gaussian process with finite moments. The term $\int_0^T |Z_{s+\epsilon}^H(a,b) - Z_{s-\epsilon}^H(a,b)|^2 ds$ is a nonnegative random variable whose moments are finite due to the Gaussian nature and fixed interval of integration.

Thus, $\left(\mathbb{E}\left[\left(\int_0^T |Z_{s+\epsilon}^H(a,b) - Z_{s-\epsilon}^H(a,b)|^2 ds\right)^2\right]\right)^{1/2}$ is finite for any fixed $\epsilon > 0$.

Consequently, for fixed ϵ, $\mathbb{E}[|LHS(u^{n,\epsilon}) - LHS(u^\epsilon)|^2]$ converges to 0 as $n \to \infty$.

Convergence of $A^\epsilon u^n$ to $A^\epsilon u$: we want to show that $\mathbb{E}[|A^\epsilon u^n - A^\epsilon u|^2] \to 0$ as $n \to \infty$.

$$|A^\epsilon u^n - A^\epsilon u| = \left|\frac{1}{2\epsilon}\int_0^T \langle D(u_s^n - u_s), \mathbf{1}_{[s-\epsilon,s+\epsilon]}\rangle_{|\mathcal{H}_{a,b}|} ds\right|.$$

By the Cauchy–Schwarz inequality for integrals and the properties of the inner product:

$$\leq \frac{1}{2\epsilon}\int_0^T \|D(u_s^n - u_s)\|_{|\mathcal{H}_{a,b}|} \|\mathbf{1}_{[s-\epsilon,s+\epsilon]}\|_{|\mathcal{H}_{a,b}|} ds.$$

From the definition of the norm in $|\mathcal{H}_{a,b}|$ and Corollary 5.1, it holds that:

$$\|\mathbf{1}_{[s-\epsilon,s+\epsilon]}\|_{|\mathcal{H}_{a,b}|}^2 \leq C\int_{s-\epsilon}^{s+\epsilon}\int_{s-\epsilon}^{s+\epsilon} |u-v|^{2H-2} du dv \leq C'(2\epsilon)^{2H}$$

for some constants C, C'. Thus, $\|\mathbf{1}_{[s-\epsilon,s+\epsilon]}\|_{|\mathcal{H}_{a,b}|} \leq C''\epsilon^H$ for some constant C''. Therefore:

$$\begin{aligned} |A^\epsilon u^n - A^\epsilon u| &\leq \frac{C''\epsilon^H}{2\epsilon}\int_0^T \|D(u_s^n - u_s)\|_{|\mathcal{H}_{a,b}|} ds \\ &= \frac{C''}{2}\epsilon^{H-1}\int_0^T \|D(u_s^n - u_s)\|_{|\mathcal{H}_{a,b}|} ds. \end{aligned}$$

For a fixed $\epsilon > 0$, by the Cauchy–Schwarz inequality for integrals:

$$\mathbb{E}[|A^\epsilon u^n - A^\epsilon u|^2] \leq \left(\frac{C''}{2}\epsilon^{H-1}\right)^2 T\mathbb{E}\left[\int_0^T \|D(u_s^n - u_s)\|^2_{|\mathcal{H}_{a,b}|} ds\right].$$

Since $u^n \to u$ in $D^{1,2}(|\mathcal{H}|)$, and given the continuous embedding $|\mathcal{H}| \subset |\mathcal{H}_{a,b}|$ (from Remark 5.1, item 1) or equivalence of norms (Remark 5.1, item 2), it follows that $u^n \to u$ in $D^{1,2}(|\mathcal{H}_{a,b}|)$.

This implies that $\mathbb{E}\left[\int_0^T \|D(u_s^n - u_s)\|^2_{|\mathcal{H}_{a,b}|} ds\right]$ converges to 0 as $n \to \infty$. Thus, for fixed ϵ, $\mathbb{E}[|A^\epsilon u^n - A^\epsilon u|^2]$ converges to 0 as $n \to \infty$.

By passing to the limit $n \to \infty$ in the equality derived for u^n, equation [5.17] is established for $u \in D^{1,2}(|\mathcal{H}|)$.

Convergence of $\delta(u^\epsilon)$ *to* $\delta(u)$: we now prove that $\delta(u^\epsilon)$ converges in $L^2(\Omega)$ to $\delta(u)$ as $\epsilon \to 0$. We use a standard approximation argument. Take a sequence $\{u^n\} \subset S_T$ such that $u^n \to u$ in the norm of the space $D^{1,2}(|\mathcal{H}|)$ as $n \to \infty$.

For every $n \geq 1$ and $\epsilon > 0$, we have the triangle inequality:

$$\begin{aligned}&\mathbb{E}\left[\delta(u^\epsilon) - \delta(u)\right]^2 \\ &\leq 3\left[\mathbb{E}\left[\delta(u^\epsilon) - \delta(u^{n,\epsilon})\right]^2 + \mathbb{E}\left[\delta(u^{n,\epsilon}) - \delta(u^n)\right]^2 + \mathbb{E}\left[\delta(u^n) - \delta(u)\right]^2\right].\end{aligned} \quad [5.18]$$

– *Term 1*: $\mathbb{E}\left[\delta(u^\epsilon) - \delta(u^{n,\epsilon})\right]^2$. The divergence operator δ is a continuous operator from $D^{1,2}(|\mathcal{H}_{a,b}|)$ to $L^2(\Omega)$. Therefore, there exists a constant $C_\delta > 0$ such that:

$$\mathbb{E}\left[\delta(u^\epsilon) - \delta(u^{n,\epsilon})\right]^2 \leq C_\delta \|u^\epsilon - u^{n,\epsilon}\|^2_{D^{1,2}(|\mathcal{H}_{a,b}|)}.$$

By Lemma 5.4, there exists a constant $e_{H,a,b}$ (or $e_{H,a}$ if $a = b$) such that $\|u^\epsilon - u^{n,\epsilon}\|^2_{D^{1,2}(|\mathcal{H}_{a,b}|)} \leq e_{H,a,b}\|u - u^n\|^2_{D^{1,2}(|\mathcal{H}_{a,b}|)}$. Since $u^n \to u$ in $D^{1,2}(|\mathcal{H}|)$ (which implies convergence in $D^{1,2}(|\mathcal{H}_{a,b}|)$), this term converges to 0 as $n \to \infty$. Thus, for any $\eta > 0$, we can choose N_η large enough such that for all $n \geq N_\eta$, this term is less than $\eta/3$.

– *Term 3*: $\mathbb{E}\left[\delta(u^n) - \delta(u)\right]^2$. Similarly, due to the continuity of δ:

$$\mathbb{E}\left[\delta(u^n) - \delta(u)\right]^2 \leq C_\delta \|u^n - u\|^2_{D^{1,2}(|\mathcal{H}_{a,b}|)}.$$

As $u^n \to u$ in $D^{1,2}(|\mathcal{H}|)$, this term also converges to 0 as $n \to \infty$. For the same N_η chosen above, this term is also less than $\eta/3$.

– *Term 2*: $\mathbb{E}\left(\delta(u^{n,\epsilon}) - \delta(u^n)\right)^2$. For a fixed n, $u^n \in S_T$ is a smooth process. It is a known property of the Skorokhod integral that the mollification $u^{n,\epsilon}$ converges to u^n in the $D^{1,2}(|\mathcal{H}|)$ norm as $\epsilon \to 0$.

Specifically, $u^{n,\epsilon} \to u^n$ in $D^{1,2}(|\mathcal{H}|)$ implies convergence in $D^{1,2}(|\mathcal{H}_{a,b}|)$ due to the continuous embedding.

By the continuity of the divergence operator δ, this implies that:

$$\mathbb{E}\left[\delta(u^{n,\epsilon}) - \delta(u^n)\right]^2 \to 0 \; as \; \epsilon \to 0.$$

Therefore, for the chosen N_η, there exists an $\epsilon_0 > 0$ such that for all $\epsilon < \epsilon_0$, this term is less than $\eta/3$.

Combining these results, for every $\eta > 0$, we can first choose N_η large enough. Then, for this fixed N_η, we choose ϵ_0 small enough. Thus, for $\epsilon < \epsilon_0$:

$$\mathbb{E}\left[\delta(u^\epsilon) - \delta(u)\right]^2 \leq 3\left[\eta/3 + \eta/3 + \eta/3\right] = \eta.$$

Since η is arbitrary, this implies that:

$$\lim_{\epsilon \to 0} \mathbb{E}\left[\delta(u^\epsilon) - \delta(u)\right]^2 = 0.$$

This completes the proof. □

Proof of Theorem 5.1. By the definition of the symmetric integral [5.10] and the result of Lemma 5.5, the symmetric integral is defined as the limit in probability of $\frac{1}{2\epsilon}\int_0^T u_s(Z^H_{s+\epsilon}(a,b) - Z^H_{s-\epsilon}(a,b))ds$. From Lemma 5.5, we know that this expression equals $\delta(u^\epsilon) + A^\epsilon u$, and that $\delta(u^\epsilon)$ converges in $L^2(\Omega)$ to $\delta(u)$ as $\epsilon \to 0$.

Therefore, to complete the proof of Theorem 5.1, it suffices to show that $A^\epsilon u$ converges in probability to the second term on the right-hand side of equation [5.12], namely to:

$$\alpha_H \int_0^T \int_0^T D_s u_t \left[(a^2+b^2)|t-s|^{2H-2} - 2ab(t+s)^{2H-2}\right] dsdt.$$

This is equivalent to showing that $A^\epsilon u$ converges to

$$\int_0^T \int_0^T D_s u_t \frac{\partial^2 R^{H,a,b}(s,t)}{\partial s \partial t} dsdt.$$

Let us expand $A^\epsilon u$ using its definition and the inner product in $|\mathcal{H}_{a,b}|$. The inner product is defined as

$$\langle f, g \rangle_{|\mathcal{H}_{a,b}|} = \int_0^T \int_0^T f(r)g(x) \frac{\partial^2 R^{H,a,b}(r,x)}{\partial r \partial x} drdx.$$

In $A^\epsilon u = \frac{1}{2\epsilon}\int_0^T \langle Du_s, \mathbf{1}_{[s-\epsilon,s+\epsilon]}\rangle_{|\mathcal{H}_{a,b}|} ds$, the term Du_s represents the Malliavin derivative of u_s, which is a process indexed by the differentiation variable, say $D_r u_s$.

The indicator function $\mathbf{1}_{[s-\epsilon,s+\epsilon]}$ is a function of the second integration variable, say x. Thus,

$$\begin{aligned} A^\epsilon u &= \frac{1}{2\epsilon}\int_0^T \left(\int_0^T\int_0^T (D_r u_s)\mathbf{1}_{[s-\epsilon,s+\epsilon]}(x)\frac{\partial^2 R^{H,a,b}(r,x)}{\partial r\partial x} drdx\right) ds \\ &= \frac{1}{2\epsilon}\int_0^T\int_0^T (D_r u_s)\left(\int_0^T \mathbf{1}_{[s-\epsilon,s+\epsilon]}(x)\frac{\partial^2 R^{H,a,b}(r,x)}{\partial r\partial x} dx\right) drds \\ &= \frac{1}{2\epsilon}\int_0^T\int_0^T (D_r u_s)\left(\int_{s-\epsilon}^{s+\epsilon} \frac{\partial^2 R^{H,a,b}(r,x)}{\partial r\partial x} dx\right) drds. \end{aligned}$$

Now, let us perform a change of variables in the innermost integral: let $x = s+\sigma$, so $dx = d\sigma$. When $x = s-\epsilon$, $\sigma = -\epsilon$. When $x = s+\epsilon$, $\sigma = +\epsilon$.

$$A^\epsilon u = \frac{1}{2\epsilon}\int_0^T\int_0^T (D_r u_s)\left(\int_{-\epsilon}^{+\epsilon} \frac{\partial^2 R^{H,a,b}(r,s+\sigma)}{\partial r\partial \sigma} d\sigma\right) drds.$$

Next, we apply the mean value theorem for integrals to the inner integral:

$$\int_{-\epsilon}^{+\epsilon} \frac{\partial^2 R^{H,a,b}(r,s+\sigma)}{\partial r\partial \sigma} d\sigma.$$

For fixed r and s, the function $f(\sigma) = \dfrac{\partial^2 R^{H,a,b}(r,s+\sigma)}{\partial r\partial \sigma}$ is continuous with respect to σ (away from the diagonal where $r = s+\sigma$).

Therefore, there exists some $\sigma_\epsilon \in (-\epsilon,\epsilon)$ such that:

$$\frac{1}{2\epsilon}\int_{-\epsilon}^{+\epsilon} \frac{\partial^2 R^{H,a,b}(r,s+\sigma)}{\partial r\partial \sigma} d\sigma = \frac{\partial^2 R^{H,a,b}(r,s+\sigma_\epsilon)}{\partial r\partial \sigma}. \qquad [5.19]$$

As $\epsilon \to 0$, we have $\sigma_\epsilon \to 0$. Since the function $\sigma \mapsto \dfrac{\partial^2 R^{H,a,b}(r,s+\sigma)}{\partial r\partial \sigma}$ is continuous at $\sigma = 0$ for every fixed r, s with $r \neq s$ (see properties of $R^{H,a,b}$ and its derivatives and see Exercise 5.5), we can take the pointwise limit almost surely (since the set where $r = s$ has measure zero in the (r,s) plane):

$$\lim_{\epsilon\to 0}\frac{1}{2\epsilon}\left(\int_{-\epsilon}^{+\epsilon} \frac{\partial^2 R^{H,a,b}(r,s+\sigma)}{\partial r\partial \sigma} d\sigma\right) = \frac{\partial^2 R^{H,a,b}(r,s)}{\partial r\partial s} \quad a.s.$$

Thus, the integrand of $A^\epsilon u$ converges pointwise almost surely as $\epsilon \to 0$:

$$\lim_{\epsilon\to 0}(D_r u_s)\frac{1}{2\epsilon}\left(\int_{-\epsilon}^{+\epsilon} \frac{\partial^2 R^{H,a,b}(r,s+\sigma)}{\partial r\partial \sigma} d\sigma\right) = D_r u_s\frac{\partial^2 R^{H,a,b}(r,s)}{\partial r\partial s} \quad a.s.$$

To apply the dominated convergence theorem, we need a suitable dominating function. From Lemma 5.2 (which is assumed to hold and typically provides such a bound), there exists a positive constant C_3, such that for every $\epsilon > 0$:

$$\frac{1}{2\epsilon}\left|\int_{-\epsilon}^{\epsilon} \frac{\partial^2 R^{H,a,b}(r, s+\sigma)}{\partial r \partial \sigma} d\sigma\right| \leq C_3 |r-s|^{2H-2}. \qquad [5.20]$$

This allows us to set the dominating function for the integrand of $A^\epsilon u$ as:

$$C_3 |D_r u_s||r-s|^{2H-2}.$$

The integrability of this dominating function is guaranteed by assumption [5.11] from the theorem statement, which reads:

$$\int_0^T \int_0^T |D_s u_t||t-s|^{2H-2} dsdt < \infty \ a.s.$$

(which is equivalent to $\int_0^T \int_0^T |D_r u_s||r-s|^{2H-2} drds < \infty \ a.s.$ by simply relabeling dummy variables).

Therefore, by the dominated convergence theorem, we can interchange the limit and the integrals:

$$\lim_{\epsilon \to 0} A^\epsilon u = \int_0^T \int_0^T D_r u_s \frac{\partial^2 R^{H,a,b}(r,s)}{\partial r \partial s} drds \quad a.s.$$

By relabeling the dummy integration variables ($r \leftrightarrow s$) and ($s \leftrightarrow t$), this limit becomes:

$$\int_0^T \int_0^T D_s u_t \frac{\partial^2 R^{H,a,b}(s,t)}{\partial s \partial t} dsdt \quad a.s.$$

Substituting the explicit form of $\frac{\partial^2 R^{H,a,b}(s,t)}{\partial s \partial t}$:

$$\int_0^T \int_0^T D_s u_t \left[\alpha_H (a^2+b^2)|t-s|^{2H-2} - 2ab\alpha_H (t+s)^{2H-2}\right] dsdt$$

$$= \alpha_H \int_0^T \int_0^T D_s u_t \left[(a^2+b^2)|t-s|^{2H-2} - 2ab(t+s)^{2H-2}\right] dsdt.$$

This is exactly the second term on the right-hand side of equation [5.12].

Combining this result with Lemma 5.5, which states that the symmetric integral approximation converges to $\delta(u^\epsilon) + A^\epsilon u$ and $\delta(u^\epsilon)$ converges to $\delta(u)$ in $L^2(\Omega)$, we conclude that the symmetric integral exists and the relation [5.12] holds.

This completes the proof of Theorem 5.1. □

5.4. Itô formulas with ZgfBm

We will establish the following divergence-integration Itô formula for the ZgfBm $Z^H(a, b)$.

THEOREM 5.2.– *Let f be a function of class $C^{1,2}([0,T] \times \mathbb{R}, \mathbb{R})$, such that it satisfies the growth condition:*

$$\sup_{t\in[0,T]} \max\left\{|f(t,x)|, \left|\frac{\partial f(t,x)}{\partial x}\right|, \left|\frac{\partial^2 f(t,x)}{\partial x^2}\right|\right\} \leq c\exp(\lambda x^2), \qquad [5.21]$$

for every $x \in \mathbb{R}$, where c and λ are positive constants such that:

$$\lambda < \frac{T^{-2H}}{4C_H(a,b)}. \qquad [5.22]$$

Let $Z_t^H(a,b)$ be a Gaussian process with $Z_0^H(a,b) = 0$. $C_H(a,b)$ is the constant defined in equation [5.5]. Then, for any $t \in [0,T]$, the following Itô-like formula holds:

$$\begin{aligned} f(t, Z_t^H(a,b)) &= f(0,0) + \int_0^t \frac{\partial f}{\partial x}(s, Z_s^H(a,b))\delta(Z_s^H(a,b)) \\ &+ \int_0^t \left(\frac{\partial f}{\partial s}(s, Z_s^H(a,b)) + HC_H(a,b)\frac{\partial^2 f}{\partial x^2}(s, Z_s^H(a,b))s^{2H-1}\right) ds. \end{aligned} \qquad [5.23]$$

Proof. Let us fix $t > 0$ and consider a uniform partition of the interval $[0,t]$:

$$\sigma := \left\{t_0 = 0 < t_1 = \frac{t}{n} < \cdots < t_{n-1} = \frac{(n-1)t}{n} < t_n = t\right\}.$$

As $n \to \infty$, $\sup_j(t_j - t_{j-1}) = t/n \to 0$.

By a standard localization argument (see, for example Nualart (2006, Chapter 5)), we can assume without loss of generality that f and its derivatives have compact support. This allows us to treat the partial derivatives of f appearing in the Taylor expansion as bounded by some positive constant C. The growth condition [5.21] and Lemma 5.3 ensure that this localization is valid and that the expectations of the localized terms converge appropriately. Throughout the proof, C denotes a generic positive constant whose value may change from line to line.

Using Taylor expansion of $f(t,x)$ around $(t_{j-1}, Z_{t_{j-1}}^H(a,b))$, we get:

$$\begin{aligned} f(t, Z_t^H(a,b)) &= f(0,0) + \sum_{j=1}^n \left(f(t_j, Z_{t_j}^H(a,b)) - f(t_{j-1}, Z_{t_{j-1}}^H(a,b))\right) \\ &= f(0,0) + \sum_{j=1}^n \frac{\partial f}{\partial s}(t_{j-1}, Z_{t_{j-1}}^H(a,b))(t_j - t_{j-1}) \end{aligned}$$

$$
\begin{aligned}
&+\sum_{j=1}^{n}\frac{\partial f}{\partial x}(t_{j-1},Z^H_{t_{j-1}}(a,b))(Z^H_{t_j}(a,b)-Z^H_{t_{j-1}}(a,b))\\
&+\frac{1}{2}\sum_{j=1}^{n}\frac{\partial^2 f}{\partial s^2}(\tilde{X}_j)(t_j-t_{j-1})^2\\
&+\frac{1}{2}\sum_{j=1}^{n}\frac{\partial^2 f}{\partial x^2}(\tilde{X}_j)(Z^H_{t_j}(a,b)-Z^H_{t_{j-1}}(a,b))^2\\
&+\sum_{j=1}^{n}\frac{\partial^2 f}{\partial s\partial x}(\tilde{X}_j)(t_j-t_{j-1})(Z^H_{t_j}(a,b)-Z^H_{t_{j-1}}(a,b))\\
&=f(0,0)+I_{n,1}+I_{n,2}+I_{n,3}+I_{n,4}+I_{n,5},
\end{aligned}
$$

where $\tilde{X}_j=(1-\theta_j)(t_{j-1},Z^H_{t_{j-1}}(a,b))+\theta_j(t_j,Z^H_{t_j}(a,b))$ for some random variable $\theta_j\in(0,1)$.

Let us study the limit of each term as $n\to\infty$.

Convergence of $I_{n,3}$: we consider $I_{n,3}=\frac{1}{2}\sum_{j=1}^{n}\frac{\partial^2 f}{\partial s^2}(\tilde{X}_j)(t_j-t_{j-1})^2$. Due to the localization argument, $\frac{\partial^2 f}{\partial s^2}(\tilde{X}_j)$ is bounded by a constant C. Therefore:

$$
\mathbb{E}[|I_{n,3}|]\le\frac{1}{2}\sum_{j=1}^{n}\mathbb{E}\left[\left|\frac{\partial^2 f}{\partial s^2}(\tilde{X}_j)\right|\right](t_j-t_{j-1})^2\le\frac{1}{2}\sum_{j=1}^{n}C\left(\frac{t}{n}\right)^2=\frac{Ct^2}{2n}.
$$

As $n\to\infty$, $I_{n,3}\xrightarrow{L^1(\Omega)}0$.

Convergence of $I_{n,4}$: we have $I_{n,4}=\frac{1}{2}\sum_{j=1}^{n}\frac{\partial^2 f}{\partial x^2}(\tilde{X}_j)(Z^H_{t_j}(a,b)-Z^H_{t_{j-1}}(a,b))^2$.

Using the boundedness of $\frac{\partial^2 f}{\partial x^2}(\tilde{X}_j)$ by C and Cauchy–Schwarz inequality, along with the property of Gaussian moments $\mathbb{E}[X^4]=3(\mathbb{E}[X^2])^2$:

$$
\begin{aligned}
\mathbb{E}[|I_{n,4}|]&\le\frac{1}{2}\sum_{j=1}^{n}\sqrt{\mathbb{E}\left[\left|\frac{\partial^2 f}{\partial x^2}(\tilde{X}_j)\right|^2\right]}\sqrt{\mathbb{E}\left[(Z^H_{t_j}(a,b)-Z^H_{t_{j-1}}(a,b))^4\right]}\\
&\le\frac{1}{2}\sum_{j=1}^{n}\sqrt{C^2}\sqrt{3(\mathbb{E}[(Z^H_{t_j}(a,b)-Z^H_{t_{j-1}}(a,b))^2])^2}
\end{aligned}
$$

$$= \frac{\sqrt{3}C}{2} \sum_{j=1}^{n} \mathbb{E}[(Z_{t_j}^H(a,b) - Z_{t_{j-1}}^H(a,b))^2]$$

$$= \frac{\sqrt{3}C}{2} \sum_{j=1}^{n} C_H(a,b)(t_j - t_{j-1})^{2H}$$

$$= \frac{\sqrt{3}CC_H(a,b)}{2} \sum_{j=1}^{n} \left(\frac{t}{n}\right)^{2H} = \frac{\sqrt{3}CC_H(a,b)t^{2H}}{2} n^{1-2H}.$$

Since $H > 1/2, 1-2H < 0$, so $n^{1-2H} \to 0$ as $n \to \infty$. Thus, $I_{n,4} \xrightarrow{L^1(\Omega)} 0$. This implies that the term usually associated with quadratic variation vanishes for ZgfBm with $H > 1/2$.

Convergence of $I_{n,5}$: we consider:

$$I_{n,5} = \sum_{j=1}^{n} \frac{\partial^2 f}{\partial s \partial x}(\tilde{X}_j)(t_j - t_{j-1})(Z_{t_j}^H(a,b) - Z_{t_{j-1}}^H(a,b)).$$

Using the boundedness of $\frac{\partial^2 f}{\partial s \partial x}(\tilde{X}_j)$ by C and Cauchy–Schwarz inequality:

$$\mathbb{E}[|I_{n,5}|] \leq \sum_{j=1}^{n} \mathbb{E}\left[\left|\frac{\partial^2 f}{\partial s \partial x}(\tilde{X}_j)\right| |t_j - t_{j-1}||Z_{t_j}^H(a,b) - Z_{t_{j-1}}^H(a,b)|\right]$$

$$\leq \sum_{j=1}^{n} \sqrt{\mathbb{E}\left[\left|\frac{\partial^2 f}{\partial s \partial x}(\tilde{X}_j)\right|^2\right]} |t_j - t_{j-1}| \sqrt{\mathbb{E}\left[(Z_{t_j}^H(a,b) - Z_{t_{j-1}}^H(a,b))^2\right]}$$

$$\leq \sum_{j=1}^{n} C\left(\frac{t}{n}\right) \sqrt{C_H(a,b)\left(\frac{t}{n}\right)^{2H}}$$

$$= C\sqrt{C_H(a,b)}n\left(\frac{t}{n}\right)^{1+H} = C\sqrt{C_H(a,b)}t^{1+H}n^{-H}.$$

Since $H > 1/2 > 0$, $n^{-H} \to 0$ as $n \to \infty$. Thus, $I_{n,5} \xrightarrow{L^1(\Omega)} 0$.

Convergence of $I_{n,2}$: now, to study the limit of the sequence $(I_{n,2})$, we will apply Proposition 5.1, which states the integration by parts formula for the Skorokhod integral. First, we verify the conditions on f and Z^H. Conditions [5.21] and [5.22], together with Lemma 5.3, imply that the involved functions have finite $L^2(\Omega)$ moments:

$$\mathbb{E}\left[\sup_{0 \leq t \leq T} |f(t, Z_t^H(a,b))|^2\right] \leq c^2 \mathbb{E}\left[\exp\left(2\lambda \sup_{0 \leq t \leq T} |Z_t^H(a,b)|^2\right)\right] < \infty.$$

In a similar way, we get that:

$$\mathbb{E}\left[\sup_{0\leq t\leq T}\left|\frac{\partial f(t,Z_t^H(a,b))}{\partial x}\right|^2\right]<\infty \quad \text{and} \quad \mathbb{E}\left[\sup_{0\leq t\leq T}\left|\frac{\partial^2 f(t,Z_t^H(a,b))}{\partial x^2}\right|^2\right]<\infty.$$

Therefore, under the growth condition [5.21], the process $\frac{\partial f}{\partial x}(t,Z_t^H(a,b))$ belongs to the Malliavin Sobolev space $D^{1,2}(|\mathcal{H}_{a,b}|)$ and condition [5.11] holds.

Let $F_j=\frac{\partial f}{\partial x}(t_{j-1},Z_{t_{j-1}}^H(a,b))$ and $u_j = \mathbf{1}_{(t_{j-1},t_j]}$. The term $(Z_{t_j}^H(a,b)-Z_{t_{j-1}}^H(a,b))$ can be written as the Skorokhod integral $\delta(u_j)$. Applying Proposition 5.1 (the integration by parts formula for the Skorokhod integral),

$$F\delta(h)=\delta(Fh)+\langle DF,h\rangle_{\mathcal{H}}$$

for $F\in\mathbb{D}^{1,2}$ and $h\in\mathcal{H}$), to each term in the sum for $I_{n,2}$:

$$\begin{aligned}
I_{n,2} &= \sum_{j=1}^{n}\frac{\partial f}{\partial x}(t_{j-1},Z_{t_{j-1}}^H(a,b))(Z_{t_j}^H(a,b)-Z_{t_{j-1}}^H(a,b))\\
&= \sum_{j=1}^{n}\frac{\partial f}{\partial x}(t_{j-1},Z_{t_{j-1}}^H(a,b))\delta\left(\mathbf{1}_{(t_{j-1},t_j]}\right)\\
&= \delta\left(\sum_{j=1}^{n}\frac{\partial f}{\partial x}(t_{j-1},Z_{t_{j-1}}^H(a,b))\left(\mathbf{1}_{(t_{j-1},t_j]}\right)\right)\\
&\quad+\sum_{j=1}^{n}\left\langle D\left(\frac{\partial f}{\partial x}(t_{j-1},Z_{t_{j-1}}^H(a,b))\right),\mathbf{1}_{(t_{j-1},t_j]}\right\rangle_{\mathcal{H}_{a,b}}\\
&= I_{n,2,1}+I_{n,2,2}.
\end{aligned}$$

Here, $D\left(\frac{\partial f}{\partial x}(t_{j-1},Z_{t_{j-1}}^H(a,b))\right)$ denotes the Malliavin derivative of the random variable $\frac{\partial f}{\partial x}(t_{j-1},Z_{t_{j-1}}^H(a,b))$. By the chain rule of Malliavin calculus [5.9], this derivative is given by:

$$\begin{aligned}
D_r\left(\frac{\partial f}{\partial x}(t_{j-1},Z_{t_{j-1}}^H(a,b))\right) &= \frac{\partial^2 f}{\partial x^2}(t_{j-1},Z_{t_{j-1}}^H(a,b))D_rZ_{t_{j-1}}^H(a,b)\\
&= \frac{\partial^2 f}{\partial x^2}(t_{j-1},Z_{t_{j-1}}^H(a,b))\mathbf{1}_{[0,t_{j-1}]}(r).
\end{aligned}$$

Now, we calculate the inner product term in $I_{n,2,2}$ using its definition:

$$<\varphi,\psi>_{\mathcal{H}_{a,b}}=\int_0^T\int_0^T\varphi(u)\psi(v)\frac{\partial^2\mathcal{R}^{H,a,b}}{\partial u\partial v}(u,v)dudv:$$

$$\left\langle D\left(\frac{\partial f}{\partial x}(t_{j-1}, Z^H_{t_{j-1}}(a,b))\right), \mathbf{1}_{(t_{j-1},t_j]}\right\rangle_{\mathcal{H}_{a,b}}$$
$$= \int_0^T \int_0^T \left(\frac{\partial^2 f}{\partial x^2}(t_{j-1}, Z^H_{t_{j-1}}(a,b))\mathbf{1}_{[0,t_{j-1}]}(u)\right)\mathbf{1}_{(t_{j-1},t_j]}(v)\frac{\partial^2 \mathcal{R}^{H,a,b}}{\partial u \partial v}(u,v)dudv$$
$$= \frac{\partial^2 f}{\partial x^2}(t_{j-1}, Z^H_{t_{j-1}}(a,b)) \int_0^T \int_0^T \mathbf{1}_{[0,t_{j-1}]}(u)\mathbf{1}_{(t_{j-1},t_j]}(v)\frac{\partial^2 \mathcal{R}^{H,a,b}}{\partial u \partial v}(u,v)dudv$$
$$= \frac{\partial^2 f}{\partial x^2}(t_{j-1}, Z^H_{t_{j-1}}(a,b)) \left\langle \mathbf{1}_{[0,t_{j-1}]}, \mathbf{1}_{(t_{j-1},t_j]}\right\rangle_{\mathcal{H}_{a,b}}$$

By the definition of the inner product for indicator functions:

$$\left\langle \mathbf{1}_{[0,t]}, \mathbf{1}_{[0,s]}\right\rangle_{\mathcal{H}_{a,b}} = \mathcal{R}^{H,a,b}(t,s),$$

we can further simplify the inner product term:

$$\begin{aligned}
\left\langle \mathbf{1}_{[0,t_{j-1}]}, \mathbf{1}_{(t_{j-1},t_j]}\right\rangle_{\mathcal{H}_{a,b}} &= \left\langle \mathbf{1}_{[0,t_{j-1}]}, \mathbf{1}_{[0,t_j]} - \mathbf{1}_{[0,t_{j-1}]}\right\rangle_{\mathcal{H}_{a,b}} \\
&= \left\langle \mathbf{1}_{[0,t_{j-1}]}, \mathbf{1}_{[0,t_j]}\right\rangle_{\mathcal{H}_{a,b}} - \left\langle \mathbf{1}_{[0,t_{j-1}]}, \mathbf{1}_{[0,t_{j-1}]}\right\rangle_{\mathcal{H}_{a,b}} \\
&= \mathcal{R}^{H,a,b}(t_{j-1}, t_j) - \mathcal{R}^{H,a,b}(t_{j-1}, t_{j-1}).
\end{aligned}$$

Therefore, the term $I_{n,2,2}$ becomes:

$$I_{n,2,2} = \sum_{j=1}^n \frac{\partial^2 f}{\partial x^2}(t_{j-1}, Z^H_{t_{j-1}}(a,b)) \left(\mathcal{R}^{H,a,b}(t_{j-1}, t_j) - \mathcal{R}^{H,a,b}(t_{j-1}, t_{j-1})\right).$$

We will prove that $(I_{n,2,2})$ converges, in $L^2(\Omega)$, to:

$$\int_0^t \frac{\partial^2 f}{\partial x^2}(s, Z^H_s(a,b))HC_H(a,b)s^{2H-1}ds.$$

Let $g_s = \frac{\partial^2 f}{\partial x^2}(s, Z^H_s(a,b))$ and $k_s = HC_H(a,b)s^{2H-1}$. Our target integral is $J = \int_0^t g_s k_s ds$.

The term $I_{n,2,2}$ is given by:

$$I_{n,2,2} = \sum_{j=1}^n g_{t_{j-1}} \left(\mathcal{R}^{H,a,b}(t_{j-1}, t_j) - \mathcal{R}^{H,a,b}(t_{j-1}, t_{j-1})\right).$$

To establish the $L^2(\Omega)$ convergence of $I_{n,2,2}$ to J, we employ a two-step approach using the triangle inequality:

$$\|I_{n,2,2} - J\|_{L^2(\Omega)} \leq \|I_{n,2,2} - J_n\|_{L^2(\Omega)} + \|J_n - J\|_{L^2(\Omega)},$$

where J_n is an intermediate approximating sum defined as:

$$J_n = \sum_{j=1}^{n} g_{t_{j-1}} \left(\int_{t_{j-1}}^{t_j} k_s ds \right) = \sum_{j=1}^{n} g_{t_{j-1}} \left(C_H(a,b) \frac{t_j^{2H} - t_{j-1}^{2H}}{2} \right).$$

Let us first show that $\|J_n - J\|_{L^2(\Omega)} \to 0$ as $n \to \infty$.

$$J_n - J = \sum_{j=1}^{n} g_{t_{j-1}} \int_{t_{j-1}}^{t_j} k_s ds - \int_0^t g_s k_s ds = \sum_{j=1}^{n} \int_{t_{j-1}}^{t_j} (g_{t_{j-1}} - g_s) k_s ds.$$

By the boundedness of g_s (from the localization argument) and k_s (integrable on $[0,T]$ since $2H - 1 > -1$), and the continuity of g_s paths (a consequence of $f \in C^{1,2}$ and the continuity of $Z_s^H(a,b)$ paths), we can use the uniform continuity of g_s on compact sets.

Specifically, for any $\epsilon > 0$, there exists $\delta > 0$ such that if $|s_1 - s_2| < \delta$, then $|g_{s_1} - g_{s_2}| < \epsilon$ almost surely, for $s_1, s_2 \in [0,T]$.

Since $\Delta t_j = t/n \to 0$ as $n \to \infty$, for n large enough, $t/n < \delta$. Thus, for $s \in [t_{j-1}, t_j]$:

$$|g_{t_{j-1}} - g_s| \leq \sup_{u \in [0, T - \Delta t_j]} |g_u - g_{u + \Delta t_j}|,$$

which tends to zero almost surely. Using the boundedness of g_s by C_g and k_s by $C_k s^{2H-1}$ (for small s, k_s is not bounded but integrable), we can write:

$$|J_n - J| \leq \sum_{j=1}^{n} \int_{t_{j-1}}^{t_j} |g_{t_{j-1}} - g_s||k_s| ds \leq \left(\sup_{s \in [t_{j-1}, t_j]} |g_{t_{j-1}} - g_s| \right) \sum_{j=1}^{n} \int_{t_{j-1}}^{t_j} |k_s| ds.$$

Since $\sum_{j=1}^{n} \int_{t_{j-1}}^{t_j} |k_s| ds = \int_0^t |k_s| ds < \infty$ and $\sup_{s \in [t_{j-1}, t_j]} |g_{t_{j-1}} - g_s| \to 0$ almost surely (by uniform continuity of g_s), the term $|J_n - J|$ converges to 0 almost surely. Furthermore, because g_s is bounded and k_s is integrable, $(g_{t_{j-1}} - g_s)k_s$ is also bounded in $L^2(\Omega)$, allowing us to apply the dominated convergence theorem to establish convergence in $L^2(\Omega)$. Hence, $\|J_n - J\|_{L^2(\Omega)} \to 0$ as $n \to \infty$.

Now, we will show that $K_n = \|I_{n,2,2} - J_n\|_{L^2(\Omega)} \to 0$ as $n \to \infty$.

$$K_n = \left(\mathbb{E} \left[\sum_{j=1}^{n} g_{t_{j-1}} \left((\mathcal{R}^{H,a,b}(t_{j-1}, t_j) - \mathcal{R}^{H,a,b}(t_{j-1}, t_{j-1})) - C_H(a,b) \frac{t_j^{2H} - t_{j-1}^{2H}}{2} \right) \right]^2 \right)^{1/2}.$$

Using the boundedness of $g_{t_{j-1}} = \frac{\partial^2 f}{\partial x^2}(t_{j-1}, Z^H_{t_{j-1}}(a,b))$ by a constant C (from the localization argument), we apply the Cauchy–Schwarz inequality for sums:

$$K_n^2 \le \mathbb{E}\left[\left(\sum_{j=1}^{n} |g_{t_{j-1}}| \left|(\mathcal{R}^{H,a,b}(t_{j-1},t_j) - \mathcal{R}^{H,a,b}(t_{j-1},t_{j-1}))\right.\right.\right.$$
$$\left.\left.\left. - \left(C_H(a,b)\frac{t_j^{2H} - t_{j-1}^{2H}}{2}\right)\right|\right)^2\right]$$
$$\le C^2 \left(\sum_{j=1}^{n} \left|(\mathcal{R}^{H,a,b}(t_{j-1},t_j) - \mathcal{R}^{H,a,b}(t_{j-1},t_{j-1}))\right.\right.$$
$$\left.\left. -C_H(a,b)\frac{t_j^{2H} - t_{j-1}^{2H}}{2}\right|\right)^2.$$

Let $\Delta t_j = t_j - t_{j-1} = t/n$. We denote the term inside the absolute value as $\mathcal{E}_j$:

$$\mathcal{E}_j = \left|(R^{H,a,b}(t_{j-1},t_j) - R^{H,a,b}(t_{j-1},t_{j-1})) - C_H(a,b)\frac{t_j^{2H} - t_{j-1}^{2H}}{2}\right|.$$

We substitute the definition of:

$$R^{H,a,b}(s,t) = \frac{1}{2}(a^2+b^2)(s^{2H}+t^{2H}) - \frac{ab}{2}((s+t)^{2H} - |s-t|^{2H})$$

and recall $C_H(a,b) = (a^2+b^2) - ab\cdot 2^{2H-1}$. Expanding $\mathcal{E}_j$:

$$\mathcal{E}_j = \left|\left[\frac{1}{2}(a^2+b^2)(t_{j-1}^{2H}+t_j^{2H}) - \frac{ab}{2}((t_{j-1}+t_j)^{2H} - (\Delta t_j)^{2H})\right]\right.$$
$$\left. - \left[\frac{1}{2}(a^2+b^2)(2t_{j-1}^{2H}) - \frac{ab}{2}(2t_{j-1})^{2H}\right] - C_H(a,b)\frac{t_j^{2H} - t_{j-1}^{2H}}{2}\right|$$
$$= \left|\frac{1}{2}(a^2+b^2)(t_j^{2H} - t_{j-1}^{2H}) - \frac{ab}{2}\left[(t_{j-1}+t_j)^{2H} - (\Delta t_j)^{2H} - (2t_{j-1})^{2H}\right]\right.$$
$$\left. - \left(\frac{1}{2}(a^2+b^2)(t_j^{2H} - t_{j-1}^{2H}) - ab\cdot 2^{2H-1}\frac{t_j^{2H} - t_{j-1}^{2H}}{2}\right)\right|$$
$$= \left|-\frac{ab}{2}\left[(t_{j-1}+t_j)^{2H} - (\Delta t_j)^{2H} - (2t_{j-1})^{2H}\right] + ab\cdot 2^{2H-2}(t_j^{2H} - t_{j-1}^{2H})\right|.$$

Let $t_{j-1} = (j-1)\frac{t}{n}$ and $\Delta t_j = \frac{t}{n}$. Then $t_j = j\frac{t}{n}$.

The expression inside the absolute value can be written with a common factor of $(\Delta t_j)^{2H} = (t/n)^{2H}$:

$$\left(\frac{t}{n}\right)^{2H}\left[-\frac{ab}{2}\left((2j-1)^{2H}-1-(2j-2)^{2H}\right)+ab\cdot 2^{2H-2}(j^{2H}-(j-1)^{2H})\right]$$

$$=\frac{ab}{2}\left(\frac{t}{n}\right)^{2H}\left|-(2j-1)^{2H}+1+(2j-2)^{2H}+2^{2H-1}(j^{2H}-(j-1)^{2H})\right|.$$

Let us call the term inside the absolute value $F(j)$:

$$F(j)=1+(2j-2)^{2H}-(2j-1)^{2H}+2^{2H-1}(j^{2H}-(j-1)^{2H}).$$

By Lemma A.6, we know that $|F(j)|\leq 2^{2H-1}$.

Substituting this back into the expression for $\mathcal{E}_j$:

$$\mathcal{E}_j\leq\frac{ab}{2}\left(\frac{t}{n}\right)^{2H}|F(j)|\leq\frac{ab}{2}\left(\frac{t}{n}\right)^{2H}2^{2H-1}.$$

Let $C_1=\frac{ab}{2}2^{2H-1}$ (which is a positive constant dependent only on a, b, H). So, we have the precise bound:

$$\mathcal{E}_j\leq C_1\left(\frac{t}{n}\right)^{2H}.$$

Now, substitute this bound into the expression for K_n:

$$\begin{aligned}K_n&\leq C\sum_{j=1}^{n}\mathcal{E}_j\\&\leq C\sum_{j=1}^{n}C_1\left(\frac{t}{n}\right)^{2H}\\&=CC_1n\left(\frac{t}{n}\right)^{2H}\\&=C_2t^{2H}n^{1-2H},\end{aligned}$$

where $C_2=CC_1$ is a constant. Since $H>1/2$, the exponent $1-2H<0$. This implies that $n^{1-2H}\to 0$ as $n\to\infty$. Therefore, $K_n\to 0$ as $n\to\infty$.

This demonstrates that $(I_{n,2,2})$ converges to:

$$\int_0^t HC_H(a,b)\frac{\partial^2 f}{\partial x^2}(s,Z_s^H(a,b))s^{2H-1}ds$$

in $L^2(\Omega)$.

Convergence of $I_{n,2,1} = \delta\left(\sum_{j=1}^{n} \frac{\partial f}{\partial x}(t_{j-1}, Z^H_{t_{j-1}}(a,b))\left(\mathbf{1}_{(t_{j-1},t_j]}\right)\right)$.

We will now show that $(I_{n,2,1})$ converges to

$$\delta\left(\frac{\partial f}{\partial x}(., Z^H_{.}(a,b))\mathbf{1}_{(0,t)}(.)\right) = \int_0^t \frac{\partial f}{\partial x}(s, Z^H_s(a,b))\delta Z^H_s(a,b)$$

in $L^2(\Omega)$. For this, we will first show that the sequence of functions

$$G_n(u) = \sum_{j=1}^{n} \frac{\partial f}{\partial x}(t_{j-1}, Z^H_{t_{j-1}}(a,b))\mathbf{1}_{(t_{j-1},t_j]}(u)$$

converges to $G(u) = \frac{\partial f}{\partial x}(u, Z^H_u(a,b))\mathbf{1}_{(0,t)}(u)$ in $L^2(\Omega;\mathcal{H}_{a,b})$. We consider the expected squared norm of the difference:

$$\begin{aligned}
&\mathbb{E}\left[\left\|\sum_{j=1}^{n}\left[\frac{\partial f}{\partial x}(t_{j-1}, Z^H_{t_{j-1}}(a,b)) - \frac{\partial f}{\partial x}(u, Z^H_u(a,b))\right]\mathbf{1}_{(t_{j-1},t_j]}(u)\right\|^2_{\mathcal{H}_{a,b}}\right] \\
&= \mathbb{E}\left[\sum_{i,j=1}^{n}\int_{t_{i-1}}^{t_i}\int_{t_{j-1}}^{t_j}\left(\frac{\partial f}{\partial x}(t_{i-1}, Z^H_{t_{i-1}}(a,b)) - \frac{\partial f}{\partial x}(u, Z^H_u(a,b))\right)\right. \\
&\quad \left.\times\left(\frac{\partial f}{\partial x}(t_{j-1}, Z^H_{t_{j-1}}(a,b)) - \frac{\partial f}{\partial x}(v, Z^H_v(a,b))\right)\frac{\partial^2 R^{H,a,b}}{\partial u\partial v}(u,v)dudv\right]. \\
&= \sum_{i,j=1}^{n}\int_{t_{i-1}}^{t_i}\int_{t_{j-1}}^{t_j}\mathbb{E}\left[\left(\frac{\partial f}{\partial x}(t_{i-1}, Z^H_{t_{i-1}}(a,b)) - \frac{\partial f}{\partial x}(u, Z^H_u(a,b))\right)\right. \\
&\quad \left.\times\left(\frac{\partial f}{\partial x}(t_{j-1}, Z^H_{t_{j-1}}(a,b)) - \frac{\partial f}{\partial x}(v, Z^H_v(a,b))\right)\right]\frac{\partial^2 R^{H,a,b}}{\partial u\partial v}(u,v)dudv.
\end{aligned}$$

Applying Cauchy–Schwarz Inequality to the expectation of the product, we get:

$$\begin{aligned}
&\mathbb{E}\left[\left|\frac{\partial f}{\partial x}(t_{i-1}, Z^H_{t_{i-1}}(a,b)) - \frac{\partial f}{\partial x}(u, Z^H_u(a,b))\right|\right. \\
&\quad \left.\times\left|\frac{\partial f}{\partial x}(t_{j-1}, Z^H_{t_{j-1}}(a,b)) - \frac{\partial f}{\partial x}(v, Z^H_v(a,b))\right|\right] \\
&\leq \sqrt{\mathbb{E}\left[\left|\frac{\partial f}{\partial x}(t_{i-1}, Z^H_{t_{i-1}}(a,b)) - \frac{\partial f}{\partial x}(u, Z^H_u(a,b))\right|^2\right]} \\
&\quad \times\sqrt{\mathbb{E}\left[\left|\frac{\partial f}{\partial x}(t_{j-1}, Z^H_{t_{j-1}}(a,b)) - \frac{\partial f}{\partial x}(v, Z^H_v(a,b))\right|^2\right]}.
\end{aligned}$$

By the mean value theorem, for $\phi(t,x) = \frac{\partial f}{\partial x}(t,x)$:

$$\phi(t_{i-1}, Z^H_{t_{i-1}}(a,b)) - \phi(u, Z^H_u(a,b))$$

$$= \frac{\partial^2 f}{\partial x^2}(Y_{i-1})(Z^H_{t_{i-1}}(a,b) - Z^H_u(a,b)) + \frac{\partial^2 f}{\partial s \partial x}(Y_{i-1})(t_{i-1} - u),$$

where

$$Y_{i-1} = (1-\alpha)(t_{i-1}, Z^H_{t_{i-1}}(a,b)) + \alpha(u, Z^H_u(a,b)),$$

with α being a random variable in $(0,1)$. Since $\frac{\partial^2 f}{\partial x^2}$ and $\frac{\partial^2 f}{\partial s \partial x}$ are bounded by C (due to localization), and using $(A+B)^2 \leq 2(A^2+B^2)$:

$$\mathbb{E}\left[\left|\frac{\partial f}{\partial x}(t_{i-1}, Z^H_{t_{i-1}}(a,b)) - \frac{\partial f}{\partial x}(u, Z^H_u(a,b))\right|^2\right]$$

$$\leq 2C^2\left(\mathbb{E}\left[(Z^H_{t_{i-1}}(a,b) - Z^H_u(a,b))^2\right] + (t_{i-1}-u)^2\right)$$

$$\leq 2C^2\left(C_H(a,b)|t_{i-1}-u|^{2H} + (t_{i-1}-u)^2\right)$$

$$\leq 2C^2(C_H(a,b) + T^{2-2H})|t_{i-1}-u|^{2H}$$

for every $u \in (t_{i-1}, t_i]$, and $i \in \{1, ..., n\}$. The last inequality holds because:

$$t_{i-1} - u \leq t/n \leq T,$$

and for $H \in (1/2, 1)$, $2H < 2$, so $|t_{i-1}-u|^2 \leq T^{2-2H}|t_{i-1}-u|^{2H}$.

Therefore, let $K' = 2C^2(C_H(a,b) + T^{2-2H})$.

$$\sqrt{\mathbb{E}\left[\left|\frac{\partial f}{\partial x}(t_{i-1}, Z^H_{t_{i-1}}(a,b)) - \frac{\partial f}{\partial x}(u, Z^H_u(a,b))\right|^2\right]}$$

$$\times\sqrt{\mathbb{E}\left[\left|\frac{\partial f}{\partial x}(t_{j-1}, Z^H_{t_{j-1}}(a,b)) - \frac{\partial f}{\partial x}(v, Z^H_v(a,b))\right|^2\right]}$$

$$\leq K' \, |t_{i-1}-u|^H |t_{j-1}-v|^H$$

$$\leq K' \left(\frac{t}{n}\right)^{2H}.$$

Thus,

$$\mathbb{E}\left[\left\|\sum_{j=1}^{n}\left[\frac{\partial f}{\partial x}(t_{j-1},Z_{t_{j-1}}^{H}(a,b))-\frac{\partial f}{\partial x}(.,Z_{.}^{H}(a,b))\right]\mathbf{1}_{(t_{j-1},t_j]}(.)\right\|_{\mathcal{H}_{a,b}}^{2}\right]$$

$$\leq K'\left(\frac{t}{n}\right)^{2H}\sum_{i,j=1}^{n}\int_{t_{i-1}}^{t_i}\int_{t_{j-1}}^{t_j}\left|\frac{\partial^2 R^{H,a,b}}{\partial u\partial v}(u,v)\right|dudv$$

$$\leq K'\left(\frac{t}{n}\right)^{2H}\int_0^t\int_0^t\left|\frac{\partial^2 R^{H,a,b}}{\partial u\partial v}(u,v)\right|dudv$$

$$= K'\left(\frac{t}{n}\right)^{2H}\left((a+b)^2-2^{2H}ab\right)t^{2H}.$$

As $n\to\infty$, $(t/n)^{2H}\to 0$ because $2H>0$. Consequently,

$$\lim_{n\to\infty}\mathbb{E}\left[\left\|\sum_{j=1}^{n}\left[\frac{\partial f}{\partial x}(t_{j-1},Z_{t_{j-1}}^{H}(a,b))-\frac{\partial f}{\partial x}(.,Z_{.}^{H}(a,b))\right]\mathbf{1}_{(t_{j-1},t_j]}(.)\right\|_{\mathcal{H}_{a,b}}^{2}\right]=0.$$

By the continuity of the Skorokhod integral (as an operator from $L^2(\Omega;\mathcal{H}_{a,b})$ to $L^2(\Omega)$), this implies that $I_{n,2,1}$ converges in $L^2(\Omega)$ to:

$$\delta\left(\frac{\partial f}{\partial x}(.,Z_{.}^{H}(a,b))\mathbf{1}_{(0,t)}(.)\right)=\int_0^t\frac{\partial f}{\partial x}(s,Z_s^H(a,b))\delta Z_s^H(a,b).$$

Convergence of $I_{n,1}$: $I_{n,1}=\sum_{j=1}^{n}(t_j-t_{j-1})\frac{\partial f}{\partial s}(t_{j-1},Z_{t_{j-1}}^{H}(a,b))$. This is a Riemann sum for the integral $\int_0^t\frac{\partial f}{\partial s}(s,Z_s^H(a,b))ds$. Given that $\frac{\partial f}{\partial s}(s,Z_s^H(a,b))$ is a continuous process (by continuity of f and Z^H) and bounded (by the localization argument), its Riemann sums converge to the integral almost surely.

$$\lim_{n\to\infty}I_{n,1}=\int_0^t\frac{\partial f}{\partial s}(s,Z_s^H(a,b))ds\ a.s.$$

Combining all convergent terms, we arrive at the Itô-like formula:

$$f(t,Z_t^H(a,b))=f(0,0)+\int_0^t\frac{\partial f}{\partial x}(s,Z_s^H(a,b))\delta(Z_s^H(a,b))$$

$$+\int_0^t\left(\frac{\partial f}{\partial s}(s,Z_s^H(a,b))+HC_H(a,b)\frac{\partial^2 f}{\partial x^2}(s,Z_s^H(a,b))s^{2H-1}\right)ds.$$

This completes the proof. □

COROLLARY 5.2.– *Let f be a function of class $C^2(\mathbb{R})$ such that:*

$$\max\{|f(x)|, |f'(x)|, |f''(x)|\} \leq c\exp(\lambda x^2), \qquad [5.24]$$

for every $x \in \mathbb{R}$, where c and λ are positive constants satisfying condition [5.22], that is:

$$\lambda < \frac{T^{-2H}}{4C_H(a,b)}.$$

Let $Z_t^H(a,b)$ be a Gaussian process with $Z_0^H(a,b) = 0$. Then, for any $t \in [0,T]$, the following Itô-like formula holds:

$$f(Z_t^H(a,b)) = f(0) + \int_0^t f'(Z_s^H(a,b))\delta(Z_s^H(a,b)) + \int_0^t HC_H(a,b)f''(Z_s^H(a,b))s^{2H-1}\,ds. \qquad [5.25]$$

Proof. The proof of this corollary relies on applying Theorem 5.2 to a function that does not explicitly depend on time.

Let us define a function $g : [0,T] \times \mathbb{R} \to \mathbb{R}$ such that $g(s,x) = f(x)$. Since f is assumed to be of class $C^2(\mathbb{R})$, it follows that g is of class $C^{1,2}([0,T] \times \mathbb{R}, \mathbb{R})$.

We now compute the necessary partial derivatives of $g(s,x)$ with respect to s and x: the first partial derivative of g with respect to time s:

$$\frac{\partial g}{\partial s}(s,x) = \frac{\partial}{\partial s}(f(x)) = 0.$$

The first partial derivative of g with respect to space x:

$$\frac{\partial g}{\partial x}(s,x) = f'(x).$$

The second partial derivative of g with respect to space x:

$$\frac{\partial^2 g}{\partial x^2}(s,x) = f''(x).$$

The growth condition for $f(x)$ stated in equation [5.24] is:

$$\max\{|f(x)|, |f'(x)|, |f''(x)|\} \leq c\exp(\lambda x^2).$$

Since $\frac{\partial g}{\partial s} = 0$, the supremum over t is trivial, and the bounds for $g, \frac{\partial g}{\partial x}, \frac{\partial^2 g}{\partial x^2}$ translate directly to f, f', f''. Therefore, all conditions for applying Theorem 5.2 are satisfied.

Now, we substitute these derivatives into the Itô formula provided by Theorem 5.2:

$$g(t, Z_t^H(a,b)) = g(0,0) + \int_0^t \frac{\partial g}{\partial x}(s, Z_s^H(a,b))\delta(Z_s^H(a,b))$$
$$+ \int_0^t \left(\frac{\partial g}{\partial s}(s, Z_s^H(a,b)) + HC_H(a,b)\frac{\partial^2 g}{\partial x^2}(s, Z_s^H(a,b))s^{2H-1} \right) ds.$$

Substituting the expressions for g and its derivatives in terms of f and its derivatives:

$$f(Z_t^H(a,b)) = f(0) + \int_0^t f'(Z_s^H(a,b))\delta(Z_s^H(a,b))$$
$$+ \int_0^t \left(0 + HC_H(a,b)f''(Z_s^H(a,b))s^{2H-1}\right) ds$$
$$= f(0) + \int_0^t f'(Z_s^H(a,b))\delta(Z_s^H(a,b))$$
$$+ \int_0^t HC_H(a,b)f''(Z_s^H(a,b))s^{2H-1}\, ds.$$

This final expression is exactly the Itô formula for functions that do not explicitly depend on time, as stated in Corollary 5.2. This completes the proof. □

REMARK 5.3.– *Applying the Itô formula [5.25] from Corollary 5.2 with specific choices of a and b, we can retrieve previously known Itô formulas:*

– When $a = b = \frac{1}{\sqrt{2}}$, the ZgfBm $Z_t^H(a,b)$ specializes to the sfBm, denoted by S_t^H. In this case, the constant $C_H(a,b)$ (defined in equation [5.5]) becomes:

$$C_H\left(\frac{1}{\sqrt{2}}, \frac{1}{\sqrt{2}}\right) = \left(\frac{1}{2} + \frac{1}{2}\right) - \frac{1}{2}2^{2H-1} = 1 - 2^{2H-2}.$$

Substituting this into formula [5.25], the Itô formula for sfBm is obtained as:

$$f(S_t^H) = f(0) + \int_0^t f'(S_s^H)\delta Z(S_s^H) + \int_0^t H(1 - 2^{2H-2})f''(S_s^H)s^{2H-1}\, ds. \quad [5.26]$$

This formula can be found in Yan et al. (2011).

– When $a = 1$ and $b = 0$, the ZgfBm $Z_t^H(a,b)$ becomes the standard fBm, denoted by B_t^H, with Hurst parameter H. In this case, the constant $C_H(a,b)$ simplifies to:

$$C_H(1,0) = (1^2 + 0^2) - (1)(0)2^{2H-1} = 1.$$

Substituting $C_H(1,0) = 1$ into formula [5.25], the Itô formula with fBm is retrieved as:

$$f(B_t^H) = f(0) + \int_0^t f'(B_s^H)\delta Z(B_s^H) + \int_0^t Hf''(B_s^H)s^{2H-1}\, ds. \quad [5.27]$$

This formula can be found, for example, in Aloès and Nualart (2003).

This demonstrates the generality of the derived Itô formula for ZgfBm, encompassing known results for its special cases.

COROLLARY 5.3.– *Let f be a function of class $C^2(\mathbb{R})$ such that:*

$$\max\{|f(x)|, |f'(x)|, |f''(x)|\} \leq c\exp(\lambda x^2), \qquad [5.28]$$

for every $x \in \mathbb{R}$, where c and λ are positive constants satisfying condition [5.22], that is:

$$\lambda < \frac{T^{-2H}}{4C_H(a,b)}.$$

Assume that the process $f'(Z_s^H(a,b))$ satisfies condition [5.11] from Theorem 5.1. Then, for any $t \in [0,T]$, the following Itô-like formula holds in terms of the symmetric integral:

$$\begin{aligned} f(Z_t^H(a,b)) &= f(0) + \int_0^t f'(Z_s^H(a,b))dZ_s^H(a,b) \\ &- \int_0^t \int_0^s \alpha_H D_r f'(Z_s^H(a,b)) \left[(a^2+b^2)|s-r|^{2H-2}\right. \\ &\left.-2ab(s+r)^{2H-2}\right] drds \\ &+ \int_0^t HC_H(a,b) f''(Z_s^H(a,b)) s^{2H-1}\, ds. \end{aligned} \qquad [5.29]$$

Here, $\int_0^t \cdot dZ_s^H(a,b)$ denotes the symmetric integral.

Proof. The proof starts from the Itô formula expressed in terms of the Skorokhod integral, as given by Corollary 5.2:

$$\begin{aligned} f(Z_t^H(a,b)) &= f(0) + \int_0^t f'(Z_s^H(a,b))\delta(Z_s^H(a,b)) \\ &+ \int_0^t HC_H(a,b) f''(Z_s^H(a,b)) s^{2H-1}\, ds. \end{aligned} \qquad [5.30]$$

The objective is to reexpress the Skorokhod integral term, $\int_0^t f'(Z_s^H(a,b))\delta(Z_s^H(a,b))$, using the symmetric integral.

To achieve this, we employ Theorem 5.1, which provides the relationship between the symmetric integral and the divergence (Skorokhod) integral. Let

$u_s = f'(Z_s^H(a,b))$. Theorem 5.1 states that if u_s belongs to the space $D^{1,2}(|\mathcal{H}|)$ and satisfies condition [5.11], then:

$$\int_0^t u_s dZ_s^H(a,b) = \int_0^t u_s \delta Z_s^H(a,b) + \alpha_H \int_0^t \int_0^s D_r u_s \left[(a^2+b^2)|s-r|^{2H-2} - 2ab(s+r)^{2H-2}\right] drds. \quad [5.31]$$

Here, the double integral accounts for the difference between the symmetric and divergence integrals. The integration limits for $D_r u_s$ are $r \in [0,s]$ because $u_s = f'(Z_s^H(a,b))$ is typically adapted to the filtration generated by Z_s^H, so $D_r u_s = 0$ for $r > s$.

Rearranging equation [5.31] to isolate the Skorokhod integral term:

$$\int_0^t u_s \delta Z_s^H(a,b) = \int_0^t u_s dZ_s^H(a,b) - \alpha_H \int_0^t \int_0^s D_r u_s \left[(a^2+b^2)|s-r|^{2H-2} - 2ab(s+r)^{2H-2}\right] drds. \quad [5.32]$$

Now, we substitute this expression for $\int_0^t f'(Z_s^H(a,b))\delta(Z_s^H(a,b))$ (with $u_s = f'(Z_s^H(a,b))$) back into equation [5.30]:

$$f(Z_t^H(a,b)) = f(0) + \left(\int_0^t f'(Z_s^H(a,b))dZ_s^H(a,b) - \int_0^t \int_0^s \alpha_H D_r f'(Z_s^H(a,b)) \left[(a^2+b^2)|s-r|^{2H-2} - 2ab(s+r)^{2H-2}\right] drds\right) + \int_0^t HC_H(a,b) f''(Z_s^H(a,b)) s^{2H-1}\, ds.$$

This final expression matches precisely the formula presented in Corollary 5.3. □

5.5. Examples of application

We will now provide some illustrative examples of application of the established Itô formulas.

5.5.1. *Example 1*

Consider the function $f(x) = \frac{x^2}{2}$, for every $x \in \mathbb{R}$. The function and its first two derivatives are:

$$f(x) = \frac{x^2}{2}, \quad f'(x) = x, \quad f''(x) = 1.$$

Thus, we have:

$$\max\{|f(x)|, |f'(x)|, |f''(x)|\} = \max\left\{\frac{x^2}{2}, |x|, 1\right\}.$$

To verify condition [5.24], we need to show that this expression is bounded by $ce^{\lambda x^2}$ for some positive constants c and λ satisfying condition [5.22]. For any $\lambda > 0$, we can choose c sufficiently large. For instance, for $\lambda > 0$, we know that:

$$\frac{x^2}{2} \leq \frac{1}{4\lambda}e^{\lambda x^2}.$$

Similarly, $|x| \leq \frac{1}{\sqrt{2\lambda}}e^{\lambda x^2/2} \leq \frac{1}{\sqrt{2\lambda}}e^{\lambda x^2}$. And $1 \leq e^{\lambda x^2}$. Therefore, condition [5.24] is satisfied for any $\lambda > 0$ with a suitable c (e.g. $c = \max\left(\frac{1}{4\lambda}, \frac{1}{\sqrt{2\lambda}}, 1\right)$).

Given that condition [5.22] only requires λ to be sufficiently small, this choice of $f(x)$ is indeed valid. Applying Corollary 5.2, we have:

$$f(Z_t^H(a,b)) = f(0) + \int_0^t f'(Z_s^H(a,b))\delta(Z_s^H(a,b)) + \int_0^t HC_H(a,b)f''(Z_s^H(a,b))s^{2H-1}\,ds$$

$$\frac{1}{2}(Z_t^H(a,b))^2 = \frac{1}{2}(0)^2 + \int_0^t Z_s^H(a,b)\delta(Z_s^H(a,b)) + \int_0^t HC_H(a,b)(1)s^{2H-1}\,ds$$

$$\frac{1}{2}(Z_t^H(a,b))^2 = \int_0^t Z_s^H(a,b)\delta(Z_s^H(a,b)) + \frac{1}{2}C_H(a,b)t^{2H}.$$

Rearranging the terms to solve for the integral, we get:

$$\int_0^t Z_s^H(a,b)\delta(Z_s^H(a,b)) = \frac{1}{2}\left(Z_t^H(a,b)\right)^2 - \frac{1}{2}C_H(a,b)t^{2H}.$$

This result provides the Skorokhod integral of the process $Z_s^H(a,b)$ with respect to itself.

5.5.2. *Example 2*

Consider the function $f(t,x) = tx$, for every $t \in [0,T]$ and $x \in \mathbb{R}$. The partial derivatives of $f(t,x)$ are:

$$\frac{\partial f}{\partial t}(t,x) = x, \quad \frac{\partial f}{\partial x}(t,x) = t, \quad \frac{\partial^2 f}{\partial x^2}(t,x) = 0.$$

For the growth condition, we examine:

$$\sup_{t\in[0,T]} \max\left\{|f(t,x)|, \left|\frac{\partial f(t,x)}{\partial x}\right|, \left|\frac{\partial^2 f(t,x)}{\partial x^2}\right|\right\}$$

$$= \sup_{t\in[0,T]} \max\{|tx|, |t|, 0\} = T\max\{|x|, 1\}.$$

This expression is bounded by $ce^{\lambda x^2}$ for any $\lambda > 0$. For instance, choose $c = T\max(1, e^{1/(4\lambda)})$, which follows from the inequality $|x| \leq \sqrt{1/(4\lambda)}e^{\lambda x^2}$ (a standard result from optimizing $xe^{-\lambda x^2}$) and $1 \leq e^{\lambda x^2}$. Therefore, condition [5.21] for Theorem 5.2 is satisfied.

Applying Theorem 5.2, which is the general Itô formula for time-dependent functions with the divergence integral:

$$f(t, Z_t^H(a,b)) = f(0, Z_0^H(a,b)) + \int_0^t \frac{\partial f}{\partial x}(s, Z_s^H(a,b))\delta(Z_s^H(a,b))$$

$$+ \int_0^t \left(\frac{\partial f}{\partial s}(s, Z_s^H(a,b)) + HC_H(a,b)\frac{\partial^2 f}{\partial x^2}(s, Z_s^H(a,b))s^{2H-1}\right) ds.$$

Since $Z_0^H(a,b) = 0$, $f(0, Z_0^H(a,b)) = f(0,0) = 0\cdot 0 = 0$. Substituting the partial derivatives into the formula:

$$tZ_t^H(a,b) - 0 = \int_0^t s\delta(Z_s^H(a,b)) + \int_0^t (Z_s^H(a,b) + HC_H(a,b)(0)s^{2H-1})ds.$$

This simplifies to:

$$tZ_t^H(a,b) = \int_0^t s\delta(Z_s^H(a,b)) + \int_0^t Z_s^H(a,b)ds.$$

Rearranging the terms to express the Skorokhod integral:

$$\int_0^t s\delta(Z_s^H(a,b)) = tZ_t^H(a,b) - \int_0^t Z_s^H(a,b)ds.$$

This formula provides an integration by parts-like identity for the Skorokhod integral of s with respect to $Z_s^H(a,b)$.

5.5.3. *Example 3: generalized fractional Black–Scholes equation*

We introduce a generalized version of the fractional Black–Scholes process (ZgfBSp), defined by:

$$S_t = S_0 \exp\left(\int_0^t \left(\nu(s) - \frac{1}{2}\sigma^2 HC_H(a,b)s^{2H-1}\right) ds + \sigma Z_t^H(a,b)\right).$$

Here, $S_0 > 0$ is the initial asset price, $\nu(t)$ is a deterministic function representing the instantaneous drift and $\sigma > 0$ is a constant volatility. The constant $C_H(a,b)$ is defined in equation [5.5].

To derive the stochastic differential equation for S_t, we set $S_t = S_0 f(t, Z_t^H(a,b))$, where $f(t,x)$ is defined as:

$$f(t,x) = \exp\left(\int_0^t \left(\nu(u) - \frac{1}{2}\sigma^2 H C_H(a,b) u^{2H-1}\right) du + \sigma x\right).$$

The partial derivatives of $f(t,x)$ are:

$$\frac{\partial f}{\partial t}(t,x) = f(t,x)\left(\nu(t) - \frac{1}{2}\sigma^2 H C_H(a,b) t^{2H-1}\right)$$

$$\frac{\partial f}{\partial x}(t,x) = \sigma f(t,x)$$

$$\frac{\partial^2 f}{\partial x^2}(t,x) = \sigma^2 f(t,x)$$

To apply Theorem 5.2, we must verify condition [5.21], which requires that

$$\sup_{t\in[0,T]} \max\left\{|f(t,x)|, \left|\frac{\partial f(t,x)}{\partial x}\right|, \left|\frac{\partial^2 f(t,x)}{\partial x^2}\right|\right\}$$

is bounded by $ce^{\lambda x^2}$ for suitable positive constants c and λ.

Since $f(t,x)$ is always positive and $\sigma > 0$, the maximum of the absolute values of f and its derivatives with respect to x is given by $\sigma^2 f(t,x)$. Let

$$K_u = \nu(u) - \frac{1}{2}\sigma^2 H C_H(a,b) u^{2H-1}.$$

Then $f(t,x) = \exp\left(\int_0^t K_u du + \sigma x\right)$. We can bound $|f(t,x)|$ as follows:

$$\begin{aligned} |f(t,x)| &\leq \exp\left(\int_0^T |K_u| du + \sigma|x|\right) \\ &= \exp\left(\int_0^T \left|\nu(u) - \frac{1}{2}\sigma^2 H C_H(a,b) u^{2H-1}\right| du\right) e^{\sigma|x|}. \end{aligned}$$

Let $M_T = \exp\left(\int_0^T \left|\nu(u) - \frac{1}{2}\sigma^2 H C_H(a,b) u^{2H-1}\right| du\right)$. We assume $\nu(u)$ is such that M_T is finite. Then,

$$\sup_{t\in[0,T]} \max\left\{|f(t,x)|, \left|\frac{\partial f(t,x)}{\partial x}\right|, \left|\frac{\partial^2 f(t,x)}{\partial x^2}\right|\right\} \leq \sigma^2 M_T e^{\sigma|x|}.$$

As shown in previous examples, the exponential function $e^{\sigma|x|}$ can be bounded by $c' e^{\lambda x^2}$ for any $\lambda > 0$ with a suitable constant c'. Specifically, for every $\lambda > 0$, we can choose $c = \sigma^2 M_T \max(1, e^{\sigma^2/(4\lambda)})$ such that $c' e^{\sigma|x|} \leq c e^{\lambda x^2}$. Therefore, condition [5.21] is satisfied.

Now, we apply Theorem 5.2 to $f(t, Z_t^H(a,b))$ in differential form:

$$df(t, Z_t^H(a,b)) = \frac{\partial f}{\partial x}(t, Z_t^H(a,b))\delta(Z_t^H(a,b))$$
$$+ \left(\frac{\partial f}{\partial t}(t, Z_t^H(a,b)) + HC_H(a,b)\frac{\partial^2 f}{\partial x^2}(t, Z_t^H(a,b))t^{2H-1}\right) dt.$$

Substitute the partial derivatives we calculated:

$$df(t, Z_t^H(a,b)) = \sigma f(t, Z_t^H(a,b))\delta(Z_t^H(a,b))$$
$$+ \left(f(t, Z_t^H(a,b))\left(\nu(t) - \frac{1}{2}\sigma^2 HC_H(a,b)t^{2H-1}\right)\right.$$
$$\left. + HC_H(a,b)\sigma^2 f(t, Z_t^H(a,b))t^{2H-1}\right) dt.$$

Factor out $f(t, Z_t^H(a,b))$ from the dt term:

$$df(t, Z_t^H(a,b)) = \sigma f(t, Z_t^H(a,b))\delta(Z_t^H(a,b))$$
$$+ f(t, Z_t^H(a,b))\left(\nu(t) + \frac{1}{2}\sigma^2 HC_H(a,b)t^{2H-1}\right) dt.$$

Now, recall that $S_t = S_0 f(t, Z_t^H(a,b))$. Therefore, $f(t, Z_t^H(a,b)) = S_t/S_0$. Substituting this back and multiplying by S_0:

$$\delta S_t = \sigma S_t \delta Z_t^H(a,b) + S_t\left(\nu(t) + \frac{1}{2}\sigma^2 HC_H(a,b)t^{2H-1}\right) dt.$$

REMARK 5.4.– *The generalized framework presented here offers a consistent and rigorous extension of fractional Black–Scholes models. In the particular case where $a = 1$ and $b = 0$, the ZgfBm $Z_t^H(1,0)$ precisely reduces to the standard fBm B_t^H and the constant $C_H(1,0)$ becomes unity. Substituting these specific parameters into the derived SDE, we obtain:*

$$\delta S_t = \sigma S_t \delta B_t^H + S_t\left(\nu(t) + \frac{1}{2}\sigma^2 H t^{2H-1}\right) dt.$$

This form, which includes the term $\frac{1}{2}\sigma^2 H t^{2H-1}$ in the drift, represents the mathematically correct and complete stochastic differential equation for a log-normal asset price driven by fBm, derived using a generalized Itô-type calculus *(specifically, Theorem 5.2). The presence of this "fractional Itô correction" term is crucial. It explicitly accounts for the non-semimartingale property of fBm when* $H \neq 1/2$. *This result highlights the ability of our generalized framework to accurately capture and extend the dynamics of classical fractional models under a robust stochastic calculus.*

5.6. Exercises

EXERCISE 5.1.– *Let* $H \in (1/2, 1)$. *This exercise guides you through a rigorous proof that the first distributional derivative of the function* $f(x) = |x|^{2H}$ *is given by:*

$$f'(x) = 2H|x|^{2H-1}sgn(x).$$

Recall that the first distributional derivative of a locally integrable function T *(viewed as a distribution) is defined for any test function* $\psi \in C_c^\infty(\mathbb{R})$ *by:*

$$\langle T', \psi \rangle = - \langle T, \psi' \rangle = - \int_{\mathbb{R}} T(x)\psi'(x)\, dx.$$

1) Start with the definition for $\langle f', \psi \rangle$:

$$\langle f', \psi \rangle = - \int_{\mathbb{R}} |x|^{2H} \psi'(x)\, dx.$$

Since $f(x) = |x|^{2H}$ *is not classically differentiable at* $x = 0$, *we must split the integral around this point to apply classical integration by parts rigorously. Let the support of* ψ *be contained within* $[-M, M]$ *for some* $M > 0$. *Show that this integral can be written as:*

$$- \int_{\mathbb{R}} |x|^{2H} \psi'(x)\, dx = - \lim_{\epsilon \to 0^+} \left(\int_{-M}^{-\epsilon} |x|^{2H} \psi'(x)\, dx + \int_{\epsilon}^{M} |x|^{2H} \psi'(x)\, dx \right).$$

2) Apply classical integration by parts to each of the two integrals over the intervals $[-M, -\epsilon]$ *and* $[\epsilon, M]$. *Show that all resulting boundary terms (at* $\pm M$ *and* $\pm\epsilon$*) sum to zero as* $\epsilon \to 0^+$.

3) Conclude that $f'(x) = 2H|x|^{2H-1}sgn(x)$ *in the sense of distributions (note that* $f'(x)$ *is a continuous and locally integrable function, despite not being classically differentiable at* $x = 0$*).*

EXERCISE 5.2.– *Let* $H \in (1/2, 1)$. *Building on the result from Exercise 5.1, this exercise guides you through a rigorous proof that the second distributional derivative of the function* $f(x) = |x|^{2H}$ *is given by:*

$$f''(x) = 2H(2H - 1)|x|^{2H-2}.$$

The second distributional derivative $f''(x)$ is the first distributional derivative of $f'(x)$. Thus, for any test function $\psi \in C_c^\infty(\mathbb{R})$, we have:

$$\langle f'', \psi \rangle = -\langle f', \psi' \rangle = -\int_{\mathbb{R}} g(x)\psi'(x)\,dx.$$

Our goal is to show that $\langle f'', \psi \rangle = \displaystyle\int_{\mathbb{R}} 2H(2H-1)|x|^{2H-2}\psi(x)\,dx.$

1) Given that $g(x)$ is continuous but not classically differentiable at $x = 0$, we must evaluate the integral $\int_{\mathbb{R}} g(x)\psi'(x)\,dx$ by considering a limit around the singularity. Let the support of ψ be contained within $[-M, M]$. Show that by applying classical integration by parts to each part of the split integral, all boundary terms vanish as $\epsilon \to 0^+$, leading to:

$$\begin{aligned}\int_{\mathbb{R}} g(x)\psi'(x)\,dx &= \lim_{\epsilon \to 0+}\left(\int_{-M}^{-\epsilon} g(x)\psi'(x)\,dx + \int_{\epsilon}^{M} g(x)\psi'(x)\,dx\right)\\ &= -\int_{\mathbb{R}} h(x)\psi(x)\,dx\end{aligned}$$

where $h(x)$ is the classical derivative of $g(x)$ for $x \neq 0$.

2) Conclude that the second distributional derivative of the function $f(x) = |x|^{2H}$ is given by:

$$f''(x) = 2H(2H-1)|x|^{2H-2}.$$

EXERCISE 5.3.– *Let $\frac{1}{2} < H < 1$ and $(a, b) \in \mathbb{R}^2 \setminus \{(0,0)\}$. Consider the space of measurable functions f on $[0, T]$ equipped with the following two norms:*

$$\|f\|^2_{|\mathcal{H}_{a,b}|} = \int_0^T \int_0^T |f(u)|\,|f(v)| \left|\frac{\partial^2 R^{H,a,b}(u,v)}{\partial u \partial v}\right| dudv$$

and

$$\|f\|^2_{|\mathcal{H}|} = \int_0^T \int_0^T |f(u)|\,|f(v)|\,|u-v|^{2H-2}dudv.$$

1) Use the upper bound from Corollary 5.1 to establish the existence of a positive constant C_2 such that for any measurable function f:

$$\|f\|_{|\mathcal{H}_{a,b}|} \leq C_2 \|f\|_{|\mathcal{H}|}.$$

What is the explicit value of C_2 in terms of H, a, b?

2) Based on this inequality, explain why $|\mathcal{H}| \subset |\mathcal{H}_{a,b}|$ (continuous embedding).

3) What implication does this continuous embedding have for the corresponding Malliavin Sobolev spaces $D^{1,2}(|\mathcal{H}|)$ and $D^{1,2}(\mathcal{H}_{a,b})$?

EXERCISE 5.4.– *Let $\frac{1}{2} < H < 1$. We continue to explore the properties of the spaces $|\mathcal{H}_{a,b}|$ and $|\mathcal{H}|$. Recall the definitions of the norms for a measurable function f on $[0,T]$:*

$$\|f\|^2_{|\mathcal{H}_{a,b}|} = \int_0^T \int_0^T |f(u)|\,|f(v)| \left| \frac{\partial^2 R^{H,a,b}(u,v)}{\partial u \partial v} \right| dudv$$

and

$$\|f\|^2_{|\mathcal{H}|} = \int_0^T \int_0^T |f(u)|\,|f(v)|\,|u-v|^{2H-2} dudv.$$

For all $(a,b) \in \mathbb{R}^2 \setminus \{(0,0)\}$ with $a \neq b$, we want to demonstrate the equivalent norm inequalities:

$$C_1 \|f\|_{|\mathcal{H}|} \leq \|f\|_{|\mathcal{H}_{a,b}|} \leq C_2 \|f\|_{|\mathcal{H}|},$$

where $C_1 = \alpha_H C_{1,H}^{\min}$ and $C_2 = \alpha_H C_{2,H}$.

1) Utilize both the lower and upper bounds for $\left| \frac{\partial^2 R^{H,a,b}(u,v)}{\partial u \partial v} \right|$ provided by Corollary 5.1, and apply them to the definition of $\|f\|^2_{|\mathcal{H}_{a,b}|}$ to derive the stated double inequality.

2) Explain how these equivalent norm inequalities imply that $|\mathcal{H}| = |\mathcal{H}_{a,b}|$ as sets.

3) What is the consequence of this equality of sets for their associated Malliavin Sobolev spaces, $D^{1,2}(|\mathcal{H}|)$ and $D^{1,2}(\mathcal{H}_{a,b})$?

EXERCISE 5.5.– *Consider the covariance function $R^{H,a,b}(t,s)$ given by:*

$$R^{H,a,b}(t,s) = (a^2+b^2)\frac{1}{2}(t^{2H}+s^{2H}-|t-s|^{2H})+ab(t^{2H}+s^{2H}-(t+s)^{2H}),$$

with $H \in (1/2, 1)$, and recall that:

$$\frac{\partial^2 R^{H,a,b}(t,s)}{\partial t \partial s} = \alpha_H \left[(a^2+b^2)|t-s|^{2H-2} - 2ab(t+s)^{2H-2} \right].$$

Let $f_{r,s}(\sigma) = \dfrac{\partial^2 R^{H,a,b}(r, s+\sigma)}{\partial r \partial \sigma}$.

1) Write out the explicit form of $f_{r,s}(\sigma)$ by substituting $t = r$ and $s = s+\sigma$ into the given formula for $\dfrac{\partial^2 R^{H,a,b}(t,s)}{\partial t \partial s}$.

2) For which fixed pairs (r,s) is the function $f_{r,s}(\sigma)$ continuous at $\sigma = 0$?

3) Apply the mean value theorem to the inner integral $\dfrac{1}{2\epsilon} \displaystyle\int_{-\epsilon}^{+\epsilon} f_{r,s}(\sigma) d\sigma$.

4) Deduce that for fixed (r, s) where $r \neq s$:

$$\lim_{\epsilon \to 0} \frac{1}{2\epsilon} \left(\int_{-\epsilon}^{+\epsilon} \frac{\partial^2 R^{H,a,b}(r, s + \sigma)}{\partial r \partial \sigma} d\sigma \right) = \frac{\partial^2 R^{H,a,b}(r, s)}{\partial r \partial s}.$$

EXERCISE 5.6.– *Consider a stochastic process X_t defined by:*

$$X_t = X_0 \exp\left(\mu t + \beta Z_t^H(a, b)\right),$$

where $X_0 > 0$ is a constant, μ and β are real constants, $Z_t^H(a, b)$ is a ZgfBm with Hurst parameter $H > \frac{1}{2}$ and coefficients a, b. Assume $C_H(a, b) = (a + b)^2 - 2^{2H}ab$.

1) Identify the function $f(t, x)$ such that $X_t = X_0 f(t, Z_t^H(a, b))$.

2) Calculate the partial derivatives $\frac{\partial f}{\partial t}(t, x)$, $\frac{\partial f}{\partial x}(t, x)$, and $\frac{\partial^2 f}{\partial x^2}(t, x)$.

3) Using Theorem 5.2, derive the stochastic differential equation (SDE) for X_t in the form:

$$\delta X_t = A_t X_t dt + B_t X_t \delta Z_t^H(a, b).$$

Identify the drift coefficient A_t and the diffusion coefficient B_t.

4) Compare your result with the standard geometric Brownian motion SDE. How does the fractional nature (through H and $C_H(a, b)$) influence the drift?

EXERCISE 5.7.– *Let $Y_t = (Z_t^H(a, b))^n$ for an integer $n \geq 2$, where $Z_t^H(a, b)$ is a ZgfBm with Hurst parameter $H > \frac{1}{2}$ and coefficients a, b. Assume $Z_0^H(a, b) = 0$.*

1) Identify the function $f(x)$ such that $Y_t = f(Z_t^H(a, b))$. Note that this function does not explicitly depend on time.

2) Calculate the first and second derivatives of $f(x)$ with respect to x.

3) Apply Theorem 5.2 (or its time-independent corollary, if available and proven in your text, otherwise use the full Theorem 5.2 with $\frac{\partial f}{\partial t} = 0$) to derive the stochastic differential equation for Y_t. Express your result in terms of Y_t and $Z_t^H(a, b)$.

4) Consider the specific case where $n = 2$, that is, $Y_t = (Z_t^H(a, b))^2$. Derive its SDE. How does this compare to the Itô's lemma for $(B_t)^2$ in classical (nonfractional) stochastic calculus?

6

Stochastic Partial Differential Equations Driven by a ZgfBm Noise

6.1. Introduction

Stochastic partial differential equations (SPDEs) provide a powerful framework for modeling physical phenomena subject to random fluctuations. The introduction of a stochastic term into a partial differential equation allows us to capture the inherent randomness often observed in real-world processes. A key challenge in this field lies in the selection of an appropriate noise term, as this choice must be carefully considered based on the specific equation of motion and its underlying physical interpretation.

In recent years, stochastic heat equations driven by a variety of noise processes have received significant attention, leading to a wealth of valuable insights (see, for example, Balan and Tudor (2008), Tudor (2013) and Tudor and Zili (2014)). Notably, Tudor (2013) investigated solutions to stochastic heat equations driven by fractional-white noise, a Gaussian noise process that exhibits Brownian behavior in space and fractional Brownian motion (fBm) behavior in time.

The purpose of this chapter is to investigate a new SPDE featuring a Laplacian operator, driven by a white-colored noise. This noise process behaves as a Wiener process in the spatial variable and as a generalized fractional Brownian motion (ZgfBm) in time with Hurst parameter $H \in (1/2, 1)$. Our study offers a significant generalization of existing works that focus on more particular Gaussian noises, such as fBm or sfBm. By considering the more general ZgfBm, we provide a unified framework for the analysis of SPDEs. This approach not only allows for the study of more varied and complex physical phenomena but also ensures that readers interested in SPDEs driven by fBm or sfBm will find their specific cases encompassed within our more general analysis.

6.2. The stochastic partial differential equation

For a given dimension $d \geq 1$, this chapter aims to provide a rigorous analysis of the following SPDE:

$$\begin{cases} \dfrac{\partial u_{a,b,H}}{\partial t} = \dfrac{1}{2}\Delta u_{a,b,H} + \dot{W}_{a,b,H},\ t \in (0,T],\ x \in \mathbb{R}^d \\ u_{a,b,H}(x,0) = 0. \end{cases} \quad [6.1]$$

In this equation, $u_{a,b,H}(t,x)$ is the unknown solution, Δ is the Laplacian operator with respect to the spatial variable $x \in \mathbb{R}^d$, and $\dot{W}_{a,b,H}$ is the formal derivative of the centered random noise $W_{a,b,H}$, which we will define precisely in the next section.

6.2.1. *The noise* $W_{a,b,H}$

The forcing term $W_{a,b,H}$ in equation [6.1] is a centered Gaussian random field (as introduced in Chapter 1, Definition 1.15). It is defined as a collection of random variables:

$$\{W_{a,b,H}(t,A); t \in [0,T], A \in \mathcal{B}_b(\mathbb{R}^d)\},$$

where $\mathcal{B}_b(\mathbb{R}^d)$ is the set of bounded Borel measurable sets in $\mathbb{R}^d$. Its distinctive covariance structure is given by:

$$\mathbb{E}\left[W_{a,b,H}(t,A)W_{a,b,H}(s,B)\right] = R^{H,a,b}(t,s)\lambda(A \cap B) \quad [6.2]$$

where λ represents the Lebesgue measure on $\mathbb{R}^d$. The function $R^{H,a,b}(t,s)$ is the covariance function of the ZgfBm $Z^H(a,b)$, with Hurst parameter $H \in (1/2,1)$, defined by:

$$\begin{aligned} R^{H,a,b}(t,s) &= \mathrm{Cov}\Big(Z_t^H(a,b), Z_s^H(a,b)\Big) \\ &= \frac{1}{2}(a+b)^2\left(s^{2H} + t^{2H}\right) - ab(t+s)^{2H} - \frac{a^2+b^2}{2} \mid t-s \mid^{2H}. \end{aligned}$$

This specific covariance structure is the defining characteristic of $W_{a,b,H}$ as a Gaussian random field. It indicates two crucial properties of the noise:

– *Colored in time*: the presence of $R^{H,a,b}(t,s)$ dictates the temporal correlation. This means that noise increments at different time points t and s are correlated according to the properties of the ZgfBm, leading to a "colored" temporal behavior.

– *White in space*: the term $\lambda(A \cap B)$ signifies that the noise is spatially white. As discussed in Chapter 1, section 1.4.3, this implies that for fixed time points, the random variables $W_{a,b,H}(t,A)$ and $W_{a,b,H}(t,B)$ are uncorrelated (and thus independent, due to Gaussianity) whenever the spatial sets A and B are disjoint. This is precisely the defining characteristic of a spatially white noise component, ensuring maximal randomness across distinct spatial regions.

Therefore, $W_{a,b,H}$ is a noise that exhibits temporal long-range dependence inherited from the ZgfBm, combined with a lack of spatial correlation.

6.2.2. *Wiener integrals with respect to* $W_{a,b,H}$ *and the transfer formula*

To properly define and work with stochastic integrals with respect to the noise $W_{a,b,H}$, we first introduce the associated canonical Hilbert space, and then detail the construction of the integral.

Let $\mathcal{E}$ be the set of finite linear combinations of elementary functions of the form $\mathbf{1}_{[0,t]} \times A$, where $t \in [0,T]$ and $A \in \mathcal{B}_b(\mathbb{R}^d)$. The Hilbert space $\mathcal{H}_{a,b,H}$ is then defined as the completion of $\mathcal{E}$ with respect to the inner product:

$$< \mathbf{1}_{[0,t]} \times A, \mathbf{1}_{[0,s]} \times B >_{\mathcal{H}_{a,b,H}} := \mathbb{E}\left[W_{a,b,H}(t,A)W_{a,b,H}(s,B)\right].$$

For sufficiently smooth functions $g, h \in \mathcal{H}_{a,b,H}$, the inner product can be expressed in an integral form, leveraging the covariance structure from equation [6.2]:

$$< g, h >_{\mathcal{H}_{a,b,H}} = \int_0^T \int_0^T \frac{\partial^2 R^{a,b,H}}{\partial u \partial v}(u,v)dudv \int_{\mathbb{R}^d} g(u,y)h(v,y)dy. \quad [6.3]$$

Note that the partial derivatives of $R^{a,b,H}$ are taken with respect to its time arguments.

Building upon the well-established frameworks for stochastic integration, which typically involve defining the integral first for simple, step-like functions and then extending it by density to a broader class of integrands (as described in section 1.4.5 and, for example, in Tudor (2013) and Tudor and Zili (2014)), it is possible to define *Wiener integrals* with respect to the specific noise $W_{a,b,H}$.

The construction proceeds in a standard manner:

1) For an elementary function $\varphi(u,y)$ of the form:

$$\varphi(u,y) = \sum_{i=1}^{n} c_i \mathbf{1}_{[t_i,t_{i+1})}(u)\mathbf{1}_{A_i}(y),$$

where $c_i \in \mathbb{R}$, $0 \leq t_i < t_{i+1} \leq T$, and $A_i \in \mathcal{B}_b(\mathbb{R}^d)$, the integral is defined directly as a finite sum of random variables:

$$\int_0^T \int_{\mathbb{R}^d} \varphi(u,y)W_{a,b,H}(du,dy) := \sum_{i=1}^{n} c_i W_{a,b,H}([t_i,t_{i+1}), A_i).$$

The term $W_{a,b,H}([t_i,t_{i+1}), A_i)$ here denotes the random variable:

$$W_{a,b,H}(t_{i+1}, A_i) - W_{a,b,H}(t_i, A_i)$$

when integrating over a time interval.

2) This initial definition is then extended by linearity to all functions in the space $\mathcal{E}$.

3) The crucial step for defining the integral for a broader class of functions is the extension by density. The integral is extended to all functions φ belonging to the completion of $\mathcal{E}$ with respect to the inner product defined in equation [6.3]. This completion is precisely the Hilbert space $\mathcal{H}_{a,b,H}$. Therefore, the rigorous domain of the integrand φ for which the Wiener integral is defined is the Hilbert space $\mathcal{H}_{a,b,H}$.

This Wiener integral serves as a crucial isometry, that is, it is a norm-preserving linear mapping between the deterministic Hilbert space $\mathcal{H}_{a,b,H}$ (of integrands) and the space of square-integrable random variables $L^2(\Omega)$ (where the integral resides). This isometric property ensures that the integral is a well-defined random variable and, crucially, that the inner product in $\mathcal{H}_{a,b,H}$ directly corresponds to the covariance (or expected product) of the resulting random variables in $L^2(\Omega)$. Specifically, for any $\varphi, \psi \in \mathcal{H}_{a,b,H}$, we have:

$$\mathbb{E}\left[\left(\int_0^T \int_{\mathbb{R}^d} \varphi(u,y) W_{a,b,H}(du,dy)\right)\left(\int_0^T \int_{\mathbb{R}^d} \psi(v,y) W_{a,b,H}(dv,dy)\right)\right]$$

$$= \int_0^T \int_0^T \mu_{a,b,H}(du,dv) \int_{\mathbb{R}^d} \varphi(u,y)\psi(v,y)dy. \qquad [6.4]$$

This holds for functions φ and ψ such that the following integrability conditions are met:

$$\int_0^T \int_0^T d \mid \mu_{a,b,H} \mid (u,v) \int_{\mathbb{R}^d} \mid \varphi(u,y) \mid \mid \varphi(v,y) \mid dy < \infty$$

and

$$\int_0^T \int_0^T d \mid \mu_{a,b,H} \mid (u,v) \int_{\mathbb{R}^d} \mid \psi(u,y) \mid \mid \psi(v,y) \mid dy < \infty,$$

where $\mu_{a,b,H}$ is the signed measure defined by its Radon–Nikodym derivative with respect to the Lebesgue measure on $[0,T]^2$:

$$d\mu_{a,b,H}(u,v) = \frac{\partial^2 R^{a,b,H}}{\partial u \partial v}(u,v)dudv. \qquad [6.5]$$

Here, $|\mu_{a,b,H}|$ denotes the total variation measure associated with $\mu_{a,b,H}$. This careful definition of the Wiener integral is crucial for establishing the well-posedness and solution properties of the SPDE [6.1].

6.2.2.1. *The transfer formula*

A powerful consequence of the construction of the noise $W_{a,b,H}$ and its associated Wiener integral, and a key tool in our analysis, is the transfer formula

(Proposition 6.1). This formula allows us to express a Wiener integral with respect to the colored noise $W_{a,b,H}$ in terms of an integral with respect to a standard space–time white noise $W(s,y)$. This transformation is invaluable, as it permits us to leverage the well-established theory and computational techniques available for standard white noise integrals, simplifying the analysis of the SPDE driven by the ZgfBm noise. The kernel $K_{a,b,H}(u,s)$ within the transfer formula effectively embeds the temporal correlation structure of $W_{a,b,H}$ into the standard white noise framework.

PROPOSITION 6.1 (transfer formula).– *For every function $g \in \mathcal{H}_{a,b,H}$, the Wiener integral with respect to the ZgfBm noise $W_{a,b,H}$ can be expressed as:*

$$\begin{aligned}&\int_0^T \int_{\mathbb{R}^d} g(s,y)dW_{a,bH}(s,y)\\ &= d_H \int_{\mathbb{R}}\int_{\mathbb{R}^d}\int_{\mathbb{R}} \mathbf{1}_{(0,T)}(u)g(u,y)K_{a,b,H}(u,s)dudW(s,y),\end{aligned} \qquad [6.6]$$

where $W = \{W(s,y); s \in \mathbb{R}, y \in \mathbb{R}^d\}$ is a standard space–time white noise with covariance given by:

$$\mathbb{E}\left[W(s,A)W(t,B)\right] = (t \wedge s)\lambda(A \cap B),$$

and the constants and kernel functions are defined as:

$$d_H = \frac{(H-1/2)}{2C(H)},$$

and the kernel $K_{a,b,H}(u,s)$ is given by:

$$\begin{aligned}K_{a,b,H}(u,s) &= (a+b)\left(((u-s)_+)^{H-3/2} - ((u+s)_-)^{H-3/2}\right)\\ &+(a-b)\left(((u-s)_+)^{H-3/2} + ((u+s)_-)^{H-3/2}\right).\end{aligned}$$

Proof. Let $I_1(g) := \int_0^T \int_{\mathbb{R}^d} g(u,y)dW_{a,bH}(u,y)$ denote the left-hand side of equation [6.6]. Let $I_2(g) := d_H \int_{\mathbb{R}}\int_{\mathbb{R}^d}\left(\int_0^T g(u,y)K_{a,b,H}(u,s)du\right)dW(s,y)$ denote the right-hand side. Note that we implicitly use the indicator function $\mathbf{1}_{(0,T)}(u)$ by setting the integration limits for u from 0 to T.

Both $I_1(g)$ and $I_2(g)$ are centered Gaussian random variables (since g is a deterministic function and the integrals are linear with respect to Gaussian measures). Therefore, to prove their equality in $L^2(\Omega)$, it suffices to show that their covariances are identical for any two functions $\varphi, \psi \in \mathcal{H}_{a,b,H}$.

From the isometry property of the Wiener integral with respect to $W_{a,b,H}$ (equation [6.4]) and the definition of the signed measure $\mu_{a,b,H}$ (equation [6.5]), we have:

$$\begin{aligned}\mathbb{E}[I_1(\varphi)I_1(\psi)] &= \int_0^T \int_0^T \mu_{a,b,H}(du,dv) \int_{\mathbb{R}^d} \varphi(u,y)\psi(v,y)dy \\ &= \int_0^T \int_0^T \frac{\partial^2 R^{a,b,H}}{\partial u \partial v}(u,v)dudv \int_{\mathbb{R}^d} \varphi(u,y)\psi(v,y)dy\end{aligned}$$

Let

$$F_\varphi(s,y) := d_H \int_0^T \varphi(u,y)K_{a,b,H}(u,s)du$$

and

$$F_\psi(s,y) := d_H \int_0^T \psi(v,y)K_{a,b,H}(v,s)dv.$$

The integral $I_2(g)$ is a Wiener integral with respect to the standard space–time white noise $W(s,y)$. Its isometry property (see Theorem 1.7) states:

$$\mathbb{E}[I_2(\varphi)I_2(\psi)] = \int_{\mathbb{R}} \int_{\mathbb{R}^d} F_\varphi(s,y)F_\psi(s,y)dsdy$$

$$= \int_{\mathbb{R}} \int_{\mathbb{R}^d} \left(d_H \int_0^T \varphi(u,y)K_{a,b,H}(u,s)du \right) \left(d_H \int_0^T \psi(v,y)K_{a,b,H}(v,s)dv \right) dsdy$$

$$= d_H^2 \int_0^T \int_0^T \left(\int_{\mathbb{R}} K_{a,b,H}(u,s)K_{a,b,H}(v,s)ds \right) \int_{\mathbb{R}^d} \varphi(u,y)\psi(v,y)dy\, dudv$$

We rearranged the order of integration, which is permissible for L^2 integrands by Fubini's theorem.

For $\mathbb{E}[I_1(\varphi)I_1(\psi)]$ and $\mathbb{E}[I_2(\varphi)I_2(\psi)]$ to be equal for all $\varphi, \psi \in \mathcal{H}_{a,b,H}$, it is sufficient to prove that the temporal kernel functions multiplying the inner spatial integral $\int_{\mathbb{R}^d} \varphi(u,y)\psi(v,y)dy$ are identical almost everywhere on $[0,T]^2$:

$$\frac{\partial^2 R^{a,b,H}}{\partial u \partial v}(u,v) = d_H^2 \int_{\mathbb{R}} K_{a,b,H}(u,s)K_{a,b,H}(v,s)ds$$

Let us verify this identity. Recall the integral representation of $Z_t^H(a,b)$ from Lemma 3.3:

$$Z_t^H(a,b) = d_H \int_{\mathbb{R}} \left(\int_0^t K_{a,b,H}(x,s)dx \right) dW(s).$$

The covariance function of $Z_t^H(a,b)$ is $R^{H,a,b}(t,t')$. By the isometry property of the 1D Wiener integral with respect to $W(s)$:

$$\begin{aligned} R^{H,a,b}(t,t') &= \mathbb{E}[Z_t^H(a,b)Z_{t'}^H(a,b)] \\ &= d_H^2 \int_{\mathbb{R}} \left(\int_0^t K_{a,b,H}(x,s)dx\right)\left(\int_0^{t'} K_{a,b,H}(x',s)dx'\right) ds. \end{aligned}$$

This equality holds for all $t, t' \in [0,T]$.

Now, we compute the second partial derivative with respect to t and t'. Let

$$G(t,s) := \int_0^t K_{a,b,H}(x,s)dx.$$

For a fixed s, the function $x \mapsto K_{a,b,H}(x,s)$ is integrable over $[0,t]$, since $H - 3/2 > -1$. While $K_{a,b,H}(x,s)$ has singularities at $x = s$ and $x = -s$, its integral $G(t,s)$ is a well-defined function of t and is differentiable almost everywhere on $[0,T]$. Specifically, by the fundamental theorem of calculus, its derivative with respect to t is given by:

$$G'(t,s) = K_{a,b,H}(t,s) \; for \; any \; t \in [0,T] \setminus \{s\}.$$

Therefore, for $t \in [0,T] \setminus \{s\}$:

$$\frac{\partial}{\partial t} R^{H,a,b}(t,t') = d_H^2 \int_{\mathbb{R}} K_{a,b,H}(t,s)\left(\int_0^{t'} K_{a,b,H}(x',s)dx'\right) ds.$$

Similarly, taking the partial derivative with respect to t' yields $K_{a,b,H}(t',s)$ from the inner integral term for $t' \in [0,T] \setminus \{s\}$. This gives:

$$\frac{\partial^2}{\partial t \partial t'} R^{H,a,b}(t,t') = d_H^2 \int_{\mathbb{R}} K_{a,b,H}(t,s)K_{a,b,H}(t',s)ds.$$

Since equality almost everywhere is sufficient for L^2 equivalence in this context, the result is valid.

Replacing t with u and t' with v (as in the general expression), we obtain precisely the required identity almost everywhere:

$$\frac{\partial^2 R^{a,b,H}}{\partial u \partial v}(u,v) = d_H^2 \int_{\mathbb{R}} K_{a,b,H}(u,s)K_{a,b,H}(v,s)ds.$$

Consequently, since the covariance functions $\mathbb{E}[I_1(\varphi)I_1(\psi)]$ and $\mathbb{E}[I_2(\varphi)I_2(\psi)]$ are found to be identical through their kernels (which are equal almost everywhere on $[0,T]^2$), and both $I_1(g)$ and $I_2(g)$ are centered Gaussian random variables, they are stochastically equivalent (i.e. they have the same distribution) and thus equal in $L^2(\Omega)$. □

6.3. Analysis of solution existence and characterization

In this section, we rigorously examine different solution concepts for the SPDE [6.1]. We begin by introducing and characterizing the *mild solution*, which is a widely used approach for SPDEs driven by stochastic noise that may not allow for classical solutions. To establish a rigorous connection between the mild solution and the underlying partial differential equation, we will also introduce the concept of a *weak solution*.

6.3.1. *Mild solution*

For SPDEs, especially those driven by noise with low regularity (like white noise), classical solutions in the sense of continuously differentiable functions often do not exist. In such cases, the concept of a mild solution provides a more appropriate and widely applicable framework. A mild solution transforms the original SPDE into an equivalent integral equation, leveraging the fundamental solution (Green's kernel) of the differential operator.

DEFINITION 6.1.– *A stochastic process* $\{u_{a,b,H}(t,x); (t,x) \in [0,T] \times \mathbb{R}^d\}$ *is called a mild solution of the SPDE [6.1] if it satisfies the following integral equation for all* $(t,x) \in [0,T] \times \mathbb{R}^d$*:*

$$u_{a,b,H}(t,x) := \int_0^t \int_{\mathbb{R}^d} G(t-v, x-y) W_{a,b,H}(dv, dy), \qquad [6.7]$$

where the integral on the right-hand side is a stochastic integral with respect to the driving noise $W_{a,b,H}$*, and the function* $G(t,x)$ *is the Green's kernel (also known as the heat kernel) of the heat equation, given by:*

$$G(t,x) = \begin{cases} \dfrac{1}{(2\pi t)^{\frac{d}{2}}} \exp\left(-\dfrac{|x|^2}{2t}\right) \text{ if } t > 0, x \in \mathbb{R}^d, \\ 0 \text{ if } t \le 0, x \in \mathbb{R}^d. \end{cases} \qquad [6.8]$$

PROPOSITION 6.2 (existence of mild solution).– *The mild solution* $u_{a,b,H}(t,x)$ *to the generalized fractional heat equation [6.1], as defined in [6.7], exists as an* $L^2(\Omega)$*-valued centered Gaussian random variable for all* $(t,x) \in [0,T] \times \mathbb{R}^d$ *if and only if* $d < 4H$.

Proof. For a Wiener integral of a deterministic function to be well defined in the space of square-integrable random variables $L^2(\Omega)$, its integrand must belong to the associated canonical Hilbert space of integrands. In our context, this is the Hilbert space $\mathcal{H}_{a,b,H}$, whose construction was detailed in section 6.2.2. This space consists of all measurable functions $\varphi : [0,T] \times \mathbb{R}^d \to \mathbb{R}$ such that their norm, defined by:

$$\|\varphi\|^2_{\mathcal{H}_{a,b,H}} := \int_0^T \int_0^T \mu_{a,b,H}(du,dv) \int_{\mathbb{R}^d} \varphi(u,y)\varphi(v,y)dy,$$

is finite. Here, $\mu_{a,b,H}(du,dv)$ is the signed measure defined in equation [6.5] as:

$$\frac{\partial^2 R^{a,b,H}}{\partial u \partial v}(u,v)dudv.$$

By the Itô isometry (section 6.2.2), the existence of the mild solution $u_{a,b,H}(t,x)$ in $L^2(\Omega)$ is equivalent to verifying that its integrand $\varphi(v,y) = G(t-v, x-y)$ belongs to $\mathcal{H}_{a,b,H}$, which in turn means demonstrating that its $\mathcal{H}_{a,b,H}$-norm is finite.

Furthermore, by the construction of the Wiener integral, $u_{a,b,H}(t,x)$ is a centered Gaussian random variable.

Thus, we need to show that $\|\varphi\|^2_{\mathcal{H}_{a,b,H}}$ is finite, where φ and the integral bounds are taken up to t (as the Green's function is zero for $v > t$).

Applying the definition of the norm for φ, we have:

$$\mathbb{E}\left[u_{a,b,H}(t,x)^2\right] = \|\varphi\|^2_{\mathcal{H}_{a,b,H}}$$
$$= \int_0^t \int_0^t \frac{\partial^2 R^{a,b,H}}{\partial u \partial v}(u,v)dudv \int_{\mathbb{R}^d} G(t-u, x-y)G(t-v, x-y)dy.$$

Let us analyze the inner spatial integral:

$$J(t,u,v,x) := \int_{\mathbb{R}^d} G(t-u, x-y)G(t-v, x-y)dy.$$

Using the explicit form of the Green's kernel $G(t,x)$ from equation [6.8]:

$$G(t-u, x-y) = \frac{1}{(2\pi(t-u))^{d/2}} \exp\left(-\frac{|x-y|^2}{2(t-u)}\right).$$

Substituting this into $J(t,u,v,x)$, we get:

$$J(t,u,v,x)$$
$$= \frac{1}{(2\pi)^d} \frac{1}{(t-u)^{d/2}(t-v)^{d/2}} \int_{\mathbb{R}^d} \exp\left(-|x-y|^2\left(\frac{1}{2(t-u)} + \frac{1}{2(t-v)}\right)\right) dy.$$

This integral is a d-dimensional Gaussian integral of the form

$$\int_{\mathbb{R}^d} \exp(-A|\mathbf{z}|^2)d\mathbf{z} = (\pi/A)^{d/2}.$$

Here, the integration variable is $\mathbf{z} = x - y$ (a simple shift does not change the integral value over $\mathbb{R}^d$), and the coefficient A is given by $A = \dfrac{1}{2(t-u)} + \dfrac{1}{2(t-v)}$. Applying this formula to the integral term

$$\begin{aligned} := & \frac{1}{(2\pi)^d} \frac{1}{(t-u)^{d/2}(t-v)^{d/2}} \left(\frac{\pi}{\frac{1}{2(t-u)} + \frac{1}{2(t-v)}} \right)^{d/2} \\ = & \frac{1}{(2\pi)^d} \frac{1}{(t-u)^{d/2}(t-v)^{d/2}} \left(\frac{2\pi(t-u)(t-v)}{2t-u-v} \right)^{d/2} \\ = & \frac{(2\pi)^{d/2}}{(2\pi)^d(2t-u-v)^{d/2}} = \frac{1}{(2\pi)^{d/2}(2t-u-v)^{d/2}}. \end{aligned}$$

Substituting this back into the expectation:

$$\mathbb{E}\left[u_{a,b,H}(t,x)^2\right] = C_0 \int_0^t \int_0^t \frac{\partial^2 R^{a,b,H}}{\partial u \partial v}(u,v)(2t-u-v)^{-d/2} du dv,$$

where $C_0 = (2\pi)^{-d/2}$ is a constant.

From Corollary 5.1, for $u \neq v$, we have the following bounds:

$$C_1|u-v|^{2H-2} \leq \frac{\partial^2 R^{H,a,b}(u,v)}{\partial u \partial v}(u,v) \leq C_2|u-v|^{2H-2}, \qquad [6.9]$$

for every $u, v \in [0,T]; u \neq v$, with:

$$C_1 = \alpha_H \min((a-b)^2, a^2+b^2) \quad \text{and} \quad C_2 = \alpha_H(|a|+|b|)^2.$$

Therefore, we can bound the expectation:

$$C_0 C_1 \, I(t) \leq \mathbb{E}\left[u_{a,b,H}(t,x)^2\right] \leq C_0 C_2 \, I(t).$$

with

$$I(t) = \int_0^t \int_0^t |u-v|^{2H-2}(2t-u-v)^{-d/2} du dv.$$

The convergence of the integral $I_1(t)$ is rigorously analyzed in section A.7 of the appendix. Lemma A.7 establishes that this integral is finite if and only if $H > 1/2$ and $d < 4H$.

Since C_0, C_1, C_2 are positive constants, the finiteness of $\mathbb{E}\left[u_{a,b,H}(t,x)^2\right]$ is equivalent to the finiteness of this double integral. Thus, we conclude that the mild solution exists in $L^2(\Omega)$ if and only if $H > 1/2$ and $d < 4H$. This completes the proof. □

REMARK 6.1.– *While the mild solution provides an explicit formula in terms of integral operators, its direct connection to the differential form of the SPDE is not immediately apparent, especially given the low regularity of the noise term* $W_{a,b,H}$. *To establish a more rigorous link and to extend the notion of solution to less regular functions, we will now introduce a second type of solution, known as the weak solution. This formulation will allow us to interpret the SPDE in a distributional sense, where derivatives are transferred to smooth test functions.*

6.3.2. *Weak solution*

To define a weak solution for the generalized fractional heat equation [6.1], we consider its integral formulation.

DEFINITION 6.2.– *Let* $u_{a,b,H}(t,x)$ *be a function. We say that* $u_{a,b,H}$ *is a weak solution to [6.1] if, for every test function* $\varphi \in C_c^\infty([0,T)\times\mathbb{R}^d)$, *the following integral identity holds:*

$$\begin{aligned}&\int_0^T\int_{\mathbb{R}^d} u_{a,b,H}(t,x)\left(\frac{\partial\varphi}{\partial t}(t,x)+\frac{1}{2}\Delta\varphi(t,x)\right)dxdt\\&=-\int_0^T\int_{\mathbb{R}^d}\varphi(t,x)W_{a,b,H}(dt,dx).\end{aligned} \qquad [6.10]$$

Let us now explain how this integral identity is obtained. To do so, we assume for a moment that $u_{a,b,H}(t,x)$ is a classical solution, meaning it is continuously differentiable with respect to t (i.e. $u_{a,b,H}\in C^1([0,T];\mathbb{R}^d)$) and twice continuously differentiable with respect to x (i.e. $u_{a,b,H}\in C^2(\mathbb{R}^d)$ for each t).

Multiply the first equation in [6.1] by a test function $\varphi\in C_c^\infty([0,T)\times\mathbb{R}^d)$ and integrate over $[0,T]\times\mathbb{R}^d$:

$$\begin{aligned}&\int_0^T\int_{\mathbb{R}^d}\frac{\partial u_{a,b,H}}{\partial t}(t,x)\varphi(t,x)dxdt\\&=\int_0^T\int_{\mathbb{R}^d}\frac{1}{2}\Delta u_{a,b,H}(t,x)\varphi(t,x)dxdt+\int_0^T\int_{\mathbb{R}^d}\varphi(t,x)W_{a,b,H}(dt,dx).\end{aligned}$$

Consider the term involving the time derivative. Using integration by parts with respect to t:

$$\begin{aligned}&\int_0^T\int_{\mathbb{R}^d}\frac{\partial u_{a,b,H}}{\partial t}(t,x)\varphi(t,x)dxdt\\&=\left[\int_{\mathbb{R}^d}u_{a,b,H}(t,x)\varphi(t,x)dx\right]_0^T-\int_0^T\int_{\mathbb{R}^d}u_{a,b,H}(t,x)\frac{\partial\varphi}{\partial t}(t,x)dxdt.\end{aligned}$$

Since φ has compact support in $[0, T)$, it vanishes at $t = T$ (i.e. $\varphi(T, x) = 0$). Also, the initial condition is $u_{a,b,H}(0, x) = 0$. Therefore, the boundary term at $t = 0$ and $t = T$ vanishes:

$$\left[\int_{\mathbb{R}^d} u_{a,b,H}(t,x)\varphi(t,x)dx\right]_0^T = 0.$$

So, the time derivative term simplifies to:

$$\int_0^T \int_{\mathbb{R}^d} \frac{\partial u_{a,b,H}}{\partial t}(t,x)\varphi(t,x)dxdt = -\int_0^T \int_{\mathbb{R}^d} u_{a,b,H}(t,x)\frac{\partial \varphi}{\partial t}(t,x)dxdt.$$

Next, consider the Laplacian term. Using integration by parts twice with respect to x, and noting that φ has compact support in $\mathbb{R}^d$ so boundary terms at infinity vanish:

$$\int_{\mathbb{R}^d} \Delta u_{a,b,H}(t,x)\varphi(t,x)dx = \int_{\mathbb{R}^d} u_{a,b,H}(t,x)\Delta\varphi(t,x)dx.$$

Substituting these back into the integrated PDE:

$$\begin{aligned}
&-\int_0^T \int_{\mathbb{R}^d} u_{a,b,H}(t,x)\frac{\partial \varphi}{\partial t}(t,x)dxdt \\
&= \int_0^T \int_{\mathbb{R}^d} \frac{1}{2}u_{a,b,H}(t,x)\Delta\varphi(t,x)dxdt + \int_0^T \int_{\mathbb{R}^d} \varphi(t,x)W_{a,b,H}(dt,dx).
\end{aligned}$$

Rearranging the terms, we arrive at the weak formulation [6.10].

6.3.3. *Relationship between mild and weak solutions*

In this section, we will establish that every mild solution of [6.1] is also a weak solution. This demonstrates the consistency between these two notions of solutions.

PROPOSITION 6.3.– *If $u_{a,b,H}(t,x)$ is a mild solution to [6.1], then it is also a weak solution in the sense of definition in equation [6.10].*

Proof. Let $u_{a,b,H}(t,x)$ be the mild solution to [6.1], given by the formula:

$$u_{a,b,H}(t,x) = \int_0^t \int_{\mathbb{R}^d} G(t-s, x-y)W_{a,b,H}(ds,dy),$$

where $G(t,x)$ is the heat kernel given by [6.8].

We need to show that $u_{a,b,H}(t,x)$ satisfies the weak formulation [6.10] for any test function $\varphi \in C_c^\infty([0,T) \times \mathbb{R}^d)$.

Let us substitute the mild solution formula into the left-hand side (LHS) of the weak formulation:

$$\text{LHS} = \int_0^T \int_{\mathbb{R}^d} \left(\int_0^t \int_{\mathbb{R}^d} G(t-s, x-y) W_{a,b,H}(ds, dy) \right)$$
$$\times \left(\frac{\partial \varphi}{\partial t}(t,x) + \frac{1}{2}\Delta\varphi(t,x) \right) dxdt.$$

To justify the interchange of integration order, we apply a stochastic Fubini theorem. Let

$$F(t,x;s,y) = G(t-s, x-y) \left(\frac{\partial \varphi}{\partial t}(t,x) + \frac{1}{2}\Delta\varphi(t,x) \right).$$

The conditions for applying stochastic Fubini's theorem (e.g. from Walsh (1984)) require that the integrand satisfies certain integrability conditions, often related to the finiteness of its $L^2(\Omega)$-norm integrated over the product measure spaces.

Specifically, since $G(t,x)$ is the heat kernel (smooth for $t > 0$) and $\varphi(t,x) \in C_c^\infty([0,T) \times \mathbb{R}^d)$ (smooth with compact support), the deterministic function $F(t,x;s,y)$ is smooth in all its arguments and compactly supported with respect to (t,x). The well-posedness of the mild solution in $L^2(\Omega)$ (as established in Proposition 6.2) implies the necessary integrability of $G(t-s, x-y)$ with respect to the covariance measure of $W_{a,b,H}(ds, dy)$. The compact support of φ ensures the spatial and temporal integrals are over bounded domains. These properties together guarantee that the conditions for applying stochastic Fubini's theorem are met, allowing us to interchange the order of integration.

We swap the $dsdy$ integral with the $dtdx$ integral. Note that the inner integral with respect to (t,x) must now range from s to T for t, because $G(t-s, x-y) = 0$ for $t < s$.

$$\text{LHS} = \int_0^T \int_{\mathbb{R}^d} I(s,y) W_{a,b,H}(ds, dy),$$

with

$$I(s,y) = \int_s^T \int_{\mathbb{R}^d} G(t-s, x-y) \left(\frac{\partial \varphi}{\partial t}(t,x) + \frac{1}{2}\Delta\varphi(t,x) \right) dxdt.$$

We now evaluate $I(s,y)$ by considering each term separately:

For the first term involving $\frac{\partial \varphi}{\partial t}$:

$$\int_s^T \int_{\mathbb{R}^d} G(t-s, x-y) \frac{\partial \varphi}{\partial t}(t,x) dxdt.$$

We apply integration by parts with respect to t:

$$= \left[\int_{\mathbb{R}^d} G(t-s,x-y)\varphi(t,x)dx\right]_{t=s}^{t=T} - \int_s^T \int_{\mathbb{R}^d} \frac{\partial G}{\partial t}(t-s,x-y)\varphi(t,x)dxdt.$$

Since $\varphi \in C_c^\infty([0,T) \times \mathbb{R}^d)$, it follows that $\varphi(T,x) = 0$ for all $x \in \mathbb{R}^d$. Thus, the term at the upper limit ($t = T$) vanishes.

$$= -\lim_{t \to s^+} \int_{\mathbb{R}^d} G(t-s,x-y)\varphi(t,x)dx - \int_s^T \int_{\mathbb{R}^d} \frac{\partial G}{\partial t}(t-s,x-y)\varphi(t,x)dxdt.$$

As $t \to s^+$, the heat kernel $G(t-s,x-y)$ acts as a Dirac delta function $\delta(x-y)$ in the distributional sense. Therefore, the limit evaluates to:

$$\lim_{t \to s^+} \int_{\mathbb{R}^d} G(t-s,x-y)\varphi(t,x)dx = \int_{\mathbb{R}^d} \delta(x-y)\varphi(s,x)dx = \varphi(s,y).$$

So, the first part of $I(s,y)$ becomes:

$$-\varphi(s,y) - \int_s^T \int_{\mathbb{R}^d} \frac{\partial G}{\partial t}(t-s,x-y)\varphi(t,x)dxdt.$$

For the second term involving $\Delta\varphi$:

$$\frac{1}{2}\int_s^T \int_{\mathbb{R}^d} G(t-s,x-y)\Delta_x\varphi(t,x)dxdt.$$

We apply integration by parts twice with respect to x (Green's identity). Due to the compact support of φ in $\mathbb{R}^d$, all boundary terms at infinity vanish.

$$= \frac{1}{2}\int_s^T \int_{\mathbb{R}^d} \Delta_x G(t-s,x-y)\varphi(t,x)dxdt.$$

Now, we utilize the fundamental property of the heat kernel: for $t > s$, $G(t-s,x-y)$ is the solution to the homogeneous heat equation, that is:

$$\frac{\partial G}{\partial t}(t-s,x-y) = \frac{1}{2}\Delta_x G(t-s,x-y).$$

We substitute this property into the second term:

$$\frac{1}{2}\int_s^T \int_{\mathbb{R}^d} \Delta_x G(t-s,x-y)\varphi(t,x)dxdt = \int_s^T \int_{\mathbb{R}^d} \frac{\partial G}{\partial t}(t-s,x-y)\varphi(t,x)dxdt.$$

Finally, we combine the two evaluated parts to obtain $I(s,y)$:

$$\begin{aligned} I(s,y) &= \left(-\varphi(s,y) - \int_s^T \int_{\mathbb{R}^d} \frac{\partial G}{\partial t}(t-s,x-y)\varphi(t,x)dxdt\right) \\ &\quad + \left(\int_s^T \int_{\mathbb{R}^d} \frac{\partial G}{\partial t}(t-s,x-y)\varphi(t,x)dxdt\right) \\ &= -\varphi(s,y). \end{aligned}$$

The two integral terms exactly cancel each other out, leaving only $-\varphi(s,y)$.

Substitute this result for $I(s,y)$ back into the expression for the LHS of the weak formulation:

$$\text{LHS} = \int_0^T \int_{\mathbb{R}^d} (-\varphi(s,y)) W_{a,b,H}(ds,dy) = -\int_0^T \int_{\mathbb{R}^d} \varphi(s,y) W_{a,b,H}(ds,dy).$$

This is precisely the right-hand side (RHS) of the weak formulation [6.10], where (s,y) are dummy variables for the integration.

Thus, we have rigorously shown that every mild solution $u_{a,b,H}(t,x)$ satisfies the weak formulation. □

6.3.4. *Uniqueness of the weak solution*

Having established that every mild solution is also a weak solution, the next natural question is whether the weak solution is unique. If it is, then for this linear SPDE with additive noise, the mild and weak solutions are equivalent.

PROPOSITION 6.4 (uniqueness of weak solutions).– *Let $u_1(t,x)$ and $u_2(t,x)$ be two weak solutions to the SPDE [6.1] with the same initial condition $u(0,x) = 0$. Then, $u_1(t,x) = u_2(t,x)$ almost surely for all $(t,x) \in [0,T] \times \mathbb{R}^d$.*

Proof. Let u_1 and u_2 be two weak solutions to [6.1]. By definition, for any test function $\varphi \in C_c^\infty([0,T) \times \mathbb{R}^d)$, they both satisfy the weak formulation:

$$\int_0^T \int_{\mathbb{R}^d} u_1(t,x) \left(\frac{\partial \varphi}{\partial t}(t,x) + \frac{1}{2}\Delta\varphi(t,x) \right) dxdt$$

$$= -\int_0^T \int_{\mathbb{R}^d} \varphi(t,x) W_{a,b,H}(dt,dx)$$

$$\int_0^T \int_{\mathbb{R}^d} u_2(t,x) \left(\frac{\partial \varphi}{\partial t}(t,x) + \frac{1}{2}\Delta\varphi(t,x) \right) dxdt$$

$$= -\int_0^T \int_{\mathbb{R}^d} \varphi(t,x) W_{a,b,H}(dt,dx)$$

Subtracting the second equation from the first, we find that their difference:

$$u_D(t,x) = u_1(t,x) - u_2(t,x)$$

must satisfy:

$$\int_0^T \int_{\mathbb{R}^d} u_D(t,x) \left(\frac{\partial \varphi}{\partial t}(t,x) + \frac{1}{2}\Delta\varphi(t,x) \right) dxdt = 0 \qquad [6.11]$$

for all $\varphi \in C_c^\infty([0,T) \times \mathbb{R}^d)$. Additionally, since $u_1(0,x) = u_2(0,x) = 0$, we have $u_D(0,x) = 0$.

Equation [6.11] implies that $u_D(t,x)$ is a weak solution to the homogeneous heat equation with zero initial condition:

$$\begin{cases} \dfrac{\partial u_D}{\partial t} = \dfrac{1}{2}\Delta u_D, \quad t \in [0,T], \quad x \in \mathbb{R}^d \\ u_D(t,0) = 0. \end{cases} \qquad [6.12]$$

It is a well-known result in the theory of parabolic partial differential equations that the homogeneous heat equation with zero initial data in $L^2(\mathbb{R}^d)$ (or distributionally) admits only the trivial solution. Therefore, we can conclude that $u_D(t,x) = 0$ for all $(t,x) \in [0,T] \times \mathbb{R}^d$ almost surely (see, for instance, Evans 2010, Theorem 2, section 7.1.2)).

Therefore, $u_1(t,x) = u_2(t,x)$ almost surely, proving the uniqueness of the weak solution. □

6.3.4.1. *Equivalence of mild and weak solutions*

With the proofs of Proposition 6.3 (mild solution implies weak solution) and Proposition 6.4 (uniqueness of weak solutions), we can now formally state the equivalence for the SPDE [6.1] with zero initial conditions:

THEOREM 6.1 (equivalence of mild and weak solutions).– *For the SPDE [6.1] with initial condition $u_{a,b,H}(0,x) = 0$, a process $u_{a,b,H}(t,x)$ is a mild solution if and only if it is a weak solution.*

Proof.

1) $\Longrightarrow$ If $u_{a,b,H}(t,x)$ is a mild solution, then by Proposition 6.3, it is also a weak solution.

2) $\Longleftarrow$ If $u_{a,b,H}(t,x)$ is a weak solution, we know that a mild solution to [6.1] exists (Proposition 6.2). Let u_{mild} be this mild solution. By Proposition 6.3, u_{mild} is also a weak solution. Since weak solutions are unique by Proposition 6.4, and both $u_{a,b,H}$ and u_{mild} are weak solutions to the same SPDE with the same initial condition, it must be that $u_{a,b,H}(t,x) = u_{mild}(t,x)$ almost surely. Therefore, $u_{a,b,H}(t,x)$ must be the mild solution.

This concludes the proof of the equivalence between mild and weak solutions for this specific SPDE. □

REMARK 6.2.– *Owing to this equivalence, and for a more explicit characterization of the solution, we will henceforth focus on the mild solution formulation throughout the rest of this chapter.*

6.4. Further characterization of the mild solution

To ensure the existence and well-definedness of the mild solution as established in Proposition 6.2, we will assume throughout the rest of this chapter that the dimension d and Hurst parameter H satisfy:

$$d < 4H. \tag{6.13}$$

In the following proposition, we will give an explicit expression of the covariance of the mild solution.

PROPOSITION 6.5.– *For fixed $x \in \mathbb{R}^d$, and for $s \leq t$:*

$$\begin{aligned} \mathbb{E}\left[u_{a,b,H}(t,x)u_{a,b,H}(s,x)\right] &= \left(\frac{\pi}{2}\right)^{d/2} H(2H-1) \\ &\times \int_0^t \int_0^s dudv \frac{(a^2+b^2)|u-v|^{2H-2} - 2ab(u+v)^{2H-2}}{(t+s-u-v)^{d/2}}. \end{aligned} \tag{6.14}$$

Proof. By applying the Itô isometry identity [6.4], we have:

$$\begin{aligned} &\mathbb{E}\left[u_{a,b,H}(t,x)u_{a,b,H}(s,x)\right] \\ &= \int_0^t \int_0^s \mu_{a,b,H}(du,dv) \int_{\mathbb{R}^d} G(t-u,x-y)G(s-v,x-y)dy \\ &= \int_0^t \int_0^s dudv \frac{\partial^2 R^{a,b,H}(u,v)}{\partial u \partial v} \left(\frac{1}{(2\pi)^{d/2}(t+s-u-v)^{d/2}}\right). \end{aligned}$$

The last step utilizes the result of the spatial integral of two Green's kernels, as derived in the proof of Proposition 6.2:

$$\int_{\mathbb{R}^d} G(t-u,x-y)G(s-v,x-y)dy = \frac{1}{(2\pi)^{d/2}(t+s-u-v)^{d/2}}. \tag{6.15}$$

Finally, we substitute the explicit expression for the second mixed partial derivative of the ZgfBm covariance function given by:

$$\frac{\partial^2 R^{H,a,b}(u,v)}{\partial u \partial v} = H(2H-1)\left((a^2+b^2)|u-v|^{2H-2} - 2ab(u+v)^{2H-2}\right).$$

This directly leads to the expression in [6.14], completing the proof. □

The following proposition deals with the mixed-self-similarity of the solution sample paths.

PROPOSITION 6.6.– *For every $h > 0$, the process $u_{a,b,H}(t,x)$ is mixed-self-similar of order $H - \frac{d}{4}$, in the sense that, for fixed $x \in \mathbb{R}^d$, the processes $(u_{a,b,H}(ht,x))_{t\in\mathbb{R}_+}$ and $\left(u_{ah^{H-\frac{d}{4}},bh^{H-\frac{d}{4}},H}(t,x)\right)_{t\in\mathbb{R}_+}$ have the same law.*

Proof. For a fixed $h > 0$, both of the following processes $(u_{a,b,H}(ht,x))_{t\in\mathbb{R}_+}$ and $\left(u_{ah^{H-\frac{d}{4}},bh^{H-\frac{d}{4}},H}(t,x)\right)_{t\in\mathbb{R}_+}$ are centered Gaussian processes.

Consequently, to prove that they have the same law, it is sufficient to demonstrate that they possess the same covariance function.

First, we compute the covariance of the scaled process $u_{a,b,H}(ht,x)$. As a direct consequence of Proposition 6.5 (which provides the explicit form of the covariance in equation [6.14]), we replace t with ht and s with hs in its expression:

$$\mathbb{E}\left[u_{a,b,H}(ht,x)u_{a,b,H}(hs,x)\right] = \frac{1}{(2\pi)^{d/2}}H(2H-1) \times \int_0^{ht}\int_0^{hs} dudv\frac{(a^2+b^2)|u-v|^{2H-2} - 2ab(u+v)^{2H-2}}{(ht+hs-u-v)^{d/2}}. \qquad [6.16]$$

Next, we apply a change of variables in the integral. Let $u' = \frac{u}{h}$ and $v' = \frac{v}{h}$.

This implies $u = hu'$, $v = hv'$, and the Jacobian of the transformation is $dudv = h^2du'dv'$. The integration limits transform from $[0, ht] \times [0, hs]$ to $[0, t] \times [0, s]$.

We substitute these new variables into the integrand terms:

– The difference term: $|u-v|^{2H-2} = |hu' - hv'|^{2H-2} = h^{2H-2}|u'-v'|^{2H-2}$.

– The sum term: $(u+v)^{2H-2} = (hu'+hv')^{2H-2} = h^{2H-2}(u'+v')^{2H-2}$.

– The Green's function denominator term:

$$(ht+hs-u-v)^{d/2} = (h(t+s) - hu' - hv')^{d/2} = h^{d/2}(t+s-u'-v')^{d/2}.$$

Combining these transformations, the integral becomes:

$$\int_0^{ht}\int_0^{hs}\frac{(a^2+b^2)|u-v|^{2H-2} - 2ab(u+v)^{2H-2}}{(ht+hs-u-v)^{d/2}}dudv$$

$$= \int_0^t\int_0^s \frac{h^{2H-2}\left[(a^2+b^2)|u'-v'|^{2H-2} - 2ab(u'+v')^{2H-2}\right]}{h^{d/2}(t+s-u'-v')^{d/2}}h^2du'dv'$$

$$= h^{2H-d/2}\int_0^t\int_0^s \frac{(a^2+b^2)|u'-v'|^{2H-2} - 2ab(u'+v')^{2H-2}}{(t+s-u'-v')^{d/2}}du'dv'.$$

Substituting this back into [6.16], we obtain:

$$\begin{aligned}
&\mathbb{E}\left[u_{a,b,H}(ht,x)u_{a,b,H}(hs,x)\right] \\
&= \frac{1}{(2\pi)^{d/2}}H(2H-1)h^{2H-d/2} \\
&\times \int_0^t\int_0^s dudv\frac{(a^2+b^2)|u-v|^{2H-2}-2ab(u+v)^{2H-2}}{(t+s-u-v)^{d/2}} \\
&= h^{2H-d/2}\mathbb{E}\left[u_{a,b,H}(t,x)u_{a,b,H}(s,x)\right].
\end{aligned} \quad [6.17]$$

Next, we consider the covariance of the second process, $u_{ah^{H-\frac{d}{4}},bh^{H-\frac{d}{4}},H}(t,x)$. Let $\tilde{a}=ah^{H-\frac{d}{4}}$ and $\tilde{b}=bh^{H-\frac{d}{4}}$.

Using Proposition 6.5 with these new parameters $\tilde{a}$ and $\tilde{b}$:

$$\begin{aligned}
&\mathbb{E}\left[u_{\tilde{a},\tilde{b},H}(t,x)u_{\tilde{a},\tilde{b},H}(s,x)\right] \\
&= \frac{1}{(2\pi)^{d/2}}H(2H-1)\int_0^t\int_0^s dudv\frac{(\tilde{a}^2+\tilde{b}^2)|u-v|^{2H-2}-2\tilde{a}\tilde{b}(u+v)^{2H-2}}{(t+s-u-v)^{d/2}}.
\end{aligned}$$

Substituting the expressions for $\tilde{a}$ and $\tilde{b}$:

$$\tilde{a}^2+\tilde{b}^2=a^2h^{2(H-d/4)}+b^2h^{2(H-d/4)}=(a^2+b^2)h^{2H-d/2}.$$

$$2\tilde{a}\tilde{b}=2(ah^{H-d/4})(bh^{H-d/4})=2abh^{2(H-d/4)}=2abh^{2H-d/2}.$$

Thus, the expression becomes:

$$\begin{aligned}
&\mathbb{E}\left[u_{ah^{H-\frac{d}{4}},bh^{H-\frac{d}{4}},H}(t,x)\,u_{ah^{H-\frac{d}{4}},bh^{H-\frac{d}{4}},H}(s,x)\right] \\
&= \frac{H(2H-1)}{(2\pi)^{d/2}}\int_0^t\int_0^s dudv\frac{h^{2H-d/2}\left[(a^2+b^2)|u-v|^{2H-2}-2ab(u+v)^{2H-2}\right]}{(t+s-u-v)^{d/2}} \\
&= h^{2H-d/2}\left(\frac{H(2H-1)}{(2\pi)^{d/2}}\int_0^t\int_0^s dudv\frac{(a^2+b^2)|u-v|^{2H-2}-2ab(u+v)^{2H-2}}{(t+s-u-v)^{d/2}}\right) \\
&= h^{2H-d/2}\mathbb{E}\left[u_{a,b,H}(t,x)u_{a,b,H}(s,x)\right].
\end{aligned}$$

Comparing the results from [6.17], we observe that:

$$\mathbb{E}\left[u_{a,b,H}(ht,x)u_{a,b,H}(hs,x)\right]=\mathbb{E}\left[u_{ah^{H-\frac{d}{4}},bh^{H-\frac{d}{4}},H}(t,x)\,u_{ah^{H-\frac{d}{4}},bh^{H-\frac{d}{4}},H}(s,x)\right].$$

Since their covariances are equal and both processes are centered Gaussian, they have the same law. This completes the proof of Proposition 6.6. □

6.5. Regularity properties

6.5.1. *Study of the regularity of the solution in time*

In this section, we will examine the behavior of the increments of the solution $u_{a,b,H}(t,x)$ to [6.1] with respect to the time variable t. We will establish sharp upper and lower bounds for the L^2-norm of these increments.

The key result of this section is the following theorem.

THEOREM 6.2.– *There exist two positive constants C_3 and C_4 such that, for any $s,t \in [0,T]$ and for any $x \in \mathbb{R}^d$:*

$$C_3|t-s|^{2H-\frac{d}{2}} \leq \mathbb{E}\left[|u_{a,b,H}(t,x) - u_{a,b,H}(s,x)|^2\right] \leq C_4|t-s|^{2H-\frac{d}{2}}.$$

Proof. We have:

$$\mathbb{E}\left[|u_{a,b,H}(t,x) - u_{a,b,H}(s,x)|^2\right] = R_u(t,t) - 2R_u(t,s) + R_u(s,s),$$

where R_u denotes the covariance of the process $u_{a,b,H}$ with respect to the time variable for fixed $x \in \mathbb{R}^d$:

$$\begin{aligned} R_u(t,s) &= \mathbb{E}\left[u_{a,b,H}(t,x)u_{a,b,H}(s,x)\right] \\ &= \left(\frac{\pi}{2}\right)^{d/2} \int_0^t \int_0^s dudv \frac{\partial^2 R^{a,b,H}(u,v)}{\partial u \partial v}(t+s-u-v)^{-d/2} \end{aligned}$$

for every $s,t \in [0,T]$. So,

$$\begin{aligned} &\mathbb{E}\left[|u_{a,b,H}(t,x) - u_{a,b,H}(s,x)|^2\right] \\ &= \left(\frac{\pi}{2}\right)^{d/2} \int_0^t \int_0^t dudv \frac{\partial^2 R^{a,b,H}(u,v)}{\partial u \partial v}(2t-u-v)^{-d/2} \\ &\quad -2\left(\frac{\pi}{2}\right)^{d/2} \int_0^t \int_0^s dudv \frac{\partial^2 R^{a,b,H}(u,v)}{\partial u \partial v}(t+s-u-v)^{-d/2} \\ &\quad +\left(\frac{\pi}{2}\right)^{d/2} \int_0^s \int_0^s dudv \frac{\partial^2 R^{a,b,H}(u,v)}{\partial u \partial v}(2s-u-v)^{-d/2}, \end{aligned}$$

which can also be written as:

$$\mathbb{E}\left[|u_{a,b,H}(t,x) - u_{a,b,H}(s,x)|^2\right] = A_{a,b,H}(t,s) + B_{a,b,H}(t,s) + C_{a,b,H}(t,s),$$

$$A_{a,b,H}(t,s) = \left(\frac{\pi}{2}\right)^{d/2} \int_s^t \int_s^t dudv \frac{\partial^2 R^{a,b,H}(u,v)}{\partial u \partial v}(2t-u-v)^{-d/2},$$

$$B_{a,b,H}(t,s) = \left(\frac{\pi}{2}\right)^{d/2} \int_0^s \int_0^s dudv \frac{\partial^2 R^{a,b,H}(u,v)}{\partial u \partial v} (2t-u-v)^{-d/2}$$

$$-2(t+s-u-v)^{-d/2} + (2s-u-v)^{-d/2}\Bigg],$$

and

$$C_{a,b,H}(t,s) = 2\left(\frac{\pi}{2}\right)^{d/2} \int_s^t \int_0^s dudv \frac{\partial^2 R^{a,b,H}(u,v)}{\partial u \partial v}$$

$$\times \left[(2t-u-v)^{-d/2} - 2(t+s-u-v)^{-d/2}\right].$$

Since, $C_{a,b,H}(t,s) \leq 0$ and since:

$$\frac{\partial^2 R^{a,b,H}(u,v)}{\partial u \partial v} \leq C_2 \mid u-v \mid^{2H-2},$$

$$\mathbb{E}\left[|u_{a,b,H}(t,x) - u_{a,b,H}(s,x)|^2\right]$$

$$\leq c\Bigg[\int_s^t \int_s^t dudv \mid u-v \mid^{2H-2} (2t-u-v)^{-d/2} + \int_0^s \int_0^s dudv \mid u-v \mid^{2H-2}$$

$$\times\Big((2t-u-v)^{-d/2} - 2(t+s-u-v)^{-d/2} + (2s-u-v)^{-d/2}\Big)\Bigg],$$

[6.18]

where c denote a positive constant.

All of this with Lemma A.8 and Lemma A.9 allow us to get the upper bound.

Let us now prove the lower bound. The mild solution has the form given in equation [6.7]. So, for every $x \in \mathbb{R}^d$ and $(s,t) \in [0,T]^2$:

$$u_{a,b,H}(t,x) - u_{a,b,H}(s,x)$$
$$= \int_0^T \int_{\mathbb{R}^d} \Bigg(G(t-\sigma, x-y)\mathbf{1}_{(0,t)}(\sigma) - G(s-\sigma, x-y)\mathbf{1}_{(0,s)}(\sigma)\Bigg) dW_{a,b,H}(\sigma,y).$$

[6.19]

By the transfer formula [6.6], we get:

$$u_{a,b,H}(t,x) - u_{a,b,H}(s,x) = d_H \int_{\mathbb{R}} \int_{\mathbb{R}^d} dW(\sigma,y) F_{t,s}(\sigma,y), \qquad [6.20]$$

where

$$F_{t,s}(\sigma,y) = \int_{\mathbb{R}} du K_{a,b,H}(u,\sigma)\left[G(t-u,x-y)\mathbf{1}_{(0,t)}(u)\right.$$
$$\left. -G(s-u,x-y)\mathbf{1}_{(0,s)}(u)\right]$$

with

$$K_{a,b,H}(u,\sigma) = (a+b)\left(((u-\sigma)_+)^{H-3/2} - ((u+\sigma)_-)^{H-3/2}\right)$$
$$+(a-b)\left(((u-\sigma)_+)^{H-3/2} + ((u+\sigma)_-)^{H-3/2}\right).$$

Equation [6.20] implies that:

$$\mathbb{E}\left[u_{a,b,H}(t,x) - u_{a,b,H}(s,x)\right]^2 = d_H^2 \mathbb{E}\left[\int_{\mathbb{R}}\int_{\mathbb{R}^d} dW(\sigma,y)F_{t,s}(\sigma,y)\right]^2. \quad [6.21]$$

By the isometry of the Wiener process W, we get:

$$\mathbb{E}\left[\int_{\mathbb{R}}\int_{\mathbb{R}^d} dW(\sigma,y)F_{t,s}(\sigma,y)\right]^2 = \int_{\mathbb{R}}\int_{\mathbb{R}^d} d\sigma dy F_{t,s}^2(\sigma,y) \geq \int_s^t \int_{\mathbb{R}^d} d\sigma dy F_{t,s}^2(\sigma,y).$$
$$[6.22]$$

Therefore, for every $\sigma \in [s,t]$ and $u \geq 0$, we have $(u+\sigma)_- = 0$:

$$K_{a,b,H}(u,\sigma) = (a+b)(u-\sigma)_+^{H-3/2} + (a-b)(u-\sigma)_+^{H-3/2} = 2a(u-\sigma)_+^{H-3/2},$$

and as consequence:

$$F_{t,s}(\sigma,y) = \int_0^t duG(t-u,x-y)K_{a,b,H}(u,\sigma)$$
$$- \int_0^s duG(s-u,x-y)K_{a,b,H}(u,\sigma)$$
$$= 2a\left[\int_0^t duG(t-u,x-y)(u-\sigma)_+^{H-3/2} - \int_0^s duG(s-u,x-y)(u-\sigma)_+^{H-3/2}\right].$$

As $\sigma \in [s,t]$, the term $(u-\sigma)_+^{H-3/2}$ is only non-zero when $u > \sigma$. This means the second integral, $\int_0^s duG(s-u,x-y)(u-\sigma)_+^{H-3/2}$, is zero. So, the second integral vanishes, and the first integral simplifies to an integral over the interval $[\sigma,t]$. Therefore:

$$F_{t,s}(\sigma,y) = 2a\int_\sigma^t duG(t-u,x-y)(u-\sigma)^{H-3/2}.$$

Hence:

$$\mathbb{E}\left[\int_{\mathbb{R}}\int_{\mathbb{R}^d} dW(\sigma,y)F_{t,s}(\sigma,y)\right]^2$$

$$\geq 4a^2\int_s^t\int_{\mathbb{R}^d} d\sigma dy\left(\int_\sigma^t duG(t-u,x-y)(u-\sigma)^{H-3/2}\right)^2. \qquad [6.23]$$

So, for every $s,t\in[0,T];\ s\leq t$:

$$\mathbb{E}\left[\int_{\mathbb{R}}\int_{\mathbb{R}^d} dW(\sigma,y)F_{t,s}(\sigma,y)\right]^2$$

$$\geq 4a^2\int_s^t d\sigma\int_{\mathbb{R}^d} dy\int_\sigma^t\int_\sigma^t dvduG(t-u,x-y)(u-\sigma)^{H-3/2}$$

$$\times G(t-v,x-y)(v-\sigma)^{H-3/2}$$

$$=4a^2\int_s^t du\int_s^t dv\int_{\mathbb{R}^d} dyG(t-u,x-y)G(t-v,x-y) \qquad [6.24]$$

$$\times\int_s^{u\wedge v}(u-\sigma)_+^{H-3/2}(v-\sigma)_+^{H-3/2}d\sigma$$

where in the last equality, we have used the fact that

$$(s\leq\sigma\leq t,\ \sigma\leq u\leq t,\ \sigma\leq v\leq t)\Longleftrightarrow(s\leq u\leq t,\ s\leq v\leq t,\ s\leq\sigma\leq u\wedge v).$$

Equations [6.24] and [6.15] imply:

$$\mathbb{E}\left[\int_{\mathbb{R}}\int_{\mathbb{R}^d} dW(\sigma,y)F_{t,s}(\sigma,y)\right]^2$$

$$\geq 4a^2\left(\frac{\pi}{2}\right)^{d/2}\int_s^t du\int_s^t dv(2t-u-v)^{-d/2}\int_s^{u\wedge v}(u-\sigma)_+^{H-3/2}(v-\sigma)_+^{H-3/2}d\sigma. \qquad [6.25]$$

By the change of variable $z=\dfrac{u\wedge v-\sigma}{u\vee v-\sigma}$ and by an easy calculus, we get:

$$\int_s^{u\wedge v}(u-\sigma)^{H-3/2}(v-\sigma)^{H-3/2}d\sigma=\mid u-v\mid^{2H-2}\int_0^{\frac{u\wedge v-s}{u\vee v-s}}(1-z)^{1-2H}z^{H-3/2}dz. \qquad [6.26]$$

Then, by equations [6.25] and [6.26]:

$$\mathbb{E}\left[\int_{\mathbb{R}}\int_{\mathbb{R}^d} dW(\sigma,y)F_{t,s}(\sigma,y)\right]^2$$

$$\geq 4a^2\left(\frac{\pi}{2}\right)^{d/2}\int_s^t du\int_s^t dv(2t-u-v)^{-d/2}\mid u-v\mid^{2H-2} \quad [6.27]$$

$$\times\int_0^{\frac{u\wedge v-s}{u\vee v-s}}(1-z)^{1-2H}z^{H-3/2}dz.$$

Now, by the change of variable $u-s=u'$ and $v-s=v'$, we obtain:

$$\mathbb{E}\left[\int_{\mathbb{R}}\int_{\mathbb{R}^d} dW(\sigma,y)F_{t,s}(\sigma,y)\right]^2$$

$$\geq 4a^2\left(\frac{\pi}{2}\right)^{d/2}\int_0^{t-s} du\int_0^{t-s} dv(2(t-s)-u-v)^{-d/2}\mid u-v\mid^{2H-2} \quad [6.28]$$

$$\times\int_0^{\frac{u\wedge v}{u\vee v}}(1-z)^{1-2H}z^{H-3/2}dz.$$

Finally, by the change of variable $\tilde{u}=\frac{u}{t-s}$ and $\tilde{v}=\frac{v}{t-s}$, we get:

$$\mathbb{E}\left[\int_{\mathbb{R}}\int_{\mathbb{R}^d} dW(\sigma,y)F_{t,s}(\sigma,y)\right]^2 \geq D(d,H)(t-s)^{2H-\frac{d}{2}}, \quad [6.29]$$

where $D(d,H)$ is the constant defined by:

$$D(d,H)=Const\int_0^1 du\int_0^1 dv(2-u-v)^{-d/2}\mid u-v\mid^{2H-2}$$

$$\times\int_0^{\frac{u\wedge v}{u\vee v}}(1-z)^{1-2H}z^{H-3/2}dz.$$

The constant $D(d,H)$ is clearly finite since $H>\frac{1}{2}$. □

REMARK 6.3.– *Theorem 6.2 establishes that the process* $(u_{a,b,H}(.,x))$ *is an infinite-dimensional quasi-helix (in the sense of Kahane (1993)) with index* $I=H-\frac{d}{4}$. *Quasi-helices possess a range of well-studied properties, as detailed in Kahane (1993).*

In particular, as an immediate consequence of Theorem 6.2, we get:

COROLLARY 6.1.– *For any* $x\in\mathbb{R}^d$, *the process* $t\mapsto u_{a,b,H}(t,x)$ *is Hölder continuous of order* $\delta\in(0,H-\frac{d}{4})$.

Proof. Theorem 6.2 yields the following inequality:

$$\mathbb{E}\left[|u_{a,b,H}(t,x) - u_{a,b,H}(s,x)|^2\right] \leq C_4|t-s|^{2H-\frac{d}{2}},$$

for every $(s,t) \in \mathbb{R}_+^2$ with $s \leq t$, and $(a,b) \in \mathbb{R}^2 \setminus \{(0,0)\}$. In conjunction with the Kolmogorov continuity theorem (see, e.g. Theorem 1.2 in Mishura and Zili (2018, p. 14)), this inequality yields Corollary 6.1. □

As a second consequence of Theorem 6.2, and by proceeding as in the proof of Corollary 3.3, we establish the following:

COROLLARY 6.2.– *For every $x \in \mathbb{R}^d$ and $t_0 \in \mathbb{R}$:*

$$\lim_{\epsilon \to 0} \sup_{t \in [t_0-\epsilon, t_0+\epsilon]} \left| \frac{u_{a,b,H}(t,x) - u_{a,b,H}(t_0,x)}{t-t_0} \right| = +\infty \quad a.s.$$

Consequently, the trajectories of the process $u_{a,b,H}(\cdot,x)$ are nowhere differentiable.

6.5.2. *Sharp regularity of the solution in space*

In the spirit of Tudor and Xiao (2015), in this section, we fix $t > 0$ and analyze the spatial regularity of the solution $\{u_{a,b,H}(t,x), x \in \mathbb{R}^d\}$. We will first prove the following lemma.

LEMMA 6.1.– *The Gaussian random field $\{u_{a,b,H}(t,x), x \in \mathbb{R}^d\}$ is stationary. Its spectral measure $\Delta(d\xi)$ is given by:*

$$\Delta(d\xi) = H(2H-1)(2\pi)^{-d}\left(\int_0^t\int_0^t \left[(a^2+b^2)|u-v|^{2H-2} - 2ab(u+v)^{2H-2}\right]\right.$$
$$\left.\times \exp\left(-\frac{(2t-u-v)}{2}|\xi|^2\right) dudv\right) d\xi.$$

Proof. To prove stationarity, we compute the covariance function:

$$\mathbb{E}[u_{a,b,H}(t,x)u_{a,b,H}(t,y)]$$

and show it depends only on the difference $x - y$. By the definition of $u_{a,b,H}(t,x)$ (which involves a Green kernel G and an integral with respect to a spatial variable, characteristic of solutions to certain SPDEs), and applying Parseval's identity to the integral over the spatial variable z, we have:

$$\mathbb{E}\left[u_{a,b,H}(t,x)u_{a,b,H}(t,y)\right]$$
$$= \int_0^t\int_0^t dudv \frac{\partial^2 R^{a,b,H}(u,v)}{\partial u \partial v} \int_{\mathbb{R}^d} dz\, G(x-z,t,u)G(y-z,t,v)$$

Applying the Fourier transform to the Green kernel $G(\cdot,t,u)$ (denoted $\mathcal{F}[G(\cdot,t,u)](\xi)$), we express the spatial integral in the frequency domain. Specifically, recognizing that the integral over z is a convolution type integral whose Fourier transform yields a product of Fourier transforms and the term $e^{i\langle x-y,\xi\rangle}$

$$
\begin{aligned}
:= (2\pi)^{-d} &\int_0^t \int_0^t dudv \, \frac{\partial^2 R^{a,b,H}(u,v)}{\partial u \partial v} \\
\times &\int_{\mathbb{R}^d} d\xi \, \mathcal{F}[G(\cdot,t,u)](\xi)\mathcal{F}[G(\cdot,t,v)](-\xi) e^{i\langle x-y,\xi\rangle}.
\end{aligned}
$$

Next, utilizing the specific form of the Fourier transform of the Green kernel,

$$
\mathcal{F}[G(\cdot,t,\tau)](\xi) = \exp\left(-\frac{(t-\tau)}{2}|\xi|^2\right),
$$

and rearranging the integrals

$$
\begin{aligned}
:= &\int_{\mathbb{R}^d} e^{i\langle x-y,\xi\rangle} \left((2\pi)^{-d} \int_0^t \int_0^t dudv \, \frac{\partial^2 R^{a,b,H}(u,v)}{\partial u \partial v}\right. \\
&\times \exp\left.\left(-\frac{(2t-u-v)}{2}|\xi|^2\right)\right) d\xi.
\end{aligned}
$$

Finally, substituting the known expression for $\frac{\partial^2 R^{a,b,H}(u,v)}{\partial u \partial v}$ from (Lemma 5.1)

$$
\begin{aligned}
:= H(2H-1)(2\pi)^{-d} &\int_{\mathbb{R}^d} e^{i\langle x-y,\xi\rangle} \\
\times &\left(\int_0^t \int_0^t dudv \, \left[(a^2+b^2)|u-v|^{2H-2} - 2ab(u+v)^{2H-2}\right]\right. \\
\times &\exp\left.\left(-\frac{(2t-u-v)}{2}|\xi|^2\right)\right) d\xi.
\end{aligned}
$$

The resulting covariance function depends solely on $x-y$, which confirms the stationarity of the Gaussian random field $\{u_{a,b,H}(t,x), x \in \mathbb{R}^d\}$. The spectral measure $\Delta(d\xi)$ is thus identified as the bracketed term in the final line's integrand with respect to ξ. □

REMARK 6.4.– *Lemma 6.1 establishes the spatial stationarity of the solution, which simplifies the analysis of its spatial properties. Crucially, it provides the explicit form of the spectral measure $\Delta(d\xi)$, a fundamental tool for investigating the fine regularity of Gaussian fields in the frequency domain.*

Building upon the properties of the spectral measure, we now present a result concerning its bounds for large frequencies.

COROLLARY 6.3.– *There exist two positive constants* $c_1(t, H, a, b)$ *and* $c_2(t, H, a, b)$, *depending only on t, H, a and b, such that for all* $\xi \in \mathbb{R}^d$ *with* $|\xi| \geq 1$:

$$c_1(t, H, a, b)|\xi|^{-4H} d\xi \leq \Delta(d\xi) \leq c_2(t, H, a, b)|\xi|^{-4H} d\xi.$$

Proof. By Lemma 6.1, the spectral measure $\Delta(d\xi)$ is given by:

$$\Delta(d\xi) = H(2H-1)(2\pi)^{-d} \left(\int_0^t \int_0^t [(a^2+b^2)|u-v|^{2H-2} - 2ab(u+v)^{2H-2}] \right.$$
$$\left. \times \exp\left(-\frac{(2t-u-v)}{2}|\xi|^2 \right) dudv \right) d\xi.$$

Furthermore, by equation [6.9]:

$$C_1(2\pi)^{-d} \left(\int_0^t \int_0^t |u-v|^{2H-2} \exp\left(-\frac{(2t-u-v)}{2}|\xi|^2 \right) dudv \right) d\xi$$
$$\leq \Delta(d\xi)$$
$$\leq C_2(2\pi)^{-d} \left(\int_0^t \int_0^t |u-v|^{2H-2} \exp\left(-\frac{(2t-u-v)}{2}|\xi|^2 \right) dudv \right) d\xi, \qquad [6.30]$$

for some positive constants C_1 and C_2 that absorb coefficients like $H(2H-1)$ and terms from the bracket in the full spectral measure expression.

Now, consider the integral term $I(\xi)$ within these bounds:

$$I(\xi) = \int_0^t \int_0^t |u-v|^{2H-2} \exp\left(-\frac{(2t-u-v)}{2}|\xi|^2 \right) dudv.$$

Applying the change of variables $u' = t-u$ and $v' = t-v$, the integral $I(\xi)$ becomes:

$$I(\xi) = \int_0^t \int_0^t |u'-v'|^{2H-2} \exp\left(-\frac{(u'+v')}{2}|\xi|^2 \right) du'dv'.$$

This integral is precisely the form analyzed in Proposition 4.3 of Balan and Tudor (2010). According to this proposition, there exist strictly positive constants $c_{1,H}$ and $c_{2,H}$ (depending only on H) such that for all $\xi \in \mathbb{R}^d$:

$$c_{1,H}(t^{2H} \wedge 1) \left(\frac{1}{1+|\xi|^2} \right)^{2H} \leq I(\xi) \leq c_{2,H}(t^{2H} \wedge 1) \left(\frac{1}{1+|\xi|^2} \right)^{2H}. \qquad [6.31]$$

For $|\xi| \geq 1$, we note that $\left(\frac{1}{1+|\xi|^2} \right)^{2H}$ is asymptotically equivalent to $|\xi|^{-4H}$. Specifically, for $|\xi| \geq 1$, we have:

$$\frac{1}{2|\xi|^2} \leq \frac{1}{1+|\xi|^2} \leq \frac{1}{|\xi|^2},$$

which implies:

$$2^{-2H}|\xi|^{-4H} \leq \left(\frac{1}{1+|\xi|^2}\right)^{2H} \leq |\xi|^{-4H}.$$

Substituting this asymptotic behavior into equation [6.31] and then back into equation [6.30], we find that for $|\xi| \geq 1$:

$$C_1(2\pi)^{-d}c'_{1,H}(t^{2H} \wedge 1)|\xi|^{-4H}d\xi \leq \Delta(d\xi) \leq C_2(2\pi)^{-d}c'_{2,H}(t^{2H} \wedge 1)|\xi|^{-4H}d\xi,$$

where $c'_{1,H}$ and $c'_{2,H}$ are new constants arising from the asymptotic equivalence. By defining:

$$c_1(t,H,a,b) = C_1(2\pi)^{-d}c'_{1,H}(t^{2H}\wedge 1) \; and \; c_2(t,H,a,b) = C_2(2\pi)^{-d}c'_{2,H}(t^{2H}\wedge 1),$$

which are clearly positive and depend only on t, H, a, b, the desired result is obtained. This concludes the proof. □

REMARK 6.5.– *Corollary 6.3 means, among other things, that the spectral measure $\Delta(d\xi)$ is comparable with an absolutely continuous measure with density function that is comparable to $|\xi|^{-4H}$ for all $\xi \in \mathbb{R}^d$ with $|\xi| \geq 1$. This is very interesting for the study of the regularity of $\{u_{a,b,H}(t,x), x \in \mathbb{R}^d\}$. Indeed, a direct consequence of Corollary 6.3 is the following result.*

COROLLARY 6.4.– *Let $\beta = \min\{1, 2H - \frac{d}{2}\}$ and $\rho = \begin{cases} 1 \text{ if } \beta = 1 \\ 0 \text{ otherwise.}\end{cases}$ If we denote:*

$$I(x,y) = \mathbb{E}\left[|u_{a,b,H}(t,x) - u_{a,b,H}(t,y)|^2\right],$$

then, for any $M > 0$, there exist positive and finite constants c_3, c_4 such that for any $x, y \in [-M, M]^d$:

$$c_3|x-y|^{2\beta}\left(\log\frac{1}{|x-y|}\right)^{\rho} \leq I(x,y) \leq c_4|x-y|^{2\beta}\left(\log\frac{1}{|x-y|}\right)^{\rho}. \quad [6.32]$$

Proof. Take $x, y \in [-M, M]^d$ and let $z := y - x \in \mathbb{R}^d$. By the Itô isometry and the definition of the mild solution, we have:

$$\begin{aligned} I(x,y) &= \mathbb{E}\left[|u_{a,b,H}(t,y) - u_{a,b,H}(t,x)|^2\right] \\ &= \int_0^t \int_0^t \mu_{a,b,H}(du,dv) \int_{\mathbb{R}^d} dy' \Big[G(t-u, x+z-y') - G(t-u, x-y')\Big] \\ &\quad \times \Big[G(t-v, x+z-y') - G(t-v, x-y')\Big]. \end{aligned}$$

Applying Parseval's identity for the spatial integral, and recalling the Fourier transform property

$$\mathcal{F}[f(\cdot - y_0)](\xi) = e^{i\langle y_0, \xi\rangle}\mathcal{F}[f](\xi)$$

for

$$\mathcal{F}[G(T,\cdot)](\xi) = \exp(-\frac{T}{2}|\xi|^2);$$

we get:

$$\begin{aligned}
I(x,y) &= (2\pi)^{-d}\int_0^t\int_0^t \mu_{a,b,H}(du,dv)\int_{\mathbb{R}^d} d\xi \\
&\quad\times \mathcal{F}\left(G(t-u,x+z-\cdot)-G(t-u,x-\cdot)\right)(\xi) \\
&\quad\times \overline{\mathcal{F}}\left(G(t-v,x+z-\cdot)-G(t-v,x-\cdot)\right)(\xi) \\
&= (2\pi)^{-d}\int_0^t\int_0^t \mu_{a,b,H}(du,dv)\int_{\mathbb{R}^d} d\xi \exp\left(-(2t-u-v)\frac{|\xi|^2}{2}\right) \\
&\quad\times \left(e^{i\langle x+z,\xi\rangle}-e^{i\langle x,\xi\rangle}\right)\left(e^{-i\langle x+z,\xi\rangle}-e^{-i\langle x,\xi\rangle}\right) \\
&= (2\pi)^{-d}\int_0^t\int_0^t \mu_{a,b,H}(du,dv)\int_{\mathbb{R}^d} d\xi \exp\left(-(2t-u-v)\frac{|\xi|^2}{2}\right) \\
&\quad\times \left(1-e^{i\langle z,\xi\rangle}-e^{-i\langle z,\xi\rangle}+1\right) \\
&= (2\pi)^{-d}\int_0^t\int_0^t \mu_{a,b,H}(du,dv)\int_{\mathbb{R}^d} d\xi \\
&\quad\times \exp\left(-(2t-u-v)\frac{|\xi|^2}{2}\right)(2-2\cos\langle\xi,z\rangle).
\end{aligned}$$

Therefore,

$$\mathbb{E}\left[|u_{a,b,H}(t,x)-u_{a,b,H}(t,y)|^2\right] = 2(2\pi)^{-d}\int_{\mathbb{R}^d} d\xi(1-\cos\langle\xi,z\rangle)\theta(t,\xi) \qquad [6.33]$$

where

$$\theta(t,\xi) = \int_0^t\int_0^t \frac{\partial^2 R_{a,b,H}}{\partial u\partial v}(u,v)dudv\exp\left(-(2t-u-v)\frac{|\xi|^2}{2}\right).$$

By equation [6.9]:

$$\begin{aligned}
&C_1\int_0^t\int_0^t |u-v|^{2H-2}dudv\exp\left(-(2t-u-v)\frac{|\xi|^2}{2}\right) \\
&\leq \theta(t,\xi) \qquad [6.34] \\
&\leq C_2\int_0^t\int_0^t |u-v|^{2H-2}dudv\exp\left(-(2t-u-v)\frac{|\xi|^2}{2}\right),
\end{aligned}$$

where C_1 and C_2 are two positive constants. By Lemmas A.10 and A.11, there exist two strictly positive constants C_5 and C_6 such that:

$$C_5|x-y|^{2\beta}\left(\log\frac{1}{|x-y|}\right)^{\rho}$$

$$\leq \int_{\mathbb{R}^d}(1-\cos\langle\xi,z\rangle)d\xi\int_0^t\int_0^t|u-v|^{2H-2}dudv\exp\left(-(2t-u-v)\frac{|\xi|^2}{2}\right) \quad [6.35]$$

$$\leq C_6|x-y|^{2\beta}\left(\log\frac{1}{|x-y|}\right)^{\rho}.$$

Hence, the result is a straightforward consequence of equations [6.33]–[6.35]. □

REMARK 6.6.– *This corollary provides the crucial power-law bounds for the second-order increment $I(x,y)$ (often called the variogram in stochastic process theory). These bounds are fundamental for analyzing the local Hölder regularity and other fine sample path properties of the Gaussian random field $\{u_{a,b,H}(t,x), x\in\mathbb{R}^d\}$. We now state the precise Hölder continuity result as a direct consequence.*

COROLLARY 6.5.– *Suppose $2H-\frac{d}{2}>0$. Then, for every fixed $t>0$, the sample paths of the Gaussian random field $\{u_{a,b,H}(t,x), x\in\mathbb{R}^d\}$ are almost surely Hölder continuous with any exponent α, such that $0<\alpha<\beta$, where $\beta=\min\{1,2H-d/2\}$.*

Proof. Let $I(x,y)=\mathbb{E}[|u_{a,b,H}(t,x)-u_{a,b,H}(t,y)|^2]$. From the preceding corollary, for any $M>0$, there exist positive and finite constants c_3, c_4 such that for any $x,y\in[-M,M]^d$:

$$c_3|x-y|^{2\beta}\left(\log\frac{1}{|x-y|}\right)^{\rho}\leq I(x,y)\leq c_4|x-y|^{2\beta}\left(\log\frac{1}{|x-y|}\right)^{\rho},$$

with $\rho=\begin{cases}1 \text{ if } \beta=1\\ 0 \text{ otherwise.}\end{cases}$ We know that $\{u_{a,b,H}(t,x), x\in\mathbb{R}^d\}$ is a Gaussian random field, and that for a zero-mean Gaussian random variable Z, $\mathbb{E}[|Z|^p]=C_p(\mathbb{E}[Z^2])^{p/2}$ for some constant C_p. Therefore, for any $p\geq 2$:

$$\mathbb{E}[|u_{a,b,H}(t,x)-u_{a,b,H}(t,y)|^p]=C_p(I(x,y))^{p/2}.$$

Substituting the upper bound for $I(x,y)$:

$$\mathbb{E}[|u_{a,b,H}(t,x)-u_{a,b,H}(t,y)|^p]\leq C_p\left(c_4|x-y|^{2\beta}\left(\log\frac{1}{|x-y|}\right)^{\rho}\right)^{p/2}.$$

$$\mathbb{E}[|u_{a,b,H}(t,x)-u_{a,b,H}(t,y)|^p]\leq K_1|x-y|^{p\beta}\left(\log\frac{1}{|x-y|}\right)^{p\rho/2},$$

where $K_1=C_pc_4^{p/2}$ is a positive constant.

Now, we address the logarithmic term. For any $\delta > 0$, it is a standard result that for sufficiently small $r > 0$, $\log(1/r) \leq C_\delta r^{-\delta}$ for some constant C_δ. Applying this to our bound:

$$\left(\log \frac{1}{|x-y|}\right)^{p\rho/2} \leq (C_\delta |x-y|^{-\delta})^{p\rho/2} = C_\delta^{p\rho/2} |x-y|^{-p\rho\delta/2}.$$

Substituting this into the moment bound, we get:

$$\mathbb{E}[|u_{a,b,H}(t,x) - u_{a,b,H}(t,y)|^p] \leq K_2 |x-y|^{p\beta - p\rho\delta/2},$$

where $K_2 = K_1 C_\delta^{p\rho/2}$ is a positive constant.

Let α be any exponent such that $0 < \alpha < \beta$. We want to show that the sample paths are Hölder continuous with exponent α. Choose $\delta > 0$ to be sufficiently small such that $\beta - \rho\delta/2 > \alpha$ (this is always possible because $\rho\delta/2$ can be made arbitrarily small by choosing δ small, and $\alpha < \beta$). Let $\gamma = \beta - \rho\delta/2$. Then, we have $\alpha < \gamma \leq \beta$. So,

$$\mathbb{E}[|u_{a,b,H}(t,x) - u_{a,b,H}(t,y)|^p] \leq K_2 |x-y|^{p\gamma}.$$

To apply Kolmogorov's continuity theorem (Theorem 1.5), we need to ensure that the exponent $p\gamma$ is strictly greater than d (the dimension of the index set). Since $\gamma > \alpha > 0$, we can always choose an integer p sufficiently large such that $p\gamma > d$. For instance, choose $p = \lceil d/\gamma \rceil + 1$. Thus, for this chosen p, we have:

$$\mathbb{E}[|u_{a,b,H}(t,x) - u_{a,b,H}(t,y)|^p] \leq K_2 |x-y|^{d+\epsilon_0},$$

where $\epsilon_0 = p\gamma - d > 0$.

By Kolmogorov's continuity Theorem 1.5, the process $\{u_{a,b,H}(t,x), x \in \mathbb{R}^d\}$ has a modification whose sample paths are almost surely Hölder continuous with any exponent γ, such that $0 < \gamma < \epsilon_0/p$. Substituting $\epsilon_0 = p\gamma - d$:

$$\gamma < \frac{p\gamma - d}{p} = \gamma - \frac{d}{p}.$$

Since γ can be arbitrarily close to β (by choosing δ small enough), and d/p can be arbitrarily small (by choosing p large enough), the exponent $\gamma - d/p$ can be made arbitrarily close to β. Therefore, the sample paths of $\{u_{a,b,H}(t,x), x \in \mathbb{R}^d\}$ are almost surely Hölder continuous with any exponent α such that $0 < \alpha < \beta$. □

6.6. Exercises

EXERCISE 6.1.– *Consider the constant $D(d,H)$ defined by the integral:*

$$D(d,H) = \int_0^1\int_0^1 (2-u-v)^{-d/2} \mid u-v \mid^{2H-2} \left(\int_0^{\frac{u\wedge v}{u\vee v}} (1-z)^{1-2H} z^{H-3/2} dz\right) dudv.$$

where d > 0, *and* H *is a parameter satisfying* $H > 1/2$. *The aim of this exercise is to prove rigorously that* $D(d, H)$ *is finite for* $H > 1/2$.

1) Let $g(u, v) = \int_0^{\frac{u \wedge v}{u \vee v}} (1 - z)^{1-2H} z^{H-3/2} dz$. *Show that for fixed* $u, v \in (0, 1)$, $g(u, v)$ *is finite if* $H > 1/2$. *Justify the convergence at* $z = 0$ *and* $z = 1$.

2) Let $F(u, v) = (2 - u - v)^{-d/2} \mid u - v \mid^{2H-2} g(u, v)$. *Identify all potential singularities of* $F(u, v)$ *within the domain of integration* $[0, 1] \times [0, 1]$.

3) Apply a suitable change of variables (e.g. $x = u + v$, $y = u - v$*) to transform the outer double integral. Clearly state the new variables, their integration limits and the Jacobian of the transformation.*

4) Based on the transformed integral from Step 3, rigorously demonstrate that $D(d, H)$ *is finite.*

EXERCISE 6.2.– *Following the established methodology for proving the nowhere differentiability of* $Z^H(a, b)$ *as presented in the proof of Corollary 3.3, show that for every* $x \in \mathbb{R}^d$ *and* $t_0 \in \mathbb{R}$, *the process* $u_{a,b,H}(., x)$ *is nowhere differentiable with probability one.*

Specifically, demonstrate that:

$$\lim_{\epsilon \to 0} \sup_{t \in [t_0 - \epsilon, t_0 + \epsilon]} \left| \frac{u_{a,b,H}(t, x) - u_{a,b,H}(t_0, x)}{t - t_0} \right| = +\infty \quad a.s.$$

EXERCISE 6.3 (parameter influence on spatial smoothness).– *This chapter demonstrated that the sharp spatial regularity of the Gaussian random field* $\{u_{a,b,H}(t, x), x \in \mathbb{R}^d\}$ *for fixed* $t > 0$ *is critically determined by the Hurst parameter* H *and the spatial dimension d. Specifically, its spectral measure decays as* $|\xi|^{-4H}$ *for large frequencies, leading to almost sure Hölder continuity of any order* $\alpha < \beta$, *where* $\beta = \min\{1, 2H - d/2\}$. *A logarithmic correction term is present in the second-order increment* $I(x, y)$ *if and only if* $\beta = 1$. *The term* $2H - d/2$ *plays a critical role in defining the Hölder exponent* β.

1) Consider a scenario where $2H - d/2 < 1$. *What is the resulting sharp Hölder exponent* β *in this case? Will the asymptotic behavior of the second-order increment* $I(x, y)$ *involve a logarithmic correction term?*

2) Now consider a scenario where $2H - d/2 \geq 1$. *What is the resulting sharp Hölder exponent* β *in this case? Will the asymptotic behavior of the second-order increment* $I(x, y)$ *involve a logarithmic correction term?*

3) Based on your answers for (1) and (2), explain why the formula $\beta = \min\{1, 2H - d/2\}$ *precisely captures the sharp Hölder exponent across all relevant parameter ranges.*

Appendix

A.1. Appendix 1: foundations of stochastic processes and analysis

This appendix provides a concise review of fundamental concepts from stochastic analysis and calculus, which are essential for the rigorous development presented in this book.

A.1.1. *Slutsky's theorem*

THEOREM A.1 (Slutsky's theorem).– *Let* $(X_n)_{n\geq 1}$ *and* $(Y_n)_{n\geq 1}$ *be two sequences of random variables. If* X_n *converges in distribution to a random variable* X *(denoted* $X_n \xrightarrow{d} X$*), and* Y_n *converges in probability to a constant* c *(denoted* $Y_n \xrightarrow{p} c$*), then the following hold:*

1) $X_n + Y_n \xrightarrow{d} X + c$*;*

2) $X_n Y_n \xrightarrow{d} cX$*;*

3) $X_n / Y_n \xrightarrow{d} X/c$*, provided that* $c \neq 0$.

A direct consequence is that if both X_n *and* Y_n *converge in probability to constants* c_1 *and* c_2*, respectively, then their sums, products and quotients also converge in probability to the corresponding constants, provided the denominator is non-zero. For instance, if* $X_n \xrightarrow{p} c_1$ *and* $Y_n \xrightarrow{p} c_2$*, then* $X_n + Y_n \xrightarrow{p} c_1 + c_2$.

A.1.2. *Semimartingales*

DEFINITION A.1.– *Let* $(\Omega, \mathcal{F}, (\mathcal{F}_t)_{t\geq 0}, P)$ *be a filtered probability space. A stochastic process* $M = (M_t)_{t\geq 0}$ *is called a martingale with respect to the filtration* $(\mathcal{F}_t)$ *if:*

1) M is adapted to $(\mathcal{F}_t)$, that is, M_t is $\mathcal{F}_t$-measurable for all t.

2) $\mathbb{E}[|M_t|] < \infty$ for all t.

3) $\mathbb{E}[M_t|\mathcal{F}_s] = M_s$ for all $s \leq t$.

DEFINITION A.2.– *A stochastic process $M = (M_t)_{t\geq 0}$ is a local martingale if there exists a sequence of stopping times $(\tau_n)_{n\geq 1}$ such that $\tau_n \uparrow \infty$ a.s. as $n \to \infty$, and for each n, the stopped process $M^{\tau_n} = (M_{t\wedge\tau_n})_{t\geq 0}$ is a martingale.*

A.1.2.1. *Processes of finite variation*

DEFINITION A.3.– *A stochastic process $A = (A_t)_{t\geq 0}$ is predictable if it is measurable with respect to the σ-algebra generated by all adapted left-continuous processes. Intuitively, the value of a predictable process at time t is known just before t.*

DEFINITION A.4.– *A stochastic process $A = (A_t)_{t\geq 0}$ is said to be of finite variation if its total variation on any finite interval $[0, T]$ is finite almost surely, that is, Var$(A)_T = \int_0^T |dA_s| < \infty$ a.s. for all $T > 0$.*

A.1.2.2. *Semimartingales*

DEFINITION A.5.– *A stochastic process $X = (X_t)_{t\geq 0}$ is called a semimartingale if it admits a decomposition of the form:*

$$X_t = M_t + A_t$$

where $M = (M_t)_{t\geq 0}$ is a local martingale and $A = (A_t)_{t\geq 0}$ is a predictable process of finite variation.

The class of semimartingales is of paramount importance in stochastic calculus because it is the largest class of processes for which the Itô integral can be meaningfully defined.

A.2. Appendix 2: essential mathematical tools

This appendix collects several key mathematical results, ranging from matrix properties to analytical approximation techniques, which are applied throughout the book.

A.2.1. *On positive semi-definite matrices*

LEMMA A.1.– *Let M be an $n \times n$ symmetric positive semi-definite matrix. If for some vector $\mathbf{x} \in \mathbb{R}^n$, the quadratic form $\mathbf{x}'M\mathbf{x} = 0$, then it must follow that $M\mathbf{x} = \mathbf{0}$.*

Proof. Let M be an $n \times n$ symmetric positive semi-definite matrix. By definition, a matrix M is positive semi-definite if and only if for any vector $\mathbf{v} \in \mathbb{R}^n$, $\mathbf{v}'M\mathbf{v} \geq 0$. A key property of symmetric positive semi-definite matrices is that they can be factored into the form $M = P'P$ for some matrix P (this is related to the Cholesky decomposition or square root of a matrix).

Given that $\mathbf{x}'M\mathbf{x} = 0$, we can substitute the factorization $M = P'P$:

$$\mathbf{x}'(P'P)\mathbf{x} = 0.$$

This can be rewritten using the associative property of matrix multiplication:

$$(P\mathbf{x})'(P\mathbf{x}) = 0.$$

Let $\mathbf{y} = P\mathbf{x}$. Then, the equation becomes: $\mathbf{y}^T\mathbf{y} = 0$.

The term $\mathbf{y}^T\mathbf{y}$ represents the squared Euclidean norm (or squared magnitude) of the vector $\mathbf{y}$, that is, $\|\mathbf{y}\|^2 = \sum_{i=1}^{n} y_i^2$. For the squared norm of a real vector to be zero, the vector itself must be the zero vector:

$$\|\mathbf{y}\|^2 = 0 \implies \mathbf{y} = \mathbf{0}.$$

Substituting back $\mathbf{y} = P\mathbf{x}$, we get: $P\mathbf{x} = \mathbf{0}$.

Now, we want to show that $M\mathbf{x} = \mathbf{0}$. We can use the factorization $M = P'P$ again:

$$M\mathbf{x} = (P'P)\mathbf{x} = P'(P\mathbf{x}) = P'\mathbf{0} = \mathbf{0}.$$

Thus, we have proven that if $\mathbf{x}^T M\mathbf{x} = 0$ for a symmetric positive semi-definite matrix M, then $M\mathbf{x} = \mathbf{0}$. □

A.2.2. *Asymptotic analysis and Euler–Maclaurin summation*

This section introduces essential tools from asymptotic analysis, including the Euler–Maclaurin summation formula, which are instrumental for approximating sums and understanding the limiting behavior of key quantities in our proofs.

LEMMA A.2 (Euler–Maclaurin formula to $O(N^{-2})$).– *For a function $f(x)$ that is twice continuously differentiable on $[0,1]$ (i.e. $f \in C^2([0,1])$), the following approximation holds:*

$$\frac{1}{N}\sum_{i=1}^{N} f\left(\frac{i}{N}\right) = \int_0^1 f(x)dx + \frac{f(1)-f(0)}{2N} + O\left(\frac{1}{N^2}\right).$$

Proof. Let $h = 1/N$. We aim to relate the sum $\frac{1}{N}\sum_{i=1}^{N} f(ih)$ to the integral $\int_0^1 f(x)dx$. This can be achieved by considering the error of the composite trapezoidal rule.

Consider a single subinterval $[a, b]$ with length $h = b - a$. The trapezoidal rule approximates $\int_a^b f(x)dx$ by $\frac{h}{2}(f(a) + f(b))$. To find the error, we use integration by parts twice. Let $x = a + ht$, so $dx = hdt$. The integral becomes $h\int_0^1 f(a + ht)dt$. Let $g(t) = f(a + ht)$. We want to evaluate $h\int_0^1 g(t)dt$.

$$\begin{aligned}\int_0^1 g(t)dt &= \left[g(t)\left(t - \frac{1}{2}\right)\right]_0^1 - \int_0^1 g'(t)\left(t - \frac{1}{2}\right)dt \\ &= \left(g(1)\left(\frac{1}{2}\right) - g(0)\left(-\frac{1}{2}\right)\right) - \int_0^1 g'(t)\left(t - \frac{1}{2}\right)dt \\ &= \frac{g(1) + g(0)}{2} - \int_0^1 g'(t)\left(t - \frac{1}{2}\right)dt.\end{aligned}$$

Now, let us integrate the remaining integral by parts. Let $u = g'(t)$ and $dv = (t - 1/2)dt$. Then, $v = \frac{1}{2}t^2 - \frac{1}{2}t = \frac{t(t-1)}{2}$. Note that $v(0) = v(1) = 0$.

$$\begin{aligned}\int_0^1 g'(t)\left(t - \frac{1}{2}\right)dt &= \left[g'(t)\left(\frac{t^2}{2} - \frac{t}{2}\right)\right]_0^1 - \int_0^1 g''(t)\left(\frac{t^2}{2} - \frac{t}{2}\right)dt \\ &= (0 - 0) - \int_0^1 g''(t)\left(\frac{t^2}{2} - \frac{t}{2}\right)dt.\end{aligned}$$

Substituting this back into the integral for $g(t)$:

$$\int_0^1 g(t)dt = \frac{g(1) + g(0)}{2} + \int_0^1 g''(t)\left(\frac{t^2}{2} - \frac{t}{2}\right)dt.$$

Now substitute back $g(t) = f(a + ht)$. This implies $g'(t) = hf'(a + ht)$ and $g''(t) = h^2 f''(a + ht)$.

$$h\int_0^1 f(a+ht)dt = \frac{h}{2}(f(a+h)+f(a))+h\int_0^1 h^2 f''(a+ht)\left(\frac{t^2}{2} - \frac{t}{2}\right)dt.$$

$$\int_a^b f(x)dx = \frac{h}{2}(f(b) + f(a)) + h^3\int_0^1 f''(a + ht)\left(\frac{t^2}{2} - \frac{t}{2}\right)dt.$$

Let $K(t) = \frac{t^2}{2} - \frac{t}{2} = \frac{t(t-1)}{2}$. Since $K(t) \leq 0$ for $t \in [0, 1]$ and $f''(x)$ is continuous, by the mean value theorem for integrals, there exists some $\eta_{a,b} \in (a, b)$ such that:

$$\int_0^1 f''(a + ht)K(t)dt = f''(\eta_{a,b})\int_0^1 K(t)dt$$

$$= f''(\eta_{a,b}) \left[\frac{t^3}{6} - \frac{t^2}{4}\right]_0^1 = f''(\eta_{a,b}) \left(\frac{1}{6} - \frac{1}{4}\right) = f''(\eta_{a,b}) \left(-\frac{1}{12}\right).$$

So, for a single interval $[a, b]$ with $h = b - a$:

$$\int_a^b f(x)dx = \frac{h}{2}(f(a) + f(b)) - \frac{h^3}{12} f''(\eta_{a,b}).$$

Now, we divide the interval $[0, 1]$ into N subintervals of length $h = 1/N$. The subintervals are $[ih, (i+1)h]$ for $i = 0, 1, \ldots, N-1$. Summing the result from step 1 for each subinterval:

$$\int_0^1 f(x)dx = \sum_{i=0}^{N-1} \int_{ih}^{(i+1)h} f(x)dx$$

$$= \sum_{i=0}^{N-1} \left[\frac{h}{2}(f(ih) + f((i+1)h)) - \frac{h^3}{12} f''(\eta_i)\right],$$

where $\eta_i \in (ih, (i+1)h)$. The sum of the trapezoidal approximations forms the composite trapezoidal rule:

$$\sum_{i=0}^{N-1} \frac{h}{2}(f(ih) + f((i+1)h)) = \frac{h}{2}\left(f(0) + 2\sum_{i=1}^{N-1} f(ih) + f(Nh)\right).$$

Since $h = 1/N$ and $Nh = 1$:

$$= \frac{1}{2N}\left(f(0) + 2\sum_{i=1}^{N-1} f\left(\frac{i}{N}\right) + f(1)\right).$$

The sum of error terms is:

$$\sum_{i=0}^{N-1}\left(-\frac{h^3}{12} f''(\eta_i)\right) = -\frac{h^3}{12}\sum_{i=0}^{N-1} f''(\eta_i) = -\frac{1}{12N^3}\sum_{i=0}^{N-1} f''(\eta_i).$$

Since $f''(x)$ is continuous on $[0, 1]$, it is bounded. The sum $\sum_{i=0}^{N-1} f''(\eta_i)$ is a sum of N bounded terms, so it is of order N. Thus, $\frac{1}{N^3}\sum_{i=0}^{N-1} f''(\eta_i) = O(N^{-2})$. So, the total integral equation is:

$$\int_0^1 f(x)dx = \frac{1}{N}\left[\frac{1}{2}f(0) + \sum_{i=1}^{N-1} f\left(\frac{i}{N}\right) + \frac{1}{2}f(1)\right] + O\left(\frac{1}{N^2}\right).$$

We want to express $\frac{1}{N}\sum_{i=1}^{N} f(\frac{i}{N})$. Note that $\sum_{i=1}^{N} f(\frac{i}{N}) = \sum_{i=1}^{N-1} f(\frac{i}{N}) + f(1)$. From the composite trapezoidal rule equation, we can isolate $\frac{1}{N}\sum_{i=1}^{N-1} f\left(\frac{i}{N}\right)$:

$$\frac{1}{N}\sum_{i=1}^{N-1} f\left(\frac{i}{N}\right) = \int_0^1 f(x)dx - \frac{f(0) + f(1)}{2N} + O\left(\frac{1}{N^2}\right).$$

Now, substitute this into the expression for $\frac{1}{N}\sum_{i=1}^{N} f\left(\frac{i}{N}\right)$:

$$\begin{aligned}\frac{1}{N}\sum_{i=1}^{N} f\left(\frac{i}{N}\right) &= \left(\int_0^1 f(x)dx - \frac{f(0)+f(1)}{2N} + O\left(\frac{1}{N^2}\right)\right) + \frac{f(1)}{N} \\ &= \int_0^1 f(x)dx + \frac{f(1)}{N} - \frac{f(0)}{2N} - \frac{f(1)}{2N} + O\left(\frac{1}{N^2}\right) \\ &= \int_0^1 f(x)dx + \frac{2f(1)-f(0)-f(1)}{2N} + O\left(\frac{1}{N^2}\right) \\ &= \int_0^1 f(x)dx + \frac{f(1)-f(0)}{2N} + O\left(\frac{1}{N^2}\right).\end{aligned}$$

This concludes the proof. □

LEMMA A.3 (Euler–Maclaurin for midpoint sums).– *For a function $f(x)$ that is twice continuously differentiable on $[0,1]$, the following approximation holds:*

$$\frac{1}{N}\sum_{i=1}^{N} f\left(\frac{i-1/2}{N}\right) = \int_0^1 f(x)dx + O\left(\frac{1}{N^2}\right).$$

Proof. The core idea is to analyze the error of the midpoint rule over a single subinterval and then sum these errors.

Let $h = 1/N$ be the width of each subinterval. We consider the integral over a generic subinterval $[x_k, x_k+h]$, where $x_k = kh$. The midpoint of this interval is $x_k + h/2$.

The midpoint rule approximates $\int_{x_k}^{x_k+h} f(x)dx$ by $hf(x_k+h/2)$.

Let us expand $f(x)$ using a Taylor series around the midpoint $x_k + h/2$. For any $x \in [x_k, x_k+h]$, we can write $x = x_k + h/2 + t$, where $t \in [-h/2, h/2]$.

Let $x_c = x_k + h/2$. Using Taylor's theorem with a remainder term:

$$f(x_c+t) = f(x_c) + tf'(x_c) + \frac{t^2}{2}f''(\xi_t)$$

where ξ_t is some point between x_c and $x_c + t$.

Now, we integrate $f(x)$ over the interval $[x_k, x_k+h]$, which corresponds to integrating with respect to t from $-h/2$ to $h/2$:

$$\int_{x_k}^{x_k+h} f(x)dx = \int_{-h/2}^{h/2} f(x_c+t)dt$$

$$= \int_{-h/2}^{h/2} \left(f(x_c) + tf'(x_c) + \frac{t^2}{2} f''(\xi_t) \right) dt$$

$$= \left[tf(x_c) + \frac{t^2}{2} f'(x_c) \right]_{-h/2}^{h/2} + \int_{-h/2}^{h/2} \frac{t^2}{2} f''(\xi_t)dt$$

$$= \left(\frac{h}{2} f(x_c) + \frac{(h/2)^2}{2} f'(x_c) \right) - \left(-\frac{h}{2} f(x_c) + \frac{(-h/2)^2}{2} f'(x_c) \right)$$

$$+ \int_{-h/2}^{h/2} \frac{t^2}{2} f''(\xi_t)dt$$

$$= hf(x_c) + \int_{-h/2}^{h/2} \frac{t^2}{2} f''(\xi_t)dt.$$

Note that the term involving $f'(x_c)$ canceled out. This is a crucial reason why the midpoint rule has a higher order of accuracy than the trapezoidal or rectangular rules.

Let $R_k = \int_{-h/2}^{h/2} \frac{t^2}{2} f''(\xi_t)dt$. Since $f''(x)$ is continuous on $[0, 1]$ (due to $f \in C^2$), it is bounded. By the mean value theorem for integrals (since $t^2/2 \geq 0$), there exists a $\zeta_k \in [x_k, x_k + h]$ such that:

$$\int_{-h/2}^{h/2} \frac{t^2}{2} f''(\xi_t)dt = f''(\zeta_k) \int_{-h/2}^{h/2} \frac{t^2}{2} dt = f''(\zeta_k) \left[\frac{t^3}{6} \right]_{-h/2}^{h/2} = f''(\zeta_k) \frac{h^3}{24}.$$

Thus, for each subinterval:

$$\int_{x_k}^{x_k+h} f(x)dx = hf(x_k + h/2) + \frac{h^3}{24} f''(\zeta_k).$$

Rearranging to get the midpoint approximation:

$$hf(x_k + h/2) = \int_{x_k}^{x_k+h} f(x)dx - \frac{h^3}{24} f''(\zeta_k). \qquad [A.1]$$

The sum in the lemma is $\frac{1}{N} \sum_{i=1}^{N} f\left(\frac{i-1/2}{N}\right)$. Since $h = 1/N$, this can be written as $\sum_{i=1}^{N} hf\left(\frac{i-1/2}{N}\right)$. Summing the expression [A.1] for $k = 0, 1, \ldots, N-1$ and $x_k = \frac{k}{N} = \frac{i-1}{N}$:

$$\sum_{k=0}^{N-1} hf\left(x_k + h/2\right) = \sum_{i=1}^{N} \frac{1}{N} f\left(\frac{i-1/2}{N}\right) = \sum_{i=1}^{N} \int_{(i-1)/N}^{i/N} f(x)dx - \sum_{i=1}^{N} \frac{h^3}{24} f''(\zeta_i).$$

The sum of integrals on the right is simply the total integral over $[0, 1]$:

$$\sum_{i=1}^{N} \int_{(i-1)/N}^{i/N} f(x)dx = \int_0^1 f(x)dx.$$

Now, substitute $h = 1/N$ into the error sum:

$$\sum_{i=1}^{N} \frac{h^3}{24} f''(\zeta_i) = \frac{1}{24N^3} \sum_{i=1}^{N} f''(\zeta_i).$$

Since f'' is continuous on $[0, 1]$, it is bounded. Let $M = \sup_{x\in[0,1]} |f''(x)|$. Thus,

$$\left| \sum_{i=1}^{N} f''(\zeta_i) \right| \leq M.N.$$

Therefore, the error sum is:

$$\frac{1}{24N^3} \sum_{i=1}^{N} f''(\zeta_i) = \frac{1}{24N^3} O(N) = O\left(\frac{1}{N^2}\right).$$

Combining the results:

$$\frac{1}{N} \sum_{i=1}^{N} f\left(\frac{i-1/2}{N}\right) = \int_0^1 f(x)dx + O\left(\frac{1}{N^2}\right).$$

□

LEMMA A.4 (asymptotic order of a specific summation for dependent terms).– *Let $X_{i,j}(N)$ be the quantity defined, for $i, j \in \{1, \ldots, N\}$, $i \neq j$, as:*

$$X_{i,j}(N) = -ab \left(\frac{1}{N}\right)^{2H} \left[(i+j)^{2H} - 2(i+j-1)^{2H} + (i+j-2)^{2H}\right]$$

$$-\frac{(a^2+b^2)}{2} \left(\frac{1}{N}\right)^{2H} \left[|i-j+1|^{2H} - 2|i-j|^{2H} + |i-j-1|^{2H}\right],$$

where a, b are non-zero constants and $H \in (0, 1)$.

Then, the leading order asymptotic behavior of the sum of squared terms:

$$\Sigma_N = \sum_{i\neq j} (X_{i,j}(N))^2$$

for $N \to \infty$ is given by:

– *For* $H = 1/2$*:* $\Sigma_N = 0$.

– *For* $H = 3/4$*:* $\Sigma_N \sim \frac{9(a^2+b^2)^2}{128}\frac{\ln N}{N^2}$.

– *For* $H \in (0,1)$ *and* $H \neq 1/2, H \neq 3/4$*:*

$$\Sigma_N \sim \frac{(a^2+b^2)^2}{2}(2H(2H-1))^2\frac{1}{(4H-3)(4H-2)}N^{-2}.$$

Proof. Let $f(x) = x^{2H}$. The bracketed terms in $X_{i,j}(N)$ are second-order finite differences. Specifically,

$$(i+j)^{2H} - 2(i+j-1)^{2H} + (i+j-2)^{2H}$$

is a backward difference operator $\nabla^2 f(x)$, and

$$|i-j+1|^{2H} - 2|i-j|^{2H} + |i-j-1|^{2H}$$

is a central difference operator $\delta^2 f(x)$. For a function $g(x)$ that is at least four times continuously differentiable, both $\nabla^2 g(x)$ and $\delta^2 g(x)$ have the asymptotic expansion $g''(x) + O(g^{(4)}(x))$ for large x (see Exercise 4.2).

If $H = 1/2$, then $f(x) = x$. For $g(x) = x$, $g''(x) = 0$ and all higher derivatives are zero. Thus, both finite difference terms are exactly zero. Consequently, $X_{i,j}(N) = 0$ for all $i \neq j$, and $\Sigma_N = 0$.

For $H \neq 1/2$, $f(x) = x^{2H}$. Its second derivative is $f''(x) = 2H(2H-1)x^{2H-2}$. Let $C = 2H(2H-1)$, a non-zero constant. Using the asymptotic approximation $g''(x) + O(g^{(4)}(x))$, we approximate the terms in $X_{i,j}(N)$:

$$(i+j)^{2H} - 2(i+j-1)^{2H} + (i+j-2)^{2H} = C(i+j)^{2H-2} + O((i+j)^{2H-4})$$

$$|i-j+1|^{2H} - 2|i-j|^{2H} + |i-j-1|^{2H} = C|i-j|^{2H-2} + O(|i-j|^{2H-4})$$

Substituting these into $X_{i,j}(N)$ and keeping leading order terms:

$$X_{i,j}(N) = N^{-2H}\left(-abC(i+j)^{2H-2} - \frac{(a^2+b^2)}{2}C|i-j|^{2H-2}\right) + \text{l.o.t.}$$

Squaring $X_{i,j}(N)$, we obtain:

$$(X_{i,j}(N))^2 = N^{-4H}\Bigg[(abC)^2(i+j)^{4H-4} + \left(\frac{(a^2+b^2)}{2}C\right)^2|i-j|^{4H-4}$$
$$+ab(a^2+b^2)C^2(i+j)^{2H-2}|i-j|^{2H-2}\Bigg] + \text{l.o.t.}$$

Let $K_1 = (abC)^2$, $K_2 = \frac{(a^2+b^2)^2}{4}C^2$ and $K_3 = ab(a^2+b^2)C^2$. We now analyze the sum $\Sigma_N = \sum_{i\neq j}(X_{i,j}(N))^2$. Let $p = 4H - 4$.

For the subcase $H = 3/4$, $p = 4(3/4) - 4 = -1$. The sum $\sum_{i \neq j} |i-j|^{-1}$ can be precisely evaluated:

$$2\sum_{k=1}^{N-1}(N-k)k^{-1} = 2N\sum_{k=1}^{N-1}k^{-1} - 2\sum_{k=1}^{N-1}1 \sim 2N(\ln N + \gamma) - 2(N-1) \sim 2N\ln N$$

as $N \to \infty$. The terms $\sum_{i \neq j}(i+j)^{-1}$ and $\sum_{i \neq j}(i+j)^{-1}|i-j|^{-1}$ are of order $O(N)$.

Since $N \ln N$ dominates N, the leading order of Σ_N for $H = 3/4$ is determined by the K_2 term.

$$\Sigma_N \sim N^{-4H}K_2(2N\ln N) = N^{-3}\frac{(a^2+b^2)^2}{4}C^2(2N\ln N).$$

Substituting $H = 3/4$, $C = 2(3/4)(3/4-1) = -3/8$, so $C^2 = 9/64$. Therefore:

$$\Sigma_N \sim N^{-3}\frac{(a^2+b^2)^2}{4}\frac{9}{64}(2N\ln N) = \frac{9(a^2+b^2)^2}{128}\frac{\ln N}{N^2}.$$

For the subcase where $H \in (0,1)$ and $H \neq 1/2$, $H \neq 3/4$, $p = 4H - 4 \neq -1$.

The asymptotic behavior of sums $\sum_{i \neq j}(i+j)^p$ and $\sum_{i \neq j}(i+j)^{p_1}|i-j|^{p_2}$ are generally found via integral approximations. Specifically,

$$\sum_{i \neq j}(i+j)^p \sim \frac{2^{p+2}-2}{(p+1)(p+2)}N^{p+2}$$

for $p > -2$. Also,

$$\sum_{i \neq j}|i-j|^p \sim \frac{2}{(p+1)(p+2)}N^{p+2}$$

for $p > -1$.

All three of the summands, $K_1(i+j)^{4H-4}$, $K_2|i-j|^{4H-4}$ and $K_3(i+j)^{2H-2}|i-j|^{2H-2}$, contribute terms of order N^{4H-2} to the sum. However, in the context of sums arising from differences of power-law kernels, especially for the second-order differences given, the leading order behavior is typically dominated by the term associated with $|i-j|^{4H-4}$.

This implies that the sum of the contributions from the $(i+j)$ related terms (i.e. $K_1(i+j)^{4H-4}$ and $K_3(i+j)^{2H-2}|i-j|^{2H-2}$) either sum to a lower order or cancel at the leading N^{4H-2} power. Consequently, the leading term is solely determined by the K_2 component. Thus,

$$\Sigma_N \sim N^{-4H} K_2 \left(\frac{2}{(4H-3)(4H-2)} N^{4H-2} \right).$$

Substituting $K_2 = \frac{(a^2+b^2)^2}{4}(2H(2H-1))^2$:

$$\Sigma_N \sim \frac{(a^2+b^2)^2}{2}(2H(2H-1))^2 \frac{1}{(4H-3)(4H-2)} N^{-2}.$$

This completes the proof. □

A.3. Appendix 3: properties of power-law functions

This appendix focuses on the analytical properties of power-law functions, particularly those related to convexity and bounding, which are crucial for the proofs involving the ZgfBm's parameters.

LEMMA A.5 (monotonicity of convexity error for power functions).– *Let $H \in (1/2, 1)$ and consider the function $\phi(x) = x^{2H}$ for $x > 0$. For a fixed interval width $h > 0$, define the convexity error as a function of the left endpoint a:*

$$E(a) = \frac{\phi(a) + \phi(a+h)}{2} - \phi\left(a + \frac{h}{2}\right).$$

Then, the function $a \longmapsto E(a)$ is a decreasing function for $a > 0$.

Proof. The first derivative is $\phi'(x) = 2Hx^{2H-1}$. The second derivative is:

$$\phi''(x) = 2H(2H-1)x^{2H-2}.$$

Since $H \in (1/2, 1)$, we have $2H > 1$ and $2H - 1 > 0$. Also, for $x > 0$, $x^{2H-2} > 0$. Therefore, $\phi''(x) > 0$ for $x > 0$, which confirms that $\phi(x)$ is strictly convex for $x > 0$.

The exponent $2H - 2$ is negative because $2H < 2$ (since $H < 1$). Since $2 - 2H$ is a positive exponent, x^{2-2H} is an increasing function for $x > 0$.

Consequently, $\frac{1}{x^{2-2H}}$ is a decreasing function of x for $x > 0$.

Since $\phi''(x) = 2H(2H-1)x^{2H-2}$ and $2H(2H-1)$ is a positive constant, this implies that $\phi''(x)$ is a decreasing function of x for $x > 0$.

The derivative of E is:

$$E'(a) = \frac{1}{2}[\phi'(a) + \phi'(a+h)] - \phi'\left(a + \frac{h}{2}\right).$$

Since $\phi''(x)$ is a decreasing function, it means that the derivative of $\phi'(x)$ (which is $\phi''(x)$ itself) is decreasing. A function whose derivative is decreasing is a concave function. Therefore, $\phi'(x)$ is a concave function.

For a concave function $f(x)$, for any x_1, x_2 and $t \in [0,1]$:

$$f(tx_1 + (1-t)x_2) \geq tf(x_1) + (1-t)f(x_2).$$

Applying this to $\phi'(x)$ with $x_1 = a$, $x_2 = a + h$, and $t = 1/2$:

$$\phi'\left(\frac{1}{2}a + \frac{1}{2}(a+h)\right) \geq \frac{1}{2}\phi'(a) + \frac{1}{2}\phi'(a+h).$$

$$\phi'\left(a + \frac{h}{2}\right) \geq \frac{\phi'(a) + \phi'(a+h)}{2}.$$

Rearranging this inequality, we get:

$$\frac{\phi'(a) + \phi'(a+h)}{2} - \phi'\left(a + \frac{h}{2}\right) \leq 0.$$

This shows that $E'(a) \leq 0$, which implies that $E(a)$ is a decreasing function of a. □

LEMMA A.6.– *Let $H \in (1/2, 1)$, and for $j \geq 1$, define the function $F(j)$ as:*

$$F(j) = 1 + (2j-2)^{2H} - (2j-1)^{2H} + 2^{2H-1}(j^{2H} - (j-1)^{2H}).$$

Then

$$1 \leq F(j) \leq 2^{2H-1}.$$

Proof. Let us analyze the terms within $F(j)$. We first note that for any $x \geq 0$, $2^{2H-1}x^{2H} = \frac{1}{2}(2x)^{2H}$. Applying this to the last term:

$$2^{2H-1}(j^{2H} - (j-1)^{2H}) = \frac{1}{2}((2j)^{2H} - (2(j-1))^{2H}) = \frac{1}{2}((2j)^{2H} - (2j-2)^{2H}).$$

Substituting this back into the definition of $F(j)$:

$$\begin{aligned} F(j) &= 1 + (2j-2)^{2H} - (2j-1)^{2H} + \frac{1}{2}(2j)^{2H} - \frac{1}{2}(2j-2)^{2H} \\ &= 1 + \left((2j-2)^{2H} - \frac{1}{2}(2j-2)^{2H}\right) - (2j-1)^{2H} + \frac{1}{2}(2j)^{2H} \\ &= 1 + \frac{1}{2}(2j-2)^{2H} + \frac{1}{2}(2j)^{2H} - (2j-1)^{2H}. \end{aligned}$$

Rearranging the terms, we recognize a specific structure related to convexity:

$$F(j) = 1 + \left(\frac{(2j-2)^{2H} + (2j)^{2H}}{2} - (2j-1)^{2H} \right) 1 + \tilde{h}(j),$$

with

$$\tilde{h}(j) = \frac{(2j-2)^{2H} + (2j)^{2H}}{2} - (2j-1)^{2H}.$$

Now, we prove the bounds for $\tilde{h}(j)$.

For $H > 1/2$, the function $\phi(x) = x^{2H}$ is strictly convex for $x > 0$. By the definition of convexity, for any $x_1, x_2 > 0$:

$$\phi\left(\frac{x_1 + x_2}{2}\right) \leq \frac{\phi(x_1) + \phi(x_2)}{2}.$$

Applying this with $x_1 = 2j - 2$ and $x_2 = 2j$ (which are both positive for $j \geq 1$), we have:

$$\left(\frac{(2j-2) + 2j}{2}\right)^{2H} \leq \frac{(2j-2)^{2H} + (2j)^{2H}}{2}.$$

This simplifies to:

$$(2j-1)^{2H} \leq \frac{(2j-2)^{2H} + (2j)^{2H}}{2}.$$

Therefore, by the definition of $\tilde{h}(j)$:

$$\tilde{h}(j) = \frac{(2j-2)^{2H} + (2j)^{2H}}{2} - (2j-1)^{2H} \geq 0.$$

We note that the function $\tilde{h}(j)$ corresponds to the convexity error $E(a)$ introduced in Lemma A.5, with $\phi(x) = x^{2H}$, $a = 2j - 2$ and $h = 2$.

Since $j \geq 1$, the left endpoint $a = 2j - 2 \geq 0$. For $j = 1$, $a = 0$. For $j > 1$, $a > 0$. Our analysis of $\phi''(x)$ holds for $x > 0$. The case $a = 0$ for $j = 1$ needs specific handling of 0^{2H}, which we assume to be 0 for $2H > 0$.

As $E(a)$ is a decreasing function of a, $\tilde{h}(j)$ is a decreasing function of j. Therefore, the maximum value of $\tilde{h}(j)$ must occur at the smallest possible value of j, which is $j = 1$.

Let us calculate $\tilde{h}(1)$:

$$\tilde{h}(1) = \frac{(2(1)-2)^{2H} + (2(1))^{2H}}{2} - (2(1)-1)^{2H} = \frac{0^{2H} + 2^{2H}}{2} - 1^{2H}.$$

Since $H > 1/2$, $2H > 1$, so we take $0^{2H} = 0$.

$$\tilde{h}(1) = \frac{0 + 2^{2H}}{2} - 1 = \frac{2^{2H}}{2} - 1 = 2^{2H-1} - 1.$$

Since $\tilde{h}(j) \geq 0$ (as $\phi(x)$ is convex) and $\tilde{h}(j)$ is decreasing, we conclude that for all $j \geq 1$:

$$0 \leq \tilde{h}(j) \leq 2^{2H-1} - 1.$$

This completes the proof. □

A.4. Appendix 4: integral convergence and transformations

This appendix is dedicated to a collection of technical lemmas concerning integral calculus that are essential for the proofs presented in the main text, covering topics such as the convergence conditions, scaling properties and crucial bounds of key integrals.

LEMMA A.7.– *We have:*

$$I(t) = \int_0^t \int_0^t |u - v|^{2H-2}(2t - u - v)^{-d/2} dudv < \infty \Longleftrightarrow H > 1/2 \text{ and } d < 4H.$$

Proof. The integrand, $f(u, v) = |u - v|^{2H-2}(2t - u - v)^{-d/2}$, exhibits singularities at two primary locations within the integration domain $[0, t] \times [0, t]$:

– Diagonal singularity: the term $|u - v|^{2H-2}$ becomes singular when $u = v$. This requires $2H - 2 > -1$ for local integrability along the diagonal, that is, $H > 1/2$.

– Corner singularity: the term $(2t - u - v)^{-d/2}$ becomes singular when $u + v = 2t$. This occurs specifically at the point $(u, v) = (t, t)$. This requires $-d/2 > -1$ for local integrability, that is, $d < 2$.

However, these two singularities are coupled, especially at the point (t, t) where $u = v = t$, making a direct analysis challenging. A suitable change of variables is necessary to disentangle these effects.

Let $s = u + v$ and $r = u - v$. From these, we can express u and v in terms of s and r:

$$u = \frac{s + r}{2}, \quad v = \frac{s - r}{2}.$$

The Jacobian of this transformation is given by:

$$J = \left|\det\begin{pmatrix} \frac{\partial u}{\partial s} & \frac{\partial u}{\partial r} \\ \frac{\partial v}{\partial s} & \frac{\partial v}{\partial r} \end{pmatrix}\right| = \left|\det\begin{pmatrix} 1/2 & 1/2 \\ 1/2 & -1/2 \end{pmatrix}\right| = \left|-\frac{1}{4} - \frac{1}{4}\right| = \left|-\frac{1}{2}\right| = \frac{1}{2}.$$

So, $dudv = \frac{1}{2}dsdr$.

The original square domain $0 \leq u \leq t$ and $0 \leq v \leq t$ transforms into a diamond shape in the (s, r)-plane:

– From $0 \leq u \leq t$: $0 \leq \frac{s+r}{2} \leq t \implies 0 \leq s + r \leq 2t$.

– From $0 \leq v \leq t$: $0 \leq \frac{s-r}{2} \leq t \implies 0 \leq s - r \leq 2t$.

Combining these, we get:

$-t \leq r \leq t$ (since $r = u - v$, its maximum value is $t - 0 = t$ and minimum is $0 - t = -t$).

$s \geq |r|$ (from $s + r \geq 0 \implies s \geq -r$ and $s - r \geq 0 \implies s \geq r$).

$s \leq 2t - |r|$ (from $s + r \leq 2t \implies s \leq 2t - r$ and $s - r \leq 2t \implies s \leq 2t + r$).

The integral $I(t)$ thus becomes:

$$I(t) = \int_{-t}^{t} \int_{|r|}^{2t-|r|} |r|^{2H-2}(2t - s)^{-d/2}\frac{1}{2}dsdr.$$

The integral $I(t)$ converges if and only if the integral over the transformed domain converges. The singular behavior of the integrand in the (s, r)-plane is concentrated near $r = 0$ (corresponding to $u = v$) and $s = 2t$ (corresponding to $u = t, v = t$).

Let us focus on the local behavior near $(r, s) = (0, 2t)$, as this is where the most critical combined singularity occurs. Consider a small neighborhood around (t, t) in the (u, v)-plane. Let $u = t - \epsilon_1$ and $v = t - \epsilon_2$, where $\epsilon_1, \epsilon_2 > 0$ are small. Then, $u - v = \epsilon_2 - \epsilon_1$ and $2t - u - v = \epsilon_1 + \epsilon_2$. The integral locally takes the form:

$$J = \int_0^{\delta} \int_0^{\delta} |\epsilon_2 - \epsilon_1|^{2H-2}(\epsilon_1 + \epsilon_2)^{-d/2}d\epsilon_1 d\epsilon_2,$$

for some small $\delta > 0$.

To analyze the convergence of J, we switch to polar coordinates in the (ϵ_1, ϵ_2)-plane: let $\epsilon_1 = \rho\cos\theta$ and $\epsilon_2 = \rho\sin\theta$, where $\rho \geq 0$ and $0 \leq \theta \leq \pi/2$ (since $\epsilon_1, \epsilon_2 \geq 0$). The Jacobian of this transformation is ρ. Substituting these into J:

$$J = \int_0^{\delta'} \int_0^{\pi/2} |\rho\sin\theta - \rho\cos\theta|^{2H-2}(\rho\cos\theta + \rho\sin\theta)^{-d/2}\rho\, d\rho\, d\theta$$

$$= \int_0^{\delta'} \int_0^{\pi/2} \rho^{2H-2}|\sin\theta - \cos\theta|^{2H-2}\rho^{-d/2}(\cos\theta + \sin\theta)^{-d/2}\rho\, d\rho\, d\theta$$

$$= \int_0^{\delta'} \rho^{2H-2-d/2+1} d\rho \int_0^{\pi/2} |\sin\theta - \cos\theta|^{2H-2} (\cos\theta + \sin\theta)^{-d/2} d\theta$$

$$= \int_0^{\delta'} \rho^{2H-d/2-1} d\rho \cdot C_H,$$

where $C_H = \int_0^{\pi/2} |\sin\theta - \cos\theta|^{2H-2} (\cos\theta + \sin\theta)^{-d/2} d\theta$ is a constant. For C_H to be finite, we need $2H - 2 > -1$ (i.e. $H > 1/2$), which ensures the integrability around $\theta = \pi/4$ (where $\sin\theta - \cos\theta = 0$). This is consistent with the isolated diagonal singularity condition.

The convergence of J (and thus $I(t)$) depends on the integrability of the radial part $\int_0^{\delta'} \rho^{2H-d/2-1} d\rho$. This integral converges at $\rho = 0$ if and only if the exponent is greater than -1:

$$2H - d/2 - 1 > -1$$

$$2H - d/2 > 0$$

$$4H > d \implies d < 4H.$$

□

LEMMA A.8.– *For every* $\gamma \in (\sup(d/4, 1/2), 1)$, *there exists a positive constant* $c_1(d,\gamma)$, *depending only on* d *and* γ, *such that, for every* $s, t \in [0, T]$ *with* $s \le t$:

$$\int_s^t \int_s^t dudv \mid u - v \mid^{2\gamma-2} (2t - u - v)^{-d/2} = c_1(d,\gamma) \mid t - s \mid^{2\gamma - d/2}.$$

Proof. Let $T' = t - s$. We begin by performing a change of variables to shift the integration interval from $[s, t]$ to $[0, T']$. Let $u' = u - s$ and $v' = v - s$. Then, $du = du'$ and $dv = dv'$. The new integration limits are from 0 to T' for both u' and v'. The terms in the integrand transform as follows:

– The difference term: $\mid u - v \mid = \mid (u' + s) - (v' + s) \mid = \mid u' - v' \mid$.

– The sum term:

$$2t - u - v = 2t - (u' + s) - (v' + s) = 2(t - s) - (u' + v') = 2T' - (u' + v').$$

Substituting these into the integral, we obtain:

$$\int_0^{T'} \int_0^{T'} \mid u' - v' \mid^{2\gamma-2} (2T' - (u' + v'))^{-d/2} du'dv'.$$

For simplicity, we will now drop the primes and denote this integral as $\mathcal{I}(T')$:

$$\mathcal{I}(T') = \int_0^{T'} \int_0^{T'} \mid u - v \mid^{2\gamma-2} (2T' - (u+v))^{-d/2} dudv.$$

Next, we introduce a second change of variables to separate the sum and difference components. Let $x = u + v$ and $y = u - v$. From these definitions, we can express u and v as: $u = \dfrac{x+y}{2}$ and $v = \dfrac{x-y}{2}$.

The Jacobian of this transformation is given by:

$$\left| \det \begin{pmatrix} \frac{\partial u}{\partial x} & \frac{\partial u}{\partial y} \\ \frac{\partial v}{\partial x} & \frac{\partial v}{\partial y} \end{pmatrix} \right| = \left| \det \begin{pmatrix} 1/2 & 1/2 \\ 1/2 & -1/2 \end{pmatrix} \right| = \left| -\frac{1}{4} - \frac{1}{4} \right| = \left| -\frac{1}{2} \right| = \frac{1}{2}.$$

So, the differential element transforms as $dudv = \frac{1}{2} dxdy$.

The square integration domain $(u, v) \in [0, T'] \times [0, T']$ transforms into a diamond-shaped region in the (x, y)-plane defined by:

$y \in [-T', T']$ (since $y = u - v$, its minimum is $0 - T' = -T'$ and maximum is $T' - 0 = T'$).

$x \in [|y|, 2T' - |y|]$ (derived from $u \geq 0 \implies x + y \geq 0$, $v \geq 0 \implies x - y \geq 0$, $u \leq T' \implies x + y \leq 2T'$, and $v \leq T' \implies x - y \leq 2T'$).

Substituting these into the integral $\mathcal{I}(T')$:

$$\mathcal{I}(T') = \frac{1}{2} \int_{-T'}^{T'} \int_{|y|}^{2T'-|y|} |y|^{2\gamma-2} (2T' - x)^{-d/2} dxdy.$$

Finally, to extract the scaling factor T', we perform another change of variables within the (x, y) integral. Let $X = x/T'$ and $Y = y/T'$. Then, $x = T'X$, $y = T'Y$.

The differentials are $dx = T'dX$ and $dy = T'dY$. The integration limits for Y become $[-1, 1]$. The integration limits for X become $[|Y|, 2 - |Y|]$.

Substituting these into the expression for $\mathcal{I}(T')$:

$$\begin{aligned}
\mathcal{I}(T') &= \frac{1}{2} \int_{-1}^{1} \int_{|Y|}^{2-|Y|} |T'Y|^{2\gamma-2} (2T' - T'X)^{-d/2} (T'dX)(T'dY) \\
&= \frac{1}{2} \int_{-1}^{1} \int_{|Y|}^{2-|Y|} (T')^{2\gamma-2} |Y|^{2\gamma-2} (T')^{-d/2} (2 - X)^{-d/2} (T')^2 dXdY \\
&= \frac{1}{2} (T')^{(2\gamma-2)-d/2+2} \int_{-1}^{1} \int_{|Y|}^{2-|Y|} |Y|^{2\gamma-2} (2 - X)^{-d/2} dXdY
\end{aligned}$$

$$= \frac{1}{2}(T')^{2\gamma-d/2}\int_{-1}^{1}\int_{|Y|}^{2-|Y|}|Y|^{2\gamma-2}(2-X)^{-d/2}dXdY$$
$$= (T')^{2\gamma-d/2}c_1(d,\gamma),$$

with

$$c_1(d,\gamma) = \frac{1}{2}\int_{-1}^{1}|Y|^{2\gamma-2}\left(\int_{|Y|}^{2-|Y|}(2-X)^{-d/2}dX\right)dY.$$

Let $J(Y) = \displaystyle\int_{|Y|}^{2-|Y|}(2-X)^{-d/2}dX$. We consider two cases for the exponent $-d/2$:

– *Case 1*: $d/2 \neq 1$ (i.e. $d \neq 2$). In this case, the integral is:

$$J(Y) = \left[\frac{(2-X)^{-d/2+1}}{-d/2+1}\right]_{X=|Y|}^{X=2-|Y|}$$
$$= \frac{1}{1-d/2}\left[(2-(2-|Y|))^{1-d/2}-(2-|Y|)^{1-d/2}\right]$$
$$= \frac{1}{1-d/2}\left[|Y|^{1-d/2}-(2-|Y|)^{1-d/2}\right].$$

– *Case 2*: $d/2 = 1$ (i.e. $d = 2$). In this case, the integral involves a logarithm:

$$J(Y) = [-\ln(2-X)]_{X=|Y|}^{X=2-|Y|}$$
$$= -\ln(2-(2-|Y|)) + \ln(2-|Y|)$$
$$= -\ln(|Y|) + \ln(2-|Y|) = \ln\left(\frac{2-|Y|}{|Y|}\right).$$

Combining these, the constant $c_1(d,\gamma)$ can be expressed as:

$$c_1(d,\gamma) = \frac{1}{2}\int_{-1}^{1}|Y|^{2\gamma-2}J(Y)dY.$$

Substituting the expression for $J(Y)$ and noting that the integrand is an even function of Y, we can write the integral from 0 to 1 and multiply by 2:

– *Case 1*: $d \neq 2$. The constant $c_1(d,\gamma)$ is given by:

$$c_1(d,\gamma) = \frac{1}{1-d/2}\int_{0}^{1}Y^{2\gamma-2}\left[Y^{1-d/2}-(2-Y)^{1-d/2}\right]dY.$$

We need to justify the finiteness of this integral. We can split the integrand into two parts:

$$I_A = \int_0^1 Y^{2\gamma-2}Y^{1-d/2}dY = \int_0^1 Y^{2\gamma-1-d/2}dY.$$

For this integral to be finite, the exponent of Y must be greater than -1 (a standard condition for integrability near 0).

$$2\gamma - 1 - d/2 > -1 \implies 2\gamma - d/2 > 0 \implies d < 4\gamma.$$

This condition is precisely met by our assumption $\gamma \in (d/4, 1)$. Thus, I_A is finite.

The second part is:

$$I_B = -\int_0^1 Y^{2\gamma-2}(2-Y)^{1-d/2}dY.$$

For this integral, the potential singularity is at $Y = 0$ due to $Y^{2\gamma-2}$. The term $(2-Y)^{1-d/2}$ is well behaved on $[0,1]$:

If $1 - d/2 > 0$ (i.e. $d < 2$), then $(2-Y)^{1-d/2}$ is continuous and positive on $[0,1]$.

If $1 - d/2 < 0$ (i.e. $d > 2$), then $(2-Y)^{1-d/2}$ is continuous and positive on $[0,1]$ (it just means it is $1/(2-Y)^{d/2-1}$, not singular on $[0,1]$). If $d = 2$, this case is handled separately below.

So, for I_B to be finite, we only need $2\gamma - 2 > -1$, which implies $\gamma > 1/2$. This condition is also met by our assumption $\gamma \in (\sup(d/4, 1/2), 1)$. Therefore, both I_A and I_B are finite, ensuring that $c_1(d,\gamma)$ is finite when $d \neq 2$.

– *Case 2*: $d = 2$. The constant $c_1(d,\gamma)$ is given by:

$$c_1(d,\gamma) = \int_0^1 Y^{2\gamma-2}\ln\left(\frac{2-Y}{Y}\right)dY.$$

We can rewrite this as:

$$c_1(d,\gamma) = \int_0^1 Y^{2\gamma-2}\left[\ln(2-Y) - \ln(Y)\right]dY.$$

Again, we analyze each part of the integral:

$$I_C = \int_0^1 Y^{2\gamma-2}\ln(2-Y)dY.$$

The term $\ln(2-Y)$ is continuous and finite on $[0,1]$ (it evaluates to $\ln(2)$ at $Y = 0$ and $\ln(1) = 0$ at $Y = 1$). The singularity is at $Y = 0$ due to $Y^{2\gamma-2}$. For this integral to be finite, we require $2\gamma - 2 > -1$, which means $\gamma > 1/2$. This is satisfied by $\gamma \in (\sup(d/4, 1/2), 1)$.

Thus, I_C is finite.

The second part is:

$$I_D = -\int_0^1 Y^{2\gamma-2} \ln(Y) dY.$$

The singularity for this integral is at $Y = 0$. We can use the result that $\int_0^1 x^a \ln(x) dx$ converges if $a > -1$. In our case, $a = 2\gamma - 2$. So, we need $2\gamma - 2 > -1$, which implies $\gamma > 1/2$. As before, this condition is met by $\gamma \in (d/4, 1)$ with $d = 2$.

Therefore, I_D is finite.

Since both I_C and I_D are finite, $c_1(d, \gamma)$ is finite when $d = 2$.

In summary, for both cases ($d \neq 2$ and $d = 2$), the condition $\gamma \in (d/4, 1)$ guarantees the integrability of all terms, ensuring that $c_1(d, \gamma)$ is a well-defined, finite and positive constant. The positivity arises from the fact that the original integrand $|y|^{2\gamma-2}(2T' - x)^{-d/2}$ is positive within the integration domain.

Thus, we have proved the desired equality:

$$\int_s^t \int_s^t |u - v|^{2\gamma-2}(2t - u - v)^{-d/2} du dv = c_1(d, \gamma)(t - s)^{2\gamma-d/2}.$$

This is since we have assumed that $s \leq t$, $(t - s)^{2\gamma-d/2}$ is equal to $|t - s|^{2\gamma-d/2}$. □

We cite the following technical lemma, instrumental for the results presented in section 6.5, which is proven in Tudor (2013).

LEMMA A.9.– *Denoting*

$$I(t, s) = \int_0^s \int_0^s dv du \mid u - v \mid^{2\gamma-2} \Big[(2t - u - v)^{-d/2} - 2(t + s - u - v)^{-d/2} + (2s - u - v)^{-d/2}\Big],$$

for every $\gamma \in (\sup(1/2, d/4), 1)$, *there exists a positive constant* $c_2(d, \gamma)$, *depending only on* d *and* γ, *such that, for every* $s, t \in [0, T]$ *with* $s \leq t$:

$$I(t, s) \leq c_2(d, \gamma) \mid t - s \mid^{2\gamma-d/2}.$$

The following technical lemma is due to Balan and Tudor (2010).

LEMMA A.10.– *Let* $A_t(\xi)$ *be defined as:*

$$A_t(\xi) = \int_0^t \int_0^t |u-v|^{2H-2} \exp\left(-\frac{u+v}{2}|\xi|^2\right) dudv.$$

For any $t > 0$ *and* $\xi \in \mathbb{R}^d$, *there exist strictly positive constants* $C_1 = C_1(H)$ *and* $C_2 = C_2(H)$, *such that:*

$$\frac{1}{4}(t^{2H} \wedge 1)\left(\frac{1}{1+|\xi|^2}\right)^{2H} \leq A_t(\xi) \leq C_2(t^{2H}+1)\left(\frac{1}{1+|\xi|^2}\right)^{2H}.$$

Here, C_2 *(which replaces the original* C_H*) is a constant depending only on* H, *for example,* $C_2 = b_H^2(4H)^{2H}$ *(where* b_H *is a constant related to fractional Sobolev norms). The term* $1/(1+|\xi|^2)$ *can be replaced by* $1/|\xi|^2$ *for large* $|\xi|$.

The following lemma, which provides bounds for an integral crucial to our analysis, is a direct consequence of Theorem 4 in Tudor and Xiao (2015).

LEMMA A.11.– *Let* $z \in \mathbb{R}^d$ *and define:*

$$K(z) = \int_{\mathbb{R}^d} (1-\cos\langle\xi, z\rangle)\left(\frac{1}{1+|\xi|^2}\right)^{2H} d\xi,$$

where $H > 0$ *and* $d \geq 1$. *Let* $\beta = \min\left\{1, 2H - \frac{d}{2}\right\}$ *and* $\rho = \begin{cases} 1 \text{ if } \beta = 1 \\ 0 \text{ otherwise.} \end{cases}$ *For any* $M > 0$, *there exist positive and finite constants* C_L, C_U, *such that for any* $z \in \mathbb{R}^d$ *with* $|z| \in (0, M]$*:*

$$C_L|z|^{2\beta}\left(\log\frac{1}{|z|}\right)^{\rho} \leq K(z) \leq C_U|z|^{2\beta}\left(\log\frac{1}{|z|}\right)^{\rho}. \quad [A.2]$$

References

Adler, R.J. (1981). *The Geometry of Random Fields.* John Wiley & Sons, New York.

Aloès, E. and Nualart, D. (2003). Stochastic integration with respect to the fractional Brownian motion. *Stoch. and Stoch. Rep.*, 75, 129–152.

Aloès, E., Mazet, O., Nualart, D. (2001). Stochastic calculus with respect to Gaussian processes. *Ann. Probab.*, 29, 766–801.

Ayache, A. and Xiao, Y. (2005). Asymptotic properties and Hausdorff dimensions of fractional Brownian sheets. *J. Fourier Anal. Appl.*, 11, 407–439.

Balan, R. and Tudor, C.A. (2008). The stochastic heat equation with fractional-colored noise: Existence of the solution. *Latin Amer. J. Probab. Math. Stat.,* 4, 57–87.

Balan, R. and Tudor, C.A. (2010). The stochastic wave equation with fractional noise: A random field approach. *Stochastic Processes and their Applications*, 120(12), 2468–2494.

Baudoin, F. and Nualart, D. (2003). Equivalence of Volterra processes. *Stochastic Processes and their Applications*, 107(2), 327–350.

Berman, S.M. (1969). Local times and sample functions properties of stationary Gaussian processes. *Trans. Amer. Math. Soc.,* 137, 277–299.

Biagini, F., Hu, Y., Oksendal, B., Zhang, T. (2008). *Stochastic Calculus for Fractional Brownian Motion and Applications: Probability and Its Applications*. Springer, London.

Bojdecki, T., Gorostiza, L.G., Talarczyk, A. (2004a). Sub-fractional Brownian motion and its relation to occupation times. *Statistics & Probability Letters*, 69(4), 405–419.

Bojdecki, T., Gorostiza, L.G., Talarczyk, A. (2004b). Fractional Brownian density process and its self-intersection local time of order k. *Journal of Theoretical Probability*, 17(3), 717–739.

Cheridito, P. (2001). Mixed fractional Brownian motion. *Bernoulli*, 7, 913–934.

Cutland, N.J., Kopp, P.E., Willinger, W. (1995). Stock price returns and the Joseph effect: A fractional version of the Black–Scholes model. *Progress in Probability*, 36, 327–351.

Dellacherie, C. and Meyer, P.A. (1980). *Probabilités et potentiel : chapitres V à VIII*. Hermann, Paris.

Doléans-Dade, C. and Meyer, P.A. (1970). Intégrales stochastiques par rapport aux martingales locales. *Séminaire de probabilités de Strasbourg*, 124, 77–107.

Dudley, R.M. (2002). *Real Analysis and Probability*. Cambridge University Press, Cambridge.

Dzhaparidze, K. and Zanten, H.V. (2004). A series expansion of fractional Brownian motion. *Prob. Theory Related Fields*, 130(1), 39–55.

Dzhaparidze, K. and Zanten, H.V. (2005). Optimality of an explicit series expansion of the fractional Brownian sheet. *Statist. Probab. Lett.*, 71(4), 295–301.

El-Nouty, C. (2003). The fractional mixed fractional Brownian motion. *Statistics & Probability Letters*, 65(2), 111–120.

El-Nouty, C. and Zili, M. (2015). On the sub-mixed fractional Brownian motion. *Appl. Math. J. Chin. Univ.*, 30, 27–43.

Erdelyi, A., Magnus, W., Oberhettinger, F., Tricomi, F.G. (1954). *Tables of Integral Transforms*. McGraw-Hill Book Company, New York.

Evans, L.C. (2010). *Partial Differential Equations*. American Mathematical Society, Providence, RI.

Falconer, K. (1990). *Fractal Geometry: Mathematical Foundation and Applications*. John Wiley and Sons, Chichester.

Fernique, X. (1970). Intégrabilité des vecteurs Gaussiens. *C.R. Acad. Sci. Paris Sér.* A, 270(25), 1698–1699.

Golub, G.H. and Van Loan, C.F. (1989). *Matrix Computations*. Hopkins University Press, Baltimore, MA.

Houdré, C. and Villa, J. (2003). An example of infinite dimensional quasi-helix. *Contemp. Math, Am. Math. Soc.*, 336, 195–201.

Kahane, J.P. (1993). *Some Random Series of Functions*, 2nd edition. Cambridge University Press, Cambridge.

Karatzas, I. and Shreve, S.E. (1998). *Brownian Motion and Stochastic Calculus*, 2nd edition. Springer-Verlag, New York.

Kühn, T. and Linde, W. (2002). Optimal series representation of fractional Brownian sheets. *Bernoulli*, 8(5), 669–696.

Li, W.V. and Linde, W. (1999). Approximation, metric entropy and small ball estimates for Gaussian measures. *Ann. Probab.*, 27(3), 1556–1578.

Li, W.V. and Shao, Q.M. (2001). Gaussian processes: Inequalities, small ball probabilities and applications. In *Stochastic Processes: Theory and Methods*, Shanbhag, D.N. and Rao, C.R. (eds). North-Holland, Amsterdam.

Lifshits, M. (2012). *Lectures on Gaussian Processes*. Springer, Berlin, Heidelberg.

Mandelbrot, B.B. and Van Ness, J.W. (1968). Fractional Brownian motions, fractional noises and applications. *SIAM Review*, 10(4). doi: 10.1137/1010093.

Meerschaert, M.M., Wang, W., Xiao, Y. (2013). Fernique-type inequalities and moduli of continuity of anisotropic Gaussain random fields. *Trans. Amer. Math. Soc.*, 365, 1081–1107.

Miao, Y., Ren, W., Ren, Z. (2008). On the fractional mixed fractional Brownian motion. *Applied Mathematical Sciences*, 2(35), 1729–1738.

Mishura, Y. (2008). *Stochastic Calculus for Fractional Brownian Motion and Related Processes*. Springer, Berlin, Heidelberg.

Mishura, Y. and Shevchenko, G. (2017). *Theory and Statistical Applications of Stochastic Processes*. ISTE Ltd, London and John Wiley & Sons, New York.

Mishura, Y. and Zili, M. (2018). *Stochastic Analysis of Mixed Fractional Gaussian Processes*. ISTE Press, London and Elsevier, Amsterdam.

Mishura, Y. and Zili, M. (2026). Odd fractional Brownian motion and its main properties. To appear.

Nourdin, I. (2012). *Selected Aspects of Fractional Brownian Motion*. Springer, Milan.

Nualart, D. (2006). *The Malliavin Calculus and Related Topics*. Springer, Berlin, Heidelberg.

Pang, G. and Taqqu, M.S. (2019). Nonstationary self-similar Gaussian processes as scaling limits of power-law shot noise processes and generalizations of fractional Brownian motion. *High Frequency*, 2(2), 95–112.

Peltier, R.F. and Lévy Véhel, J. (1995). Multifractional Brownian motion: Definition and preleminary results. Research Report, INRIA, No. 2645.

Protter, P. (2009). *Stochastic Integration and Differential Equations*. Springer, Berlin, Heidelberg.

Revuz, D. and Yor, M. (1999). *Continuous Martingales and Brownian Motion*. Springer, Berlin, Heidelberg.

Russo, F. and Vallois, P. (1993). Forward, backward and symmetric stochastic integration. *Probab. Theory Rel. Fields*, 97, 403–421.

Samko, S., Kilbas, A., Marichev, O. (1993). *Fractional Integrals and Derivatives: Theory and Applications*. Gordon & Breach, London/New York.

Samorodnitsky, G. and Taqqu, M.S. (1994). *Stable Non-Gaussian Random Processes*. Chapman & Hall, New York.

Sghir, A. (2013). The generalized sub-fractional Brownian motion. *Communications on Stochastic Analysis*, 7(3), 373–382.

Sottinen, T. and Valkeila, E. (2003). On arbitrage and replication in the fractional Black–Scholes pricing model. *Statistics & Decisions*, 21, 137–151.

Stricker, C. (1984). Quelques remarques sur les semimartingales Gaussiennes et le problème de l'innovation. In *Filtering and Control of Random Processes*, Korezlioglu, H., Mazziotto, G., Szpirglas, J. (eds). Springer, Berlin, Heidelberg.

Talagrand, M. (1995). Hausorff measure of trajectories of multiparameter fractional Brownian motion. *The Annals of Probability*, 23(2), 767–775.

Testard, F. (1986). *Polarité, points multiples et géométrie de certain processus Gaussiens*. Publ. du Laboratoire de Statistique et Probabilités de l'U.P.S., Toulouse.

Tudor, C.A. (2007). Some properties of the sub-fractional Brownian motion. *Stochastics*, 79(5), 431–448.

Tudor, C.A. (2008). Some aspects of stochastic calculus for the sub-fractional Brownian motion. *An International Journal of Probability and Stochastic Processes*, 79(5), 199–230.

Tudor, C.A. (2013). *Analysis of Variations for Self-similar Processes*. Springer, Cham.

Tudor, C.A. (2014). Recent developments on stochastic heat equation with additive fractional-colored noise. *Fract. Calc. Appl. Anal.*, 17(1), 224–246.

Tudor, C.A. and Xiao, Y. (2015). Sample paths of the solution to the Fractional-colored stochastic heat equation. *arXiv*. arXiv: 1501.06828v1.

Tudor, C.A. and Zili, M. (2014). Covariance measure and stochastic heat equation with fractional noise. *Fract. Calc. Appl. Anal.*, 17(3), 807–826.

Van der Vaart, A.W. and Wellner, J.A. (1996). *Weak Convergence and Empirical Processes with Applications to Statistics*. Springer, New York.

Vignat, C. (2012). A generalized Isserlis theorem for location mixtures of Gaussian random vectors. *Statistics and Probability Letters*, 82(1), 67–71.

Walsh, B.J. (1984). *An Introduction to Stochastic Partial Differential Equations*. Springer, Berlin.

Watson, G.N. (1944). *A Treatise on the Theory of Bessel Functions*. Cambridge University Press, Cambridge.

Wu, D. and Xiao, Y. (2011). On local times of antisotropic Gaussian random fields. *Communications on Stochastic Analysis*, 5(1), 15–39.

Xiao, Y. (1995). Dimension results for Gaussian vector fields and index-α stable fields. *The Annals of Probability*, 23(1), 273–291.

Xiao, Y. (2007). Strong local nondeterminism and the sample path properties of Gaussian random fields. In *Asymptotic Theory in Probability and Statistics with Applications*, Lai, T.-L., Shao, Q.-M., Qian, L. (eds). Higher Education Press, Beijing.

Xiao, Y. (2009). Sample path properties of anisotropic Gaussian random fields. In *A Minicourse on Stochastic Partial Differential Equations*, Khoshnevisan, D. and Rassoul-Agha, F. (eds). Springer, New York.

Yan, L. and Shen, G. (2010). On the collision local time of sub-fractional Brownian motions. *Statistics and Probability Letters*, 80, 296–308.

Yan, L., Shen, G., He, K. (2011). Itô's formula for the subfractional Brownian motion. *Communications on Stochastic Analysis*, 5(1), 135–159.

Zili, M. (2006). On the mixed-fractional Brownian motion. *Journal of Applied Mathematics and Stochastic Analysis*, 2006(1), 32435.

Zili, M. (2013) An optimal series expansion of sub-mixed fractional Brownian motion. *Journal of Numerical Mathematics and Stochastics*, 5(1), 93–105.

Zili, M. (2014). Mixed sub-fractional Brownian motion. *Random Operators and Stochastic Equations*, 22(3), 163–178.

Zili, M. (2017). Generalized fractional Brownian motion. *Modern Stochastics: Theory and Applications*, 4, 15–24.

Zili, M. (2018). On the generalized fractional Brownian motion. *Mathematical Models and Computer Simulation*, 10(6), 1–11.

Zili, M. (2024). .Stochastic calculus with a special generalized fractional Brownian motion. *International Journal of Applied Mathematics and Simulation*, 1(1), IJAMS 2992–1708.

Zili, M. (2025). Stochastic heat equation with a special generalized fractional noise. *Filomat*, 39(31) [Online]. Available at: https://journal.pmf.ni.ac.rs/filomat/index.php/filomat/article/view/26073.

Index

A, B

C

D

E

F

G

T, U

V, W

Other titles from

ISTE

in

Mathematics and Statistics

2026

LESFARI Ahmed
Complex Manifolds and Geometric Algebraic Analysis

PEYRARD Nathalie, DE SAPORTA Benoîte
A Comprehensive Guide to HSMM: Theory, Software, and Advanced Extensions

SIMON Jacques
Integration
(Analysis for PDEs Set – Volume 4)

2025

DUPUY Jean-François
Generalized Linear Models: Problems with Censored, Missing, and Zero-inflated Data

GAIVORONSKI Alexei A., KNOPOV Pavel S., NORKIN Vladimir I., ZASLAVSKYI Volodymyr A.
Stochastic Modeling and Optimization Methods for Critical Infrastructure Protection 1: Stochastic Modeling
Stochastic Modeling and Optimization Methods for Critical Infrastructure Protection 2: Methods and Tools

PICARD Philippe, LEFÈVRE Claude
Abel-Gontcharoff Pseudopolynomials and Stochastic Applications

2024

BOURLÈS Henri
System Theory – A Modern Approach 1: Linear Ordinary and Functional Differential Equations
(New Mathematical Methods, Systems and Applications Set – Volume 1)

MAKAROV Volodymyr, MAYKO Nataliya
Traditional Functional-Discrete Methods for the Problems of Mathematical Physics: New Aspects

2023

AIT BEN HASSI El Mustapha
The Theory of Distributions: Introduction

KOROLIOUK Dmitri, SAMOILENKO Igor
Asymptotic and Analytic Methods in Stochastic Evolutionary Systems

PARROCHIA Daniel
Graphs, Orders, Infinites and Philosophy

2022

CHAKRAVARTHY Srinivas R.
Introduction to Matrix-Analytic Methods in Queues 1: Analytical and Simulation Approach – Basics
Introduction to Matrix-Analytic Methods in Queues 2: Analytical and Simulation Approach – Queues and Simulation

LESFARI Ahmed
Integrable Systems

RADCHENKO Vadym M.
General Stochastic Measures: Integration, Path Properties and Equations

DE SAPORTA Benoîte, ZILI Mounir
Martingales and Financial Mathematics in Discrete Time

SIMON Jacques
Distributions
(Analysis for PDEs Set – Volume 3)

2021

KOROLIOUK Dmitri, SAMOILENKO Igor
Random Evolutionary Systems: Asymptotic Properties and Large Deviations

MOKLYACHUK Mikhail
Convex Optimization: Introductory Course

POGORUI Anatoliy, SWISHCHUK Anatoliy, RODRÍGUEZ-DAGNINO Ramón M.
Random Motions in Markov and Semi-Markov Random Environments 1: Homogeneous Random Motions and their Applications
Random Motions in Markov and Semi-Markov Random Environments 2: High-dimensional Random Motions and Financial Applications

PROVENZI Edoardo
From Euclidean to Hilbert Spaces: Introduction to Functional Analysis and its Applications

2020

BARBU Vlad Stefan, VERGNE Nicolas
Statistical Topics and Stochastic Models for Dependent Data with Applications

CHABANYUK Yaroslav, NIKITIN Anatolii, KHIMKA Uliana
Asymptotic Analyses for Complex Evolutionary Systems with Markov and Semi-Markov Switching Using Approximation Schemes

KOROLIOUK Dmitri
Dynamics of Statistical Experiments

MANOU-ABI Solym Mawaki, DABO-NIANG Sophie, SALONE Jean-Jacques
Mathematical Modeling of Random and Deterministic Phenomena

SIMON Jacques
Continuous Functions
(Analysis for PDEs Set – Volume 2)

2019

BANNA Oksana, MISHURA Yuliya, RALCHENKO Kostiantyn, SHKLYAR Sergiy
Fractional Brownian Motion: Approximations and Projections

GANA Kamel, BROC Guillaume
Structural Equation Modeling with lavaan

KUKUSH Alexander
Gaussian Measures in Hilbert Space: Construction and Properties

LUZ Maksym, MOKLYACHUK Mikhail
Estimation of Stochastic Processes with Stationary Increments and Cointegrated Sequences

MICHELITSCH Thomas, PÉREZ RIASCOS Alejandro, COLLET Bernard, NOWAKOWSKI Andrzej, NICOLLEAU Franck
Fractional Dynamics on Networks and Lattices

VOTSI Irene, LIMNIOS Nikolaos, PAPADIMITRIOU Eleftheria, TSAKLIDIS George
Earthquake Statistical Analysis through Multi-state Modeling
(Statistical Methods for Earthquakes Set – Volume 2)

2018

AZAÏS Romain, BOUGUET Florian
Statistical Inference for Piecewise-deterministic Markov Processes

IBRAHIMI Mohammed
Mergers & Acquisitions: Theory, Strategy, Finance

PARROCHIA Daniel
Mathematics and Philosophy

2017

CARONI Chysseis
First Hitting Time Regression Models: Lifetime Data Analysis Based on Underlying Stochastic Processes
(Mathematical Models and Methods in Reliability Set – Volume 4)

CELANT Giorgio, BRONIATOWSKI Michel
Interpolation and Extrapolation Optimal Designs 2: Finite Dimensional General Models

CONSOLE Rodolfo, MURRU Maura, FALCONE Giuseppe
Earthquake Occurrence: Short- and Long-term Models and their Validation
(Statistical Methods for Earthquakes Set – Volume 1)

D'AMICO Guglielmo, DI BIASE Giuseppe, JANSSEN Jacques, MANCA Raimondo
Semi-Markov Migration Models for Credit Risk
(Stochastic Models for Insurance Set – Volume 1)

GONZÁLEZ VELASCO Miguel, del PUERTO GARCÍA Inés, YANEV George P.
Controlled Branching Processes
(Branching Processes, Branching Random Walks and Branching Particle Fields Set – Volume 2)

HARLAMOV Boris
Stochastic Analysis of Risk and Management
(Stochastic Models in Survival Analysis and Reliability Set – Volume 2)

KERSTING Götz, VATUTIN Vladimir
Discrete Time Branching Processes in Random Environment
(Branching Processes, Branching Random Walks and Branching Particle Fields Set – Volume 1)

MISHURA Yuliya, SHEVCHENKO Georgiy
Theory and Statistical Applications of Stochastic Processes

NIKULIN Mikhail, CHIMITOVA Ekaterina
Chi-squared Goodness-of-fit Tests for Censored Data
(Stochastic Models in Survival Analysis and Reliability Set – Volume 3)

SIMON Jacques
Banach, Fréchet, Hilbert and Neumann Spaces
(Analysis for PDEs Set – Volume 1)

2016

CELANT Giorgio, BRONIATOWSKI Michel
Interpolation and Extrapolation Optimal Designs 1: Polynomial Regression and Approximation Theory

CHIASSERINI Carla Fabiana, GRIBAUDO Marco, MANINI Daniele
Analytical Modeling of Wireless Communication Systems
(Stochastic Models in Computer Science and Telecommunication Networks Set – Volume 1)

GOUDON Thierry
Mathematics for Modeling and Scientific Computing

KAHLE Waltraud, MERCIER Sophie, PAROISSIN Christian
Degradation Processes in Reliability
(Mathematial Models and Methods in Reliability Set – Volume 3)

KERN Michel
Numerical Methods for Inverse Problems

RYKOV Vladimir
Reliability of Engineering Systems and Technological Risks
(Stochastic Models in Survival Analysis and Reliability Set – Volume 1)

2015

DEVOLDER Pierre, JANSSEN Jacques, MANCA Raimondo
Basic Stochastic Processes

LE GAT Yves
Recurrent Event Modeling Based on the Yule Process
(Mathematical Models and Methods in Reliability Set – Volume 2)

DE SAPORTA Benoîte, DUFOUR François, ZHANG Huilong
Numerical Methods for Simulation and Optimization of Piecewise Deterministic Markov Processes

2014

COOKE Roger M., NIEBOER Daan, MISIEWICZ Jolanta
Fat-tailed Distributions: Data, Diagnostics and Dependence
(Mathematical Models and Methods in Reliability Set – Volume 1)

MACKEVIČIUS Vigirdas
Integral and Measure: From Rather Simple to Rather Complex

PASCHOS Vangelis Th.
Combinatorial Optimization – 3-volume series – 2nd edition
Concepts of Combinatorial Optimization / Concepts and Fundamentals – Volume 1
Paradigms of Combinatorial Optimization – Volume 2
Applications of Combinatorial Optimization – Volume 3

2013

COUALLIER Vincent, GERVILLE-RÉACHE Léo, HUBER Catherine, LIMNIOS Nikolaos, MESBAH Mounir
Statistical Models and Methods for Reliability and Survival Analysis

JANSSEN Jacques, MANCA Oronzio, MANCA Raimondo
Applied Diffusion Processes from Engineering to Finance

SERICOLA Bruno
Markov Chains: Theory, Algorithms and Applications

2012

BOSQ Denis
Mathematical Statistics and Stochastic Processes

CHRISTENSEN Karl Bang, KREINER Svend, MESBAH Mounir
Rasch Models in Health

DECREUSEFOND Laurent, MOYAL Pascal
Stochastic Modeling and Analysis of Telecom Networks

DEVOLDER Pierre, JANSSEN Jacques, MANCA Raimondo
Stochastic Methods for Pension Funds

2011

MACKEVIČIUS Vigirdas
Introduction to Stochastic Analysis: Integrals and Differential Equations

MAHJOUB Ridha
Recent Progress in Combinatorial Optimization – ISCO2010

RAYNAUD Hervé, ARROW Kenneth
Managerial Logic

2010

BAGDONAVIČIUS Vilijandas, KRUOPIS Julius, NIKULIN Mikhail
Nonparametric Tests for Censored Data
Nonparametric Tests for Complete Data

IOSIFESCU Marius, LIMNIOS Nikoloas, OPRISAN Gheorghe
Introduction to Stochastic Models

VASSILIOU P.C.G.
Discrete-time Asset Pricing Models in Applied Stochastic Finance
Continuous-time Asset Pricing Models in Applied Stochastic Finance

2008

ANISIMOV Vladimir
Switching Processes in Queuing Models

FICHE Georges, HÉBUTERNE Gérard
Mathematics for Engineers

HUBER Catherine, LIMNIOS Nikolaos, MESBAH Mounir, NIKULIN Mikhail
Mathematical Methods in Survival Analysis, Reliability and Quality of Life

JANSSEN Jacques, MANCA Raimondo, VOLPE Ernesto
Mathematical Finance

2007

HARLAMOV Boris
Continuous Semi-Markov Processes

2006

CLERC Maurice
Particle Swarm Optimization

Printed and bound by CPI Group (UK) Ltd, Croydon, CR0 4YY

13/09/2026

14961048-0004